Phylogeny, Anatomy and Physiology of Ancient Fishes

Phylogeny, Anatomy and Physiology of Ancient Fishes

Editors

Giacomo Zaccone
Department of Food and Environmental Science
University of Messina
Messina
Italy

K. Dabrowski
School of Environment & Natural Resources
Ohio State University
Columbus, OH
USA

Michael S. Hedrick
Developmental Integrative Biology Research Cluster
Department of Biological Sciences
University of North Texas
Denton, TX
USA

J.M.O. Fernandes
Marine Genomics Research Group
Faculty of Biosciences and Aquaculture
University of Nordland
Bodø
Norway

José M. Icardo
Department of Anatomy and Cell Biology
Faculty of Medicine
University of Cantabria
Santander
SPAIN

CRC Press
Taylor & Francis Group
Boca Raton London New York

CRC Press is an imprint of the
Taylor & Francis Group, an **informa** business

A SCIENCE PUBLISHERS BOOK

Cover illustrations reproduced by kind courtesy of one of the editors, José M. Icardo.

CRC Press
Taylor & Francis Group
6000 Broken Sound Parkway NW, Suite 300
Boca Raton, FL 33487-2742

First issued in paperback 2019

ISBN-13: 978-1-4987-0755-8 (hbk)
ISBN-13: 978-0-367-37752-6 (pbk)

Visit the Taylor & Francis Web site at
http://www.taylorandfrancis.com

and the CRC Press Web site at
http://www.crcpress.com

Preface

Primitive fishes, including the Agnathans as the first representative of the vertebrates of today, have populated the aquatic environment for about the last 500 million years. Extant primitive species are essentially similar to their fossil counterparts, retaining a number of characteristics that no longer exist in living related species (or taxa). By studying the molecular, morphological and physiological characteristics of ancient fishes we ought to analyze relevant episodes of our own evolutionary story. This is probably what makes this kind of study so appealing both to the public imagination and to many researchers from very different fields.

In line with previous excellent compilations, like the one presented in Volume 26 of the Fish Physiology Series, year 2007, the objective of this work is to bring together the knowledge and experience of authors who are experts in phylogenetic relationships, morphology and physiology, in order to assemble a volume rich in information and new developments. In addition to excellent phylogenetic approaches, this volume combines studies on comparative anatomy, biology, biochemistry and molecular biology. The different chapters include the heart and the circulatory system, air respiration and the control of breathing, the digestive tract and gut development, the olfactory system and several other unusual features such as the presence of slime glands. Cyclostomes, lungfishes, chondrosteans and holosteans constitute the main subjects of study. Teleosts are also protagonists of this work since, inevitably, exceptional biological strategies and specific morphological and physiological features have to be confronted with related behavior or characteristics (or the lack of them) exhibited by the more derived species. In this context, evolutionary relationships among the major actinopterygian and sarcopterygian lineages have not been conclusively resolved, and a consensus phylogeny for teleost is emerging today based on analysis of various subsets of actinopterygian taxa.

All the chapters are state-of-the-art works on the topics under discussion. The list of subjects dealt with had to be pruned and some other topics had to be abandoned for various reasons. However, by engaging people from different fields and disciplines we sought to avoid overlooking interactions between the different levels of biological organization and to create a comprehensive view of the topics and animals studied. The view that emerges is that ancient fishes may be very complex organisms. They are often contemplated as "relics" of evolution or, as it is more commonly said, living fossils. This is probably true, but relics do not survive through geological times. Ancient fishes had the capability to adapt to changing environments and are able to thrive in present day conditions. The objective of research programs should not solely be for the purpose of increasing our knowledge base: our goal should be enhance the conditions for the preservation of these animals for generations to come.

The present volume is intended to be useful to professionals and researchers interested in the diverse topics of study, to educators, to advanced students and, in general, to all of those sharing a biological interest in aquatic creatures.

The editors are indebted to all the authors who generously expressed their willingness to cooperate, were able to complete their chapters and had the patience to bear with the inevitable delays inherent on a project of this kind. Additional thanks are due to the authors involved with the revision process.

The Editors

Contents

Preface   v

1. Phylogenetic Introduction   1
   Hans-Peter Schultze

2. The Occurrence and Function of the NOS/NO System in the   19
   Heart of the Eel and the African Lungfish
   Daniela Amelio, Filippo Garofalo, Sandra Imbrogno and Bruno Tota

3. Evolutionary Aspects on the Comparative Biology of Lungfishes: 38
   Emphasis on South-American Lungfish, *Lepidosiren paradoxa*
   *Vera Maria Fonseca de Almeida-Val, Luciana Mara Lopes Fé and
   Derek Felipe de Campos*

4. Developmental Physiology of the Australian Lungfish,   57
   Neoceratodus forsteri
   Casey A. Mueller

5. Aestivation in African Lungfishes: Physiology, Biochemistry and   81
   Molecular Biology
   *Shit F. Chew, Biyun Ching, You R. Chng, Jasmine L.Y. Ong,
   Kum C. Hiong, Xiu L. Chen and Yuen K. Ip*

6. Anatomy of the Heart and Circulation in Lungfishes   133
   José M. Icardo, Bruno Tota and Yuen K. Ip

7. The Cardiac Outflow Tract of Primitive Fishes   151
   Adrian C. Grimes

8. Control of Breathing in Primitive Fishes   179
   Michael S. Hedrick and Stephen L. Katz

9. The Lung-Swimbladder Issue: A Simple Case   201
   of Homology—Or Not?
   Markus Lambertz and Steven F. Perry

10. The Gut and Associated Organs in the African Lungfish   212
    Protopterus annectens
    José M. Icardo

11. Morphology, Histology, and Functional Structure of the Alimentary Canal of Sturgeon   233
Ramón Carmona Martos, Cristina E. Trenzado Romero and *Ana Sanz Rus*

12. The Structural Organization in the Olfactory System of the Teleosts and Garfishes   260
Michal Kuciel, Krystyna Żuwała, Eugenia Rita Lauriano, Leszek Satora and *Giacomo Zaccone*

13. Hagfish Slime and Slime Glands   272
Douglas S. Fudge, Timothy M. Winegard and *Julia E. Herr*

Index   291

1

Phylogenetic Introduction

Hans-Peter Schultze

The term 'Fishes' is used either widely including all vertebrates with gills (Craniata or Vertebrata = agnathans + gnathostomes) or restricted to jaw-bearing vertebrates with gills (gnathostomes). The gnathostomes include the tetrapods, which is often not recognized. Tetrapods are a lineage within the Osteichthyes (Hennig 1983 introduced the term 'Osteognathostomata' for bony fishes + tetrapods). In this book the term 'Fishes' is used in the wide sense, thus the term 'ancient fishes' refers to agnathans and different gnathostomes (excellent overview of extant and fossil gnathostomes in Janvier 1996) excluding advanced actinopterygians (specifically advanced teleosts) and sarcopterygians (specifically tetrapods).

Craniata or Vertebrata

Craniates or Vertebrates are chordates with a division of the body in the head, trunk and tail (= regionalization). The name Craniata refers to the presence of a head. The name Vertebrata indicates the presence of vertebral elements, which are missing in hagfishes. Thus earlier hagfishes, Myxinoida, were considered more primitive than Petromyzontida and fossil agnathans, which were classified together with the gnathostomes as vertebrates. A key character of Craniata and Vertebrata is the neural crest, which induces many of their features. The head includes a pair of eyes, a pair of nasal capsules, a brain, dorsal median organs (pineal- and parietal organ), neurohypophysis and adenohypophysis, two semicircular canals in the labyrinth and external

Natural History Museum University of Kansas Lawrence, Kansas, USA.
E-mail: hp1937@ku.edu

branchial openings. The trunk encloses notochord, muscles, heart and the digestive and reproductive systems, whereas only notochord and muscles form the tail region (behind the anus) as a propulsive organ. Cartilage is a basic tissue of vertebrates, it forms basidorsalia and basiventralia around the notochord. The epidermis is composed of many layers.

The vertebrates are divided in 'Agnatha' (taxa without jaws) and Gnathostomata (taxa with biting jaws). The Cyclostomata (Myxinoida + Petromyzontida) are the recent 'agnathans' (Fig. 1), they do not possess an external skeleton. The fossil record of Petromyzontida reaches back into the Late Devonian (360 Million years ago), that of the Myxinoida into the Late Carboniferous (310 Million years ago). The Cyclostomata are monophyletic after molecular (Heimberg et al. 2010) and developmental data (Oisi et al. 2013), whereas some morphological data like neuroanatomic features placed the Petromyzontida closer to the Gnathostomata (Khonsari et al. 2009). Besides molecular data the early development of the nasohypophysial region, the presence of an unpaired nasohypophysial duct, enlarged gill chambers and a broad neural cord indicate a close relationship of both taxa. The Myxinoida are characterized by a number of degenerated characters: loss of one semicircular canal, loss of pineal- and parietal organ, loss of cerebellum, loss of basidorsalia and basiventralia in the adults, loss of a typical lateral line system and other. These are all features in common to all other vertebrates (Petromyzontida and Gnathostomata).

Fossil 'Agnatha' possess a heavy external skeleton of bone and dentine in form of shields and/or scales, therefore they are also called 'ostracoderms' (shell-skinned). The 'ostracoderms' are paraphyletic like the 'agnathans'; they include taxa with mineralized exoskeleton on the line to the gnathostomes, but not an ancestor and all descendants. The first phosphatic mineralized elements occur in the latest Cambrian (490 Million years ago) and full body preserved 'agnathans' (= 'ostracoderms') in the Early Ordovician (475 Million years ago). They dominated the vertebrate fauna of Ordovician to Early Devonian times, and were replaced by gnathostomes during the Devonian. They disappeared at the end of the Late Devonian (360 Million years ago). The different groups are very well defined and easy to distinguish, but common features are rare so that the phylogenetic arrangement can differ greatly between authors and even between trees of the same author. The Osteostraci are considered commonly as sister taxon of the Gnathostomata based on the presence of a heterocercal (epicercal) caudal fin, short based pectoral fins with endoskeletal shoulder girdle, sclerotical (scleral) ring (around the eye) and bone with bone cells.

Going back to the Cambrian, naked 'agnathans' (Myllokunmingiida: *Haikouichthys*; Shu et al. 1999; Shu 2003) have been recorded from the Early Cambrian (520 Million years ago). *Haikouichthys* possessed W-shaped myomeres like all vertebrates, a pair of eyes, a pair of nasal capsules between the eyes, possibly arcualia in the anterior part of the notochord and other vertebrate features. A fin fold reaches from the anterior dorsal margin around

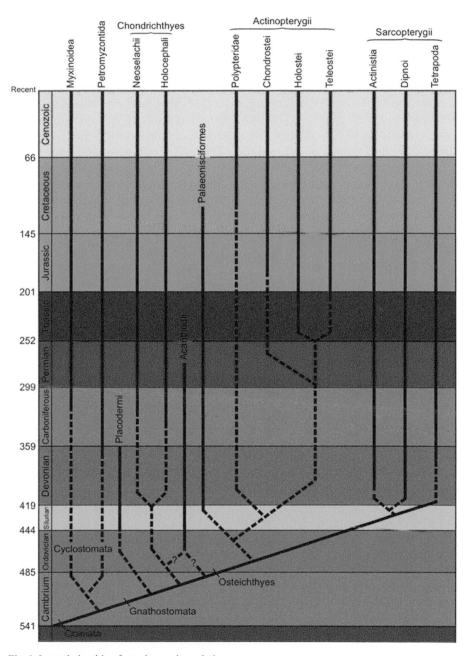

Fig. 1. Interrelationship of vertebrates through time.

the caudal tip to the anal fin and continues in front of the anal fin anteriad to the posterior gill region. The presence of a pair of nasal capsules in *Haikouichthys* indicates that this feature occurring in Pteraspidomorphi and Galeaspidida, two 'ostracoderm' groups, may be the ancestral structure in vertebrates, and that the unpaired nasal opening connected with a naso-hypophysal duct is a secondary structure occurring in cyclostomes and osteostracans (and possibly in other 'agnathans'). *Haikouichthys* has a row of gills, which appear as chambers like in *Petromyzon*. The digestive tract is a simple duct. All these structures place the Myllokunmingiida at the base of the vertebrates.

All 'agnathans' possess gill chambers and external gill openings so that the oxygen uptake is primarily over the gills. A diverticle from the anterior gut comparable with the lung or swimbladder, the topic of the book, of advanced gnathostomes is unknown in 'agnathans'. Thus the interrelationships of the gnathostomes where lung and/or swimbladder occur will be concentrated here.

Gnathostomata

Gnathostomata (jawed mouth) are characterized by the possession of the upper and lower jaws. The mandibular arch (jaws) is followed by the hyoid arch and five gill arches, which possess a medial and segmented part, the basibranchials. The ectodermal gills lie lateral to the gill arches. Further characters of the gnathostomes are intrinsic eye muscles, eyes with true cornea and the ability to accommodate, three semicircular canals (horizontal canal added to the two vertical ones of 'agnathans'), neuromasts of sensory lines in canals, and calcified exo- and endoskeleton. The body has two paired fins with endoskeletal girdles, a strongly folded trunk musculature with a septum horizontale, which separates epiaxial from hypoaxial muscle portions, four arcualia per segment around the notochord and dorsal motoric and ventral visceral spinal nerve rami united from fibers of the dorsal and ventral spinal nerve roots.

The Gnathostomata encompass Placodermi, Chondrichthyes, Acanthodii and Osteichthyes (Fig. 1). Gnathostomes extend back to the Ordovician (450 Million years ago). The latest phylogenetic analysis by Zhu et al. (2013) places the Placodermi as non-monophyletic group at the base of the gnathostomes, whereas Osteichthyes, Acanthodii and Chondrichthyes form the crown group of the gnathostomes. The acanthodians as non-monophyletic group are the sister group of the Chondrichthyes.

Placodermi

The Placodermi (Denison 1978; Goujet and Young 2004) are placed at the base of the Gnathostomata, either as a monophyletic group or paraphyletic on the stem lineage (Zhu et al. 2013) to the crown gnathostomes (Chondrichthyes +

Osteichthyes). They are characterized by a macromeric exoskeleton of plates, which cover the head and part of the trunk. The plates are named after the region, where they occur, like the central, marginal plate, etc. on the head and median dorsal plate, anterior and posterior lateral plate, etc. in the shoulder girdle. The lateral line system is clearly visible as grooves in the dermal plates. A distinct opercular apparatus is lacking; the submarginal plate may function like an operculum. The palatoquadrate is attached to the exoskeletal cheek bones (Zhu et al. 2013, Suppl. Fig. 24b,d,e), so that the adductor musculature inserts on the inside of the palatoquadrate and not on the outside as in other gnathostomes. The jaws (called gnathals) possess sharpened areas formed by semidentine, a dentine, where the dentine cell bodies are not retreated to the pulp cavity. The presence of teeth, which grow in a dentine lamina, may be present in some advanced taxa (Smith and Johanson 2003), nevertheless their presence is denied (Burrow 2003), and the growth simulation by Rücklin et al. (2012) shows a continuous enlargement starting with the tooth like margin. The occurrence of possible true teeth is shown by these authors in advanced placoderms (arthrodires). Therefore it is argued that the teeth of placoderms appear independently from those in other gnathostomes. In these studies some primitive Early Devonian placoderms like *Stensioella* and *Pseudopetalichthys* (Gross 1962) are neglected. Gross (1962, p. 56, Fig. 4) described and figured small teeth on the jaws and gill rakers on the gill arches of *Stensioella*. This could indicate that teeth and jaws are present in primitive gnathostomes and are lost in most placoderms.

The placoderms possess a double articulation between the head and body, an exoskeletal articulation between dorsolateral plates (so-called paranuchals) of the head and lateral plates of shoulder girdle in addition to an endoskeletal articulation between the endocranium and a synarcual (= fused anterior vertebral elements).

Stensiö (1963) suggested homologization of some placoderm plates with bones of osteichthyans. Zhu et al. (2013, supplementary Fig. 10) gave a wider range of homologization with different osteichthyan taxa in their description of the oldest known placoderm *Entelognathus*. These homologizations are not commonly accepted.

In some placoderms the males possess a copulatory organ, claspers, which have a different composition than those of chondrichthyans.

Except from the head with its well-ossified neurocranium the internal anatomy is little known. Stensiö (1963) gives a detailed interpretation of the canals (occupied by nerves and blood vessels) and spaces (occupied by the brain and the labyrinth) within the neurocranium. These interpretations have in general been supported by following investigations (Young 1980, 1984; Goujet 1984 and others). In contrast only the intestine with a spiral valve (Denison 1941, 1978) is generally accepted as the known internal structure of the abdomen. The presence of 'lungs' was an over-interpretation by Denison (1941; see below).

The oldest placoderms are known from the Silurian; the group became extinct at the end of the Devonian.

Crown Gnathostomes

The extant gnathostomes, Chondrichthyes and Osteichthyes, and the fossil Acanthodii are the advanced gnathostomes (Fig. 1), the crown gnathostomes. These are fishes with heterocercal (= epicercal) tail and two dorsal fins in the basal forms. In recent years endocranial features as ventral cranial and otico-occipital fissure have been shown to be supporting features for crown gnathostomes (see below), whereas they were considered earlier as features, which separate osteichthyans from chondrichthyans. They are distinct from Placodermi in many features especially in the formation of the shoulder girdle. The shoulder girdle does not form a complete ring as in placoderms.

Chondrichthyes

The Chondrichthyes have one feature in common, prismatic calcified cartilage, which distinguish them from all other fishes. The prismatic cartilage is formed by tesserae of hydroxylapatite. The main part of the endoskeleton is cartilaginous (therefore cartilaginous fishes) even though they have the capability to form perichondral bone (Peignoux-Deville et al. 1982). Some chondrichthyans possess mineralized (also hydroxylapatite of changing structure, so that it was earlier used to identify sharks, Hasse 1885) centra.

Their exoskeleton is micromeric, head and body are covered by tooth-like scales. The dentine (orthodentine) surrounds the pulp cavity and is covered by enameloid; the base is formed by acellular bone, except by cellular bone in some Paleozoic sharks. One distinguishes growing and non-growing scales. Non-growing scales start to grow from the outside towards the pulp cavity. Growing scales grow by addition of scales around a primary one. The lateral line canal runs between scales on the body. Scales are known back into Late Ordovician (450 Million years ago; Turner et al. 2004), whereas first chondrichthyan teeth appear in Early Devonian (410 Million years ago).

Permanent, lifelong exchange of teeth is a characteristic of elasmobranch chondrichthyans. The teeth are formed on the inner side of the jaws and moved into functional position to the margin where they are shed independent from their state of use. The tooth plates of holocephalans are fusion of tooth families (Didier et al. 1994; Stahl 1999).

The braincase (neurocranium) forms one unit in extant and nearly all fossil chondrichthyans. This feature was used as a main difference between chondrichthyans and osteichthyans, until fissures where discovered in the neurocranium of the Devonian *Pucapampella*. *Pucapampella* is a chondrichthyan from the Middle Devonian (390 Million years ago) of South America (Maisey 2001) and from the Lower Devonian (400 Million years ago) of South

Africa (Maisey and Anderson 2001); the genus is placed near the base of Chondrichthyes. The neurocranium of *Pucapampella* is divided in to an anterior and posterior part like in acanthodians and osteichthyans. The separation is formed by the ventral cranial fissure on the ventral side of the neurocranium, which continues laterally as an otico-occipital fissure. The ventrolateral fissure is closed in chondrichthyans above *Pucapampella*, whereas few Paleozoic chondrichthyans like the Carboniferous *Cobelodus* and *Stethacanthus* within the Symmoriida and *Orthacanthus* within the Xenacanthida still possess the otico-occipital fissure (Maisey 2007). The canals for efferent pseudobranchial artery and for lateral aorta run in cartilage. The eye stalk, a cartilaginous element, which supports the eye ball, was taken earlier as a chondrichthyan feature, but it is now known from early osteichthyans, too (Zhu et al. 2001; Basden and Young 2001).

In elasmobranch chondrichthyans each gill slit opens directly to the outside. There is no ossified opercular plate; the opercular flap of holocephalans is fleshy. The copulatory organ, clasper, is taken as a common feature for chondrichthyans. It is not known in osteichthyans nor in acanthodians.

Lung, swimbladder or divercles of esophagus indicating such structures are not known in extant or fossil chondrichthyans.

Acanthodii

Acanthodians or 'spiny sharks' possess spines in front of each fin except the caudal fin (Denison 1979). Some taxa possess pairs of additional spines between the pectoral and the pelvic fin spines. They represent a fossil taxon restricted to the Paleozoic. Their relationship within the crown gnathostomes—either closer to the chondrichthyans or closer to the osteichthyans—has been in discussion for a long time (Fig. 1).

Acanthodians have a micromeric exoskeleton like chondrichthyans, nevertheless few bony plates exist in the opercular region and in the shoulder girdle. Branchiostegal like elements cover the gill chamber behind the hyoid arch. Additional branchiostegal like gill covers can appear on the subsequent gill arches.

Characteristic for acanthodians are their concentric growing scales. Like onion skins, one layer of bone and dentine is placed around the former one. The bone can be acellular or cellular covered by orthodentine (= normal dentine) or mesodentine (interbranching dentine tubules). The base is rounded, a pulp cavity does not exist; they resemble in that way micromeric scales in some actinopterygians.

Classically the acanthodians are divided in three orders. The Climatiida have a well ossified exoskeletal shoulder girdle. Some taxa possess symphysial tooth whorls, other tooth families which seem to have followed a permanent tooth replacement like chondrichthyans. The Ischnacanthida possess dentigerous jaw bones; these grow from posterior anteriad, so that the anterior teeth are larger as the posterior ones. The last order Acanthodida possess no teeth and only one dorsal fin; endocranial and visceral structure are only known of this group.

Endocranium and visceral skeleton (palatoquadrate, Meckelian cartilage and gill arches) are known of the Permian *Acanthodes*, the geological youngest acanthodian. Nelson (1968) argued, that the backward directed epibranchial indicate the relationship to chondrichthyans; Nelson referred to Reis (1890, 1896) and Dean (1907), who proposed the relationship to chondrichthyans. Miles (1973) in contrast argued, that different features of the neurocranium are an indication for the relationship with osteichthyans. One feature, cranial fissures, cannot be used for such relationships anymore because cranial fissures have also been discovered in primitive chondrichthyans (see above; Maisey 2001). Acanthodians possess three otoliths like osteichthyans (Schultze 1990). Recently, Brazeau (2009) and Davis et al. (2012) divided the acanthodians in one group close to osteichthyans and another group close to chondrichthyans. In the analysis of Zhu et al. (2013), all three convential acanthodian groups appear between osteichthyans and chondrichthyans at the base of the chondrichthyans.

Acanthodian spines first occurred back in the Late Ordovician, the youngest acanthodian are known from the Middle Permian. No indication of lung or swimbladder are known from acanthodians.

Osteichthyes

Osteichthyans possess a macromeric bone pattern. There is one basic plan for all osteichthyans only. On the skull roof, two pair of bones, the postparietals and parietals, cover the otico-occipital (posterior) and the ethmosphenoid (anterior) region of the braincase. The parietal opening lies between the parietals, sometimes surrounded by pineal plate(s). This bone is wrongly homologized with the frontal of tetrapods in actinopterygians (see Schultze 2008 for homologization; Janvier 1996; Jollie 1962). The frontal is a bone restricted to tetrapods and their closest relatives, the Elpistostegalia. A mosaic of bones including rostrals and nasals lie in front of the parietals. A row of extrascapularia follows the postparietals, the extrascapularia carry the occipital commissure, the connection between left and right lateral line system. Maxilla and dentary carry marginal teeth. The opercular region is covered by bony plates, operculum, suboperculum and ventral branchiostegals. They possess a median gular between the lower jaws. Exoskeletal clavicle and cleithrum cover the endoskeletal scapulocoracoid. Postcleithrum (anocleithrum), supracleithrum and posttemporal form the connection to the skull roof. The body is covered by rhombic scales, which interlock vertically by peg and socket. They possess enamel, a tissue formed by the basal epidermal layer.

These are the only vertebrates with lung or swimbladder; these structures are known in extant forms and lungs also in some fossil forms (actinistians).

The earliest forms are known from the Silurian (425 Million years ago). There is discussion whether some early forms belong to sarcopterygians or actinopterygians.

Sarcopterygii

Extant sarcopterygians are coelacanths (Actinistia), lungfish (Dipnoi) and tetrapods (Fig. 1). The coelacanths are represented today by one genus (*Latimeria*) with two species, the lungfish by three genera (*Neoceratodus, Protopterus* and *Lepidosiren*) and six species. Both taxa were much more successful in the past, especially in the Paleozoic. Today the so successful tetrapods are the offspring of other piscine sarcopterygians, the Devonian Elpistostegalia and Osteolepiformes.

Basal sarcopterygians are characterized by an intracranial joint between otico-occipital and ethmosphenoidal part of the endocranium. Today only the coelacanths possess the joint within all fishes. The intracranial joint has been lost independently in lungfish and tetrapods. The intracranial joint lies between postparietals and parietals in the dermal skull roof; a parietal opening can be found between the parietals except in lungfish. Lateral to the postparietals, two (sometimes more) bones, tabular and supratemporal, are formed. They carry the extension of the main lateral line, the temporal canal, over to the intertemporal lateral to the parietal and as infraorbital canal around the orbit. There is an uneven number of extrascapularia behind the postparietals, a median and lateral extrascapularia. Dipnoans deviate from the general pattern of sarcopterygians in having a mosaic of bones, which is reduced in modern lungfish; a clear homologization of the lungfish pattern with the general pattern is not known.

Both nasal openings lie anterolateral to the orbit; the supraorbital canal runs medial to both nasal openings. The coelacanths possess an electrosensory organ, the rostral organ, between the nasal cavities. Both nasal openings are positioned within the palate of lungfish. Only the posterior nasal opening is placed in the palate of tetrapods. In the closest relatives to the tetrapods, the Elpistostegalia, the posterior nasal opening is placed close to the margin of the upper jaw. More than five (often a high number) sclerotic bones form the ring around the eye ball. The orbit is surrounded by lacrimal, jugal and postorbital. The cheek region is characterized by a squamosal, which carries the jugal canal from the preopercular canal to the infraorbital canal. The mandibular canal passes through the infradentaries (from behind: supraangular, angular, postsplenial, splenial); the dentary carries teeth but no sensory canal.

True prismatic enamel covers the dentine of the whole tooth. A thin layer of true prismatic enamel covers bones and scales pierced by pores of the pore-canal system. Enamel, dentine and the pore-canal system form the cosmine, which is found in Devonian sarcopterygians with the exception of coelacanths and the earliest forms. The function of the pore canal system is not fully explained, it is interpreted as a complex cutaneous system of vessels by Bemis and Northcutt (1992). Round scales of coelacanths, lungfish (with few exceptions) and other piscine sarcopterygian lack cosmine and enamel. Recent and most fossil lungfish possess tooth plates; small teeth occur in some early lungfish (*Uranolophus, Griphognathus* and other) and larger in the earliest lungfish *Diabolepis* (Smith and Chang 1990).

The dentine of some sarcopterygian teeth (not in teeth of Actinistia and Dipnoi) is folded as plicidentine. One can distinguish different kinds of plicidentine. Labyrinthodont plicidentine occurs in Elpistostegalia (*Panderichthys*) and early tetrapods.

In the exoskeletal shoulder girdle, a bone, the anocleithrum, is intercalated between and overlapped by supracleithrum and cleithrum. The endoskeletal shoulder girdle, the triradiate scapulocoracoid, is attached on three points to the inside of the exoskeletal shoulder girdle. An axial internal fin skeleton attaches to the articular surface of pectoral and pelvic girdle; preaxial and postaxial radials branch off, on which the fin rays sit, from the axial elements. These axial structures are homologues of the extremities of tetrapods. Sarcopterygians have primitively two dorsal fins and a heterocercal caudal fin like all crown gnathostomes. A symmetrical caudal fin of different structure appears independently in the different groups of sarcopterygians. Extant and most fossil coelacanths possess two dorsal fins and a rounded symmetrical caudal fin with an accessory middle process, whereas extant lungfish do not show any distinct dorsal or anal fin, and the caudal fin appears to form a continuous fin fold, which is interrupted at the tip of tail (Arratia et al. 2001).

The notochord forms a fluid filled tube; the fluid of *Latimeria* is considered a life prolonging elixir by some Japanese (Bruton and Stobbs 1991). Ossifications around the notochord appear in some Paleozoic dipnoans and some osteolepiform rhipidistians, some of the latter show structures typical for early tetrapods.

Lungs with pulmonary veins and arteries are characteristic for extant sarcopterygians (lungfish, tetrapods); the extant coelacanth *Latimeria* possesses a fatty organ connected to the digestive system with pulmonary veins and arteries, which represent a modified lung. The outside of the lung is covered by scale-like structures in fossil coelacanths, so that we have a record of lungs back in to the Carboniferous (325 million years ago). Of course the tetrapods occurring at the same time had lungs, even though branchial arches and external gills were recorded for some early tetrapods.

All early sarcopterygians, many of them from the Silurian and Early Devonian of China, have been found in marine deposits. Even the earliest tetrapods are from coastal marine deposits. It is therefore to be assumed that the lung originated in marine fishes.

Most fossil coelacanths are marine like the extant *Latimeria*, some may have entered fresh water. Although lungfish are originally marine, they have entered fresh water in the Carboniferous, probably already in the Devonian.

Actinopterygii

The actinopterygians (Fig. 1) are osteichthyans with only one dorsal fin (division of the one dorsal fin into two in some advanced teleosts). There is one Early Devonian genus, *Dialipina*, with two dorsal fins (Schultze and Cumbaa 2001), which is considered a stem group osteichthyan by other (Zhu et al. 2013). Schultze and Cumbaa (2001) placed *Dialipina* within actinopterygians based on typical actinopterygian rhombic scales with narrow peg in contrast to a broad peg in scales of sarcopterygians. The actinopterygians possess fringing fulcra; these are shingle-like scales in front of the first rays of their fins. The endoskeleton of pectoral and pelvic fins are broad based with many radials placed besides each other.

A unique character is the formation of the tooth tips by highly mineralized dentine, the acrodin; enamel covers the neck of the teeth. The enamel of actinopterygians (ganoine) is formed by the basal layer of the epidermis differently from the 'true' enamel of sarcopterygians. It forms multiple layers in contrast to the one-layered enamel of sarcopterygians. In the transition to round (elasmoid) scales the dentine layer between external ganoine and basal bone (isopedine) is lost first. The round scales do not possess ganoine, the superficial layer corresponds to the isopedine, the basal layer of the rhombic scale (it shows bone cells in early teleosts). The basal layer of elasmoid scales is formed by mineralization of connective tissue. There are two kinds of elasmoid scales. Amioid scales with radial structures in the covered part of the scales of halecomorphs. Cycloid scales with circuli in the covered part are unique to teleosts; ctenoid scales are cycloid scales with spines, they occur in different teleostean groups derived in parallel from cycloid scales.

The skull roof is covered by a pair of postparietals (the so-called parietals) and parietals (the so-called frontals; Schultze 2008); paired extrascapulae lie behind the postparietals. In contrast to sarcopterygians, the supraorbital canal runs between both nasal openings, where the posterior nasal opening lies at the anterior margin of the orbit. Like in sarcopterygians, the supraorbital canal joins the infraorbital canal and there is a commissure between both sides in the snout (see *Polypterus*). There is no jugal canal connecting the preopercular canal with the infraorbital canal as in sarcopterygians. The preopercular canal enters directly into the temporal canal. The mandibular canal passes not through the posterior infradentary (supraangular); it runs through the angular in to the dentary. There are no splenials like in sarcopterygians, and the dentary carries teeth on its margin and the anterior part of the mandibular canal; thus it is a dentalosplenial.

There is no squamosal like in sarcopterygians. A long preoperculum extends the whole depth from the jaw articulation to the skull roof in front of the opercular series. An extension of the posterior part of the maxilla enters the space between infraorbitals and preoperculum. In some forms so-called suborbitals appear above the posterior part of the maxilla between infraorbitals and preoperculum. A small dermal bone, the dermohyal, lies above the endoskeletal epihyal between the dorsal part of the preoperculum and the operculum.

The shoulder girdle is formed by posttemporal, supracleithrum, cleithrum and clavicula, one to three postcleithra lie behind the cleithrum. The heterocercal caudal fin is shortened in early Mesozoic forms (hemiheterocercal) and transformed to a homocercal tail in teleosts. These terms refer to the internal structure of the caudal fin, whereas the external shape of the caudal can be symmetrical or not, even with an extended lower lobe as in flying fish. The homocercal tail is characterized by reduction of number of centra and the transformation of structures above (urohyals, epurals) and below (hypurals) the vertebral column. Originally the notochord is not ossified, ossification occurs independently in different groups (palaeonisciforms, lepisosteiforms, amiiforms, teleosts).

Earliest actinopterygians are marine, some Devonian and Carboniferous forms are interpreted as freshwater forms. Lung and swimbladder occur in actinopterygians.

Distribution of Lungs and Swimbladder

'Lungs' in Placodermi

Denison (1941, 1978) interpreted 'lungs' as sediment infillings, which reach from the foremost part of the ventral side of the gill region backwards below the gill region in the Devonian antiarch placoderm *Bothriolepis canadensis*. Myers (1942) and Stensiö (1948) opposed the interpretation because of the position of the structures. McAllister (1987, 1996) left these structures out in his representation of the internal structures of *Bothriolepis* and showed only the intestine with its spiral valve. Arsenault et al. (2004) interpreted the structures as 'storage of clean water' or 'pharyngeal diverticles' (see also Janvier et al. 2007); the latter interpretation seems unlikely because of the position of the structures starting in the front of the gill region and placed below the gill arches. Despite clarifications and suggestions of other interpretations, the presence of 'lungs' in *Bothriolepis* received wide acceptance (see Goujet 2011).

The 'lung' interpretation was often connected with a paleoenvironmental interpretation of the sediments as fresh water, where *Bothriolepis canadensis* occurs. In contrast Myers (1942) argued against the existence of 'lungs' on the base that *Bothriolepis canadensis* occurs in a supposed freshwater environment. In the extreme, *Bothriolepis canadensis* was interpreted as occasional terrestrial (Wells and Dorr 1985). In contrast it has been shown clearly that the *Bothriolepis*-bearing sediments at Miguasha, Quebec, Canada, are coastal marine to estuarine deposits (Schultze 1996; Cloutier et al. 1996, 2011).

Goujet (2011) has compiled all the arguments, which oppose the existence of a 'lung' in *Bothriolepis canadensis*:

1. The structure in question is a posterior paired expansion of the oralobranchial chamber, which reaches posteriad ventral of the gill arches. It has no connection to the esophagus and can therefore not be interpreted as a 'lung'.
2. *Bothriolepis canadensis* occurs in a marginal marine environment, and there is no indication of fresh water or that the fish lived in a seasonal arid environment.
3. *Bothriolepis* occurred worldwide in the Late Devonian and in very different environments so that a coastal-marine dispersion is more plausible for *Bothriolepis* than a freshwater environment.

Janvier et al. (2007, p. 710) mentioned the possibility of a "merely fortuitous accumulation of mud that have entered the armour."

Lung and Swimbladder in Osteichthyans

Lung and swimbladder occur in extant fishes in osteichthyans only, lungs in sarcopterygians (Actinistia, Dipnoi, Tetrapoda) and in basal actinopterygians (polypterids) and a swimbladder in teleostean fishes. Chondrichthyans possess no comparable structure. Lung or swimbladder are not preserved in fossil fishes with the exception of actinistians (Brito et al. 2010), where the lung is encased in scale-like structures and thus preserved in actinistians since the Carboniferous; not yet found in Devonian actinistians. Thus we have to rely on extant forms to reconstruct basically the ancestral occurrence of lung or swimbladder from the extant distribution (Fig. 2). The sarcopterygians possess a lung with a ventral connection to the esophagus/anterior region of the stomach; its blood supply derives from the IVth gill arch artery. We find a comparable situation in the most primitive extant actinopterygians,

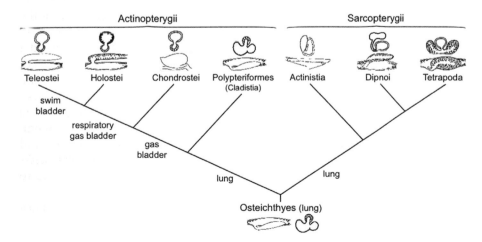

Fig. 2. Distribution of lung and swimbladder in osteichthyans.

the Polypteriformes. The distribution of these structures indicates a lung as the ancestral structure in osteichthyans (Fig. 2). This is supported by gene comparison (Cass et al. 2013). It was the accepted idea already in the second half of the 19th century after Sagemehl (1885). Wilder (1877: Fig. 7) showed a sequence of cross and longitudinal sections from lungs to swimbladder, which was repeatedly refigured, first by Dean (1895), to whom most authors refer the presentation. Liem (1988), Graham (1997), Perry et al. (2001), Longo et al. (2013) and Cass et al. (2013) presented cladograms, where the lung is the ancestral structure in osteichthyans; Liem (1988) and Graham (1997) even argued that it is a primitive structure for gnathostomes and lost in chondrichthyans based on the wrong assumption that placoderms have lungs.

Campbell and Barwick (1988) argued on functional grounds (existence of a cranial rib) and occurrence in supposedly freshwater deposits that a lung in dipnoans developed independently from that of tetrapods, because early lungfish are marine. This argument is not supported by a phylogenetic analysis of all osteichthyans and by more recent discoveries, that early choanate fishes and especially early tetrapods (Schultze 1997, 2013; Niedzwiedzki et al. 2010; George and Blieck 2011) are marine. One has to accept that lungs developed in marine environment (Packard 1974). Earliest tetrapods possess still ossified gill arches and slits, so that one has to accept, that branchial oxygen uptake occurred in early tetrapods even though they had lungs (Laurin 2010; Clack 2012). The lung is a ventral paired structure in tetrapods and lepidosirenids within lungfish like in polypteriforms, whereas *Neoceratodus* shows a derived situation with an unpaired dorsal lung and a connecting duct from the ventral esophagus running on the right side of the stomach.

The extant most primitive actinopterygians, the Polypteriformes, possess paired lungs ventral to the alimentary duct, the left is reduced (Lechleuther et al. 1989). A gas bladder, an organ dorsal to the alimentary duct, appears within actinopterygians (Fig. 2). The Chondrostei possess a gas bladder with a connection to the esophagus in its posterior part (Rauther 1937; Marinelli and Strenger 1973), otherwise there are different blood supplies in different chondrostean genera. After Longo et al. (2013), the coronary arteries in *Polyodon* and the esophageal arteries in *Acipenser* are homologous to the pulmonary arteries in *Protopterus*, *Polypterus* and *Amia*, based on their branching pattern, which corresponds to the position of Chondrostei between Polypteriformes and Holostei in the cladogram (Fig. 2). *Acipenser*, but not *Polyodon*, possess a celiacomesenteric artery as teleosts. This has to be interpreted as a convergent acquirement (Longo et al. 2013).

That means that the non-respiratory gas bladder of chondrosteans is convergent to that of teleosteans.

A respiratory (or pulmonoid) gas bladder occurs in holosteans (*Amia*, *Lepisosteus*), elopomorphs, osteoglossomorphs and some primitive euteleosts.

The transition from a pulmonary gas bladder to a swimbladder with hydrostatic, hearing or sound production function was a continuous process (Liem 1989; Graham 1997).

References

Arratia, G., H.-P. Schultze and J. Casciotta. 2001. Vertebral column and associated elements in dipnoans and comparison with other fishes: development and homology. J. Morphol. 250: 101–172.

Arsenault, M., S. Desbiens, P. Janvier and J. Kerr. 2004. New data on the soft tissues and external morphology of the antiarch *Bothriolepis canadensis* (Whiteaves, 1880), from the Upper Devonian of Miguasha, Quebec. pp. 439–454. *In*: G. Arratia, M.V.H. Wilson and R. Cloutier (eds.). Recent Advances in the Origin and Early Radiation of Vertebrates. Dr. F. Pfeil, München.

Basden, A.M. and G.C. Young. 2001. A primitive actinopterygian neurocranium from the Early Devonian of southeastern Australia. J. Vert. Paleontol. 21: 754–766.

Bemis, W.E. and G. Northcutt. 1992. Skin and blood vessels of the snout of the Australian lungfish, *Neoceratodus forsteri*, and their significance for interpreting the cosmine of Devonian lungfishes. Acta. Zool. Stockhol 73: 115–139.

Brazeau, M. 2009. The braincase and the jaws of a Devonian 'acanthodian' and modern gnathostome origins. Nature 457: 305–308.

Brito, P., F.J. Meunier, G. Clément and D. Geffard-Kuriyama. 2010. The histological structure of the calcified lung of the fossil coelacanth *Axelrodichthys araripensis* (Actinistia: Mawsonidae). Palaeontology 53(6): 1281–1290.

Bruton, M.N. and R.E. Stobbs. 1991. The ecology and conservation of the coelacanth *Latimeria chalumnae*. Environ. Biol. Fish. 32: 313–339.

Burrow, C.J. 2003. Comment on "Separate evolutionary origins of teeth from evidence in fossil jawed vertebrates." Science 300: 1661.

Campbell, K.S.W. and R.E. Barwick. 1988. Geological and paleontological information and phylogenetic hypotheses. Geol. Mag. 125: 207–277.

Clack, J.A. 2012. Gaining Ground. The Origin and Evolution of Tetrapods. Second Edition. XVI + 523 pp. Indiana University Press, Bloomington, Indiana.

Cloutier, R., S. Loboziak, A.-M. Candilier and A. Blieck. 1996. Biostratigraphy of the Upper Devonian Escuminac Formation, eastern Québec, Canada: a comparative study based on miospores and fishes. Rev. Palaeobot. Palynol. 93: 191–215.

Cloutier, R., J.-N. Proust and B. Tessier. 2011. The Miguasha fossil-fish-Lagerstätte: a consequence of the Devonian land-sea interactions. Palaeobiodiversity Palaeoenvironments 91: 293–323.

Davis, S.P., J.A. Finarelli and M.I. Coates. 2012. *Acanthodes* and shark-like conditions in the last common ancestor of modern gnathostomes. Nature 486: 247–250.

Dean, B. 1895. Fishes, Living and Fossil, an Outline of their Forms and Probable Relationships. Columbia Univ. Ser. 3: 300 pp. Macmillan & Co., New York, London.

Dean, B. 1907. Notes on acanthodian sharks. Amer. J. Anat. 7: 209–222.

Denison, R.H. 1941. The soft anatomy of *Bothriolepis*. J. Paleontol. 15(5): 553–561.

Denison, R.H. 1978. Placodermi. VI + 128 pp. *In*: H.-P. Schultze (ed.). Handbook of Paleoichthyology, Vol. 2. Gustav Fischer, Stuttgart, New York.

Denison, R.H. 1979. Acanthodii. VI + 62 pp. *In*: H.-P. Schultze (ed.). Handbook of Paleoichthyology, Vol. 5. Gustav Fischer, Stuttgart, New York.

Didier, D.A., B.J. Stahl and R. Zangerl. 1994. Development and growth of compound tooth plates in *Callorhinchus milii* (Chondrichthyes, Holocephali). J. Morphol. 222: 73–89.

George, D. and A. Blieck. 2011. Rise of earliest tetrapods: An early Devonian origin from marine environment. Plos One 6(7): e22136. doi: 10.1371/journal.pone.0022136: 7 p.

Goujet, D. 1984. Les poissons Placodermes du Spitsberg. Arthrodires Dolichothoraci de la Formation de Wood Bay (Dévonien inférieur). Cahiers Paléontol., Section Vertébrés. 284 p. CNRS, Paris.

Goujet, D. 2011. "Lungs" in Placoderms, a persistent palaeobiological myth related to environmental preconceived interpretations. Comptes Rendus Palevol. 10: 323–329.

Goujet, D. and G.C. Young. 2004. Placoderm anatomy and phylogeny: new insights. pp. 109–126. *In*: G. Arratia, M.V.H. Wilson and R. Cloutier (eds.). Recent Advances in the Origin and Early Radiation of Vertebrates. Dr. F. Pfeil, München.

Graham, J.B. 1997. Air-Breathing Fishes—Evolution, Diversity and Adaptation. Academic Press, London, 299 pp.

Gross, W. 1962. Neuuntersuchung der Stensiöellida (Arthrodira, Unterdevon). Notizbl. Hess. L.-A. Bodenforschung Wiesbaden 90: 48–86.

Hasse, C. 1885. Das natürliche System der Elasmobranchier auf Grundlage des Baues und der Entwicklung ihrer Wirbelsäule. 3 Vols.: Allgemeiner Theil. 1879. VI + 76 pp., Besonderer Theil. 1882. VI + 284 pp., Supplement. 1885. 27 pp. Gustav Fischer, Jena.

Heimberg, A.M., R. Cowper-Sallari, M. Sémon, P.C.J. Donoghue and K.J. Peterson. 2010. microRNAs reveal the interrelationships of hagfish, lampreys, and gnathostomes and the nature of the ancestral vertebrate. Proc. Natl. Acad. Sci. USA 107: 19379–19383.

Hennig, W. 1983. Stammesgeschichte der Chordaten. Fortschr. Zool. Syst. Evolut.-forsch., Heft 2. 208 p. Paul Parey Verlag, Berlin.

Janvier, P. 1996. Early Vertebrates. Oxford Monographs on Geology and Geophysics 33. XII + 393 pp. Clarendon Press, Oxford.

Janvier, P., S. Desbien and J.A. Willett. 2007. New evidence for the controversial "lungs" of the Late Devonian antiarch *Bothriolepis canadensis* (Whiteaves, 1880) (Placodermi: Antiarcha). J. Vert. Paleontol. 27(3): 709–710.

Jollie, M. 1962. Chordate Morphology. XIV + 478 pp. Reinhold Publ. Corp., New York.

Khonsari, R.H., B. Li, P. Vernier, R.G. Northcutt and P. Janvier. 2009. The anatomy of the agnathan brain and craniate phylogeny. Acta Zoologica (Stockholm) 90 (Suppl. 1): 52–68.

Laurin, M. 2010. How Vertebrates Left the Water. XV +199 pp. University of California Press, Berkeley, Los Angeles, London.

Lechleuthner, A., U. Schumacher, R.D. Negele and U. Welsch. 1989. Lungs of *Polypterus* and *Erpetoichthys*. J. Morphol. 201: 161–178.

Liem, K. 1988. Form and function of lungs: the evolution of air breathing. Amer. Zoologist 28: 739–759.

Liem, K.F. 1989. Gas bladders in teleosts: functional conservatism and morphological diversity. Amer. Zoologist 29: 333–352.

Longo, S., M. Riccio and A.R. McCune. 2013. Homology of lungs and gas bladders: Insights from arterial vasculature. J. Morphol. 274: 687–703.

Maisey, J.G. 2001. A primitive chondrichthyan braincase from the Middle Devonian of Bolivia. pp. 263–288. *In*: P.E. Ahlberg (ed.). Major Events in Early Vertebrate Evolution: Palaeontology, Phylogeny, Genetics and Development. Syst. Assoc. Spec. Vol. Ser. 61. Taylor & Francis, London, New York.

Maisey, J.G. 2007. The braincase in Paleozoic symmoriiform and cladoselachian sharks. Bull. Amer. Mus. Natur. Hist. 307: 1–122.

Maisey, J.G. and M.E. Anderson. 2001. A primitive chondrichthyan braincase from the Early Devonian of South Africa. J. Vert. Paleontol. 21: 702–713.

Marinelli, W. and A. Strenger. 1973. Vergleichende Anatomie und Morphologie der Wirbeltiere. IV. *Acipenser* ruthenus (L.). 460 pp. Deuticke, Wien.

McAllister, J.A. 1987. Phylogenetic distribution and reassessment of the intestines of fossil and modern fishes. Zool. Jb. Anat. 115: 281–294.

McAllister, J.A. 1996. Coprolitic remains from the Devonian Escuminac Formation. pp. 328–347. *In*: H.-P. Schultze and R. Cloutier (eds.). Devonian Fishes and Plants of Miguasha, Quebec, Canada. Dr. F. Pfeil, München.

Miles, R. 1973. Relationships of acanthodians. pp. 63–103. *In*: P.H. Greenwood, R.S. Miles and C. Patterson (eds.). Interrelationships of Fishes. Zool. J. Linn. Soc., 53, Suppl. 1. Academic Press, London.

Myers, G.S. 1942. The "lungs" of *Bothriolepis*. Stanford Ichthyol Bull. 2: 134–136.

Nelson, G. 1968. Gill-arch structure in *Acanthodes*. pp. 128–143. *In*: T. Ørvig (ed.). Nobel Symposium 4. Current Problems of Lower Vertebrate Phylogeny. Almqvist & Wiksell, Stockholm.

Niedzwiedzki, G., P. Szrek, K. Narkiewicz, M. Narkiewicz and P.E. Ahlberg. 2010. Tetrapod trackways from the early Middle Devonian period of Poland. Nature 463: 43–44.

Oisi, Y., K.G. Ota, S. Kuraku, S. Fujimoto and S. Kuratani. 2013. Craniofacial development of hagfishes and the evolution of vertebrates. Nature 493: 175–180.

Packard, G.C. 1974. The evolution of air-breathing in Paleozoic gnathostome fishes. Evolution 28: 320–325.

Peignoux-Deville, J., F. Lallier and B. Vidal. 1982. Evidence for the presence of osseous tissue in dogfish vertebrae. Cell Tissue Res. 222: 605–614.

Perry, S.F., R.J. Wilson, C. Straus, M.B. Harris and J.E. Remmers. 2001. Which came first, the lung or the breath? Comp. Biochem. Physiol. A 129: 37–47.

Rauther, M. 1937. VII. Der Intestinaltraktus (Fortsetzung). In Bronn's Klassen des Tierreiches. Echte Fische Teil 1 Anatomie, Physiologie und Entwicklungsgeschichte. Erste Hälfte. 679–910; Akademische Verlagsges. Leipzig.

Reis, O.M. 1890. Zur Kenntnis des Skelets der Acanthodinen. Geogn. Jh. 3: 1–43.

Reis, O.M. 1896. Ueber *Acanthodes bronni* Agassiz. Morph. Arb. 6: 143–220.

Rücklin, M., P.C.J. Donoghue, Z. Johanson, K. Trinajstic, F. Marone and M. Stampanoni. 2012. Development of teeth and jaws in the earliest jawed vertebrates. Nature 491: 748–751.

Sagemehl, M. 1885. Beiträge zur vergleichenden Anatomie der Fische III. Das Cranium der Characiden nebst allgemeinen Bemerkungen über die mit einem Weber'schen Apparat versehenen Physostomenfamilien. Morphol. Jb. 10: 1–119.

Schultze, H.-P. 1990. A new acanthodian from the Pennsylvanian of Utah, U.S.A., and the distribution of otoliths in gnathostomes. J. Vert. Paleontol. 10: 49–58.

Schultze, H.-P. 1996. Terrestrial biota in coastal marine deposits: Fossil-Lagerstätten in the Pennsylvanian of Kansas, USA. Palaeogeogr., Palaeoclimatol., Palaeoecol. 119: 255–273.

Schultze, H.-P. 1997. Umweltbedingungen beim Übergang von Fisch zu Tetrapode.—Sitzber. Ges. Naturforsch. Freunde Berlin (N.F.) 36: 59–77.

Schultze, H.-P. 2008. Nomenclature and homologization of cranial bones in actinopterygians. pp. 23–48. *In*: G. Arratia, H.-P. Schultze and M.V.H. Wilson (eds.). Mesozoic Fishes 4—Homology and Phylogeny. Dr. Friedrich Pfeil, München.

Schultze, H.-P. 2013. The paleoenvironment at the transition from piscine to tetrapod sarcopterygians. *In*: S.G. Lucas, W.A. DiMichele, J.E. Barrick, J.W. Schneider and J.A. Spielmann (eds.). The Carboniferous-Permian Transition. New Mexico Mus. Natur. Hist. Sci., Bull. 60: 373–397. April 2013.

Schultze, H.-P. and S.L. Cumbaa. 2001. *Dialipina* and the character of basal actinopterygians. pp. 315–332. *In*: P.E. Ahlberg (ed.). Major Events in Early Vertebrate Evolution. Palaeontology, Phylogeny and Development. Syst. Assoc. Spec. Vol. Ser. 61. Taylor & Francis, New York.

Shu, D. 2003. A paleontological perspective of vertebrate origins. Chinese Science Bull. 48: 725–735.

Shu, D., H.-L. Luo, S. Conway Morris, X.-L. Zhang, S.-X. Chen, J. Han, M. Zhu, Y. Li and L.-Z. Chen. 1999. Lower Cambrian vertebrates from South China. Nature 402: 42–46.

Smith, M.M. and M.-M. Chang. 1990. The dentition of *Diabolepis speratus* Chang and Yu, with further consideration of its relationships and the primitive dipnoan dentition. J. Vert. Paleontol. 10: 420–433.

Smith, M.M. and Z. Johanson. 2003. Separate evolutionary origins of teeth from evidence in fossil jawed vertebrates. Science 299: 1235–1236.

Stahl, B. 1999. Chondrichthyes III. Holocephali. 164 pp. *In*: H.-P. Schultze (ed.). Handbook of Paleoichthyology, Vol. 4. Dr. Friedrich Pfeil, München.

Stensiö, E.A:son. 1948. On the Placodermi of the Upper Devonian of East Greenland. II. Antiarchi: Subfamily Bothriolepinae. Medd. Grønland 139 (Palaeozool. Groenlandica 2): 622 pp.

Stensiö, E.A:son. 1963. Anatomical studies on the arthrodiran head. Part 1. Preface, geological and geographical distribution, the organization of the head in the Dolichothoraci, Coccosteomorphi and Pachyosteomorphi. Taxonomic appendix. Kgl. Sv. VetskapsAkad. Handl. 9: 1–419.

Turner, S., A. Blieck and G.S. Nowlan. 2004. Vertebrates (agnathans and gnathostomes). pp. 327–335. *In*: B.D. Webby, F. Paris, M.L. Droser and I.G. Percival (eds.). The Great Ordovician Biodiversification Event (IGCP 410 volume). Columbia University, New York.

Wells, N.A. and J.A. Dorr, Jr. 1985. Form and function of the fish *Bothriolepis* (Devonian; Placodermi, Antiarchi): the first terrestrial vertebrate? Michigan Academician 17: 157–170.

Wilder, B.G. 1877. Gar-pikes, old and young. Popular Science Monthly 11: 1–12.

Young, G.C. 1980. A new Early Devonian placoderm from the New South Wales, Australia, with a discussion of placoderm phylogeny. Palaeontographica A 167: 10–76.

Young, G.C. 1984. Reconstruction of the jaws and braincase in the Devonian placoderm fish *Bothriolepis*. Palaeontology 27(3): 635–661.

Zhu, M., X. Yu and P.E. Ahlberg. 2001. A primitive sarcopterygian fish with an eyestalk. Nature 410: 81–84.

Zhu, M., X. Yu, P.E. Ahlberg, B. Choo, J. Lu, T. Qiao, Q.-M. Qu, W.-J. Zhao, L.-T. Jia, H. Blom and Y. Zhu. 2013. A Silurian placoderm with osteichthyan-like marginal jaw bones. Nature 502: 188–193.

2

The Occurrence and Function of the NOS/NO System in the Heart of the Eel and the African Lungfish

Daniela Amelio,[a] Filippo Garofalo,[b] Sandra Imbrogno[c] and Bruno Tota*

Introduction

Evidence obtained in the last 15 years has shown the presence and function of the Nitric Oxide Synthase (NOS)/Nitric Oxide (NO) system in the fish heart, highlighting its importance for the cardio-circulatory homeostasis of these animals. This knowledge appears of relevant evolutionary significance since it helps to elucidate basic principles of unity and diversity in vertebrate cardiac biology and adaptations. Moreover, the fish heart is the ancestor of the mammalian heart and, therefore, its morpho-functional design can be well suited to reveal novel aspects of the pleiotropic role exerted by the NO signaling in the present-day mammalian heart.

The fish heart is a venous heart since, depending on its angio-structural patterns, is supplied only, or mostly, by venous blood (see below; Tota et al. 1983; Farrell and Jones 1992). In comparison with higher and warm-blooded

University of Calabria, Dept. of Cell Biology, Arcavacata di Rende, 87036 (Cosenza), Italy.
[a] E-mail: daniela.amelio@unical.it
[b] E-mail: filigaro@libero.it
[c] E-mail: sandra.imbrogno@unical.it
* Corresponding author: tota@unical.it

vertebrates, it is designed as a low pressure pump exposed to relatively low and variable pO_2 levels, being endowed with stretch sensors for changes in pressure and volume gradients of the venous return (Farrell and Jones 1992; Olson 1998). In most fish hearts, the myocardial and the Endocardial Endothelial (EE) cells appear particularly rich in specific secretory granules that constitute the final step in the regulated secretory pathway for a variety of cardiac hormones including endothelin and Atrial Natriuretic Peptides (ANPs) (Aardal and Helle 1991; Olson 1998; Tota et al. 2010). Accordingly, the concept of the heart as an endocrine organ can be particularly evidenced in the fish heart whose endocrine-paracrine regulation appears more relevant than in the mammalian heart in which the neural control is more developed. Therefore, the fish heart can provide a conceptual tool to explore various aspects of the NO-mediated signal-transduction pathways and integration, such as the cross-talk between different cell types, e.g., EE and myocardial cells (see below). Furthermore, the fish heart differs from the endotherm heart in myoarchitecture (trabecular or *spongiosa* vs. *compacta* type), intracardiac blood supply (lacunary vs. vascular) (Tota et al. 1983; Icardo et al. 2005), myocardial ultrastructure (lack of T-tubes) and pumping performance (e.g., sensitivity to filling pressure) (see for references Farrell and Jones 1992; Tibbits et al. 1992). Due to these features, studies addressed on spatial (e.g., compensation for ventricular heterogeneity) and temporal (e.g., developmental cardiac remodeling) paradigms of NO in the fish heart may be of particular interest. Some examples will be illustrated below. We will use the heart of an ancient teleost, the eel (*Anguilla anguilla*), to discuss different aspects of NO-mediated modulation, and the heart of the African lungfish *Protopterus* to illustrate the NOS implication in the stress response related to the aestivation.

NOS Phylogeny and Isoform Localization in the Fish Heart

The NOS/NO system is one of the oldest bioregulatory pathways (Feelisch and Martin 1995; Palumbo 2005). However, the lack of a direct correspondence between protein configuration and function has made it difficult to elucidate the basic events of NOS genes evolution and the structural/functional specialization of the different NOS isoforms.

Andreakis et al. (2011) used genomic databases and NOS protein analysis from 33 invertebrate and 63 vertebrate species to highlight a possible scenario of NOS evolution in metazoans, particularly in vertebrates (see Fig. 1). The recognition of a canonical neuronal-like identity across all taxa allowed to push its evolution back to the early metazoan history. Conceivably, NOS duplication of this early neuronal-like configuration in different animal phylogenies may have generated distinct enzyme specializations during evolution. NOSI, NOSII and NOSIII appear unambiguously grouped into three monophyletic clusters. Since there is no evidence for a NOSIII gene in agnathans, chondrichthyans and teleosts; this strongly suggests that the NOSIII isoform appeared during

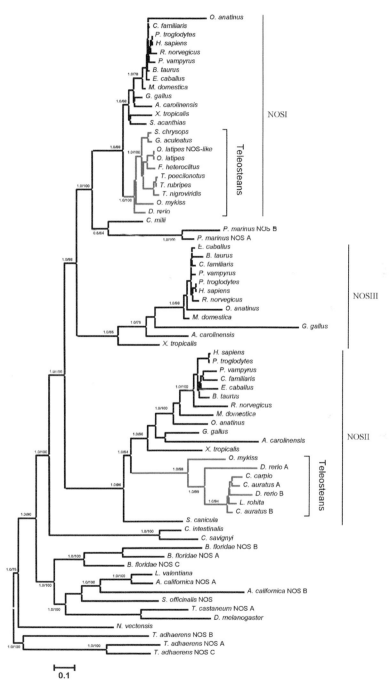

Fig. 1. Genealogical relationships among NOS protein sequences encountered in selected vertebrate and invertebrate taxa. From Andreakis et al. (2011) [The authors thank Andreakis et al. (2011) for kind permission to use their data].

tetrapod evolution, probably resulting from independent gene duplication without involving large chromosomal regions. Likewise, NOSI and NOSII, generated from a large-scale duplication event, diverged before the teleost-tetrapod split. It is important to recognize that the apparent absence of a canonical endothelial NOS (eNOS or NOSIII) in fish suggests that the efforts made to identify the different NOS isoforms in fish solely on the basis of physio-pharmacological and immuno-histochemical criteria must be taken with caution. On the other hand, the evidence obtained with both NADPH-diaphorase, immunolocalization with heterologous mammalian antibodies and physio-pharmacological analysis with NOS inhibitors indicates that eNOS-like activity is present in various fish tissues (for references see Andreakis et al. 2011), including the heart (Tota et al. 2005; Amelio et al. 2008; Garofalo et al. 2009a,b), thus prompting further studies to resolve this contrasting issue. Indeed, as emphasized by Andreakis et al. (2011), a neuronal NOS (nNOS or NOSI) isoform with an endothelial-like consensus, e.g., myristoylation, in fish could cover some functional traits of the eNOS isoform identity. Since a myristoylation consensus sequence has been identified in the nNOS from the lamprey *Petromizon marinus* and in NOSIIb from the zebrafish, the possibility exists that this isoform could function in a manner similar to the mammalian NOSIII, which is preferentially expressed in the adult heart (Lepiller et al. 2009).

Using species with distinct phylogenies and ecophysiological habitats, Tota et al. (2005) and Amelio et al. (2006, 2008) evaluated NOS localization and distribution in the fish heart (Fig. 2). Their survey included extreme stenotherms, such as the endemic inhabitants of the Antarctic waters, i.e., the hemoglobinless *Chionodraco hamatus*, the red-blooded *Trematomus bernacchii* and the naturally occurring Hb^-/Mb^- knockout *Chaenocephalus aceratus*. Conceivably, the stably icy and richly oxygenated Antarctic habitat allowed survival of these animals despite their evolutionary loss of Hb and red blood cells (Garofalo et al. 2009a). The documented widespread expression of a constitutive eNOS-like isoform also in the EE of avascular or poorly vascularized hearts (e.g., lungfish and eel) suggests the ubiquitous role of this tissue as a relevant paracrine source of bioactive NO. In all species examined, eNOS-like isoform was prevalently immunolocalized at the level of the EE lining the *trabeculae* and, to a lesser extent, on the myocardiocytes of both the atrium (Fig. 2A,E) and the ventricle (Fig. 2B,C,F,G,H). The eNOS-like isoform was mainly associated with the plasmalemma, being less evident within the cytoplasm. It has been proposed that in the fish heart the very large EE intracavitary surface remarkably contributes to eNOS-derived NO production, allowing an adequate NO gradient for the deep trabecular myocytes (Katz 1988; Pinsky et al. 1997; Petroff et al. 2001; Elrod et al. 2008; and references therein). In particular, in *Protopterus* heart (Fig. 2A,B,C,E,F,G), a high eNOS-like expression has been observed at the endocardial level. Thus, it is conceivable that the elevated intracavitary NO production might paracrinally contribute to myocardial protection against the ischemic-like conditions induced by the large pO_2 fluctuations experienced during aestivation, when Heart Rate (HR)

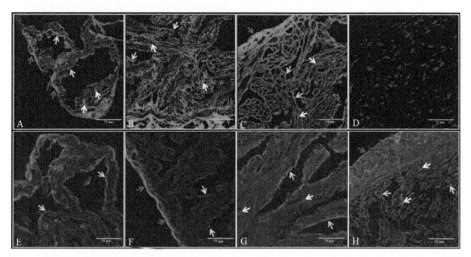

Fig. 2. Immunolocalization of eNOS-like in atrium (A, E) and ventricle (B, C, F, G, H) of FW *P. dolloi* (A–C), FW *P. annectens* (E, F, G) and *A. anguilla* (H). eNOS is localized in epicardium (blue arrows), endocardial (yellow arrows) and vascular (red arrows) endothelium, and in the cytoplasm (white arrows) of the myocardiocytes. Negative control is shown in D. Nuclei are counterstained with propidium iodide (A–D) and Hoechst (E–G). Modified from Amelio et al. 2008, 2013 and Imbrogno et al. 2013.

and intracardiac blood flow are also lowered (see below). Similarly, the relevant detection of eNOS-like and inducible NOS (iNOS or NOSII) in the ventricular myocardiocytes of both the *spongiosa* (lungfish, eel and tuna) and the *compacta* (eel and tuna) can also be consistent with an autocrine NO role. Moreover, as in mammals, the eNOS-like isoform expressed in the visceral pericardium (Fig. 2C,F,H) of all fish examined (Amelio et al. 2006, 2008, 2013a) may contribute to the autocrine-paracrine activities (e.g., ANPs in *T. bernacchii* and *C. hamatus*: Cerra et al. 1997) attributed to this tissue.

Table 1 summarizes NOS compartmentalization in the heart of various fish species, providing a morphological counterpart to the functional aspects outlined below.

NO-mediated Regulation of Cardiac Function in Fish

NO signaling is implicated both in the intrinsic and extrinsic modulation of cardiac performance.

The Cardiac Output (CO), i.e., the product of HR and Stroke Volume (SV), is regulated through both intrinsic and extrinsic (neuro-humoral) mechanisms. As well as in other vertebrates, also in fish hemodynamic loads (filling pressure or preload, and systemic aortic pressure or afterload) are primary determinants of CO. The neuro-humoral control is mainly attained through the adrenergic and cholinergic innervations, as well as a variety of humoral agents, including

Table 1. eNOS and iNOS immunolocalization pattern in the ventricle of various fish species.

	eNOS				iNOS			
	EP	EE	VE	Myo	EP	EE	VE	Myo
C. aceratus	++	++		+	−	−	−	++
C. hamatus	++	++		+	−	−	−	++
T. bernacchii	++	++	++	+	−	−	−	++
A. anguilla	++	++	++	+	−	−	−	++
P. dolloi	++	++	++	+	?	?	?	?
P. annectens	++	++	++	+	?	?	?	?
T. thynnus thynnus	++	++	++	+	?	?	?	?
C. auratus	++	++	++	+	?	?	?	?

EP = Epicardium; EE = Endocardial Endothelium; VE = Vascular Endothelium;
Myo = Myocardiocytes

angiotensin II (AngII), cardiac natriuretic peptides and catecholamines (CAs) released by both extracardiac and cardiac chromaffin cells (Randall and Perry 1992; Farrell and Jones 1992; Nilsson and Holmgren 1992; Tota et al. 2010).

Intrinsic (Frank-Starling Mechanism) Regulation

The preload-induced increases in developed force (i.e., the Frank-Starling response) play a key role in the regulation of cardiac function in fish, as documented in various species, including the eel (Tota et al. 1991; Farrell and Jones 1992; Dunmall and Schreer 2003; Icardo et al. 2005; Imbrogno et al. 2011; Amelio et al. 2013a, 2013b). This has been partly attributed to the remarkable myocardial extensibility of the highly trabeculated heart (Shiels and White 2008). The increased HR has also been indicated as a major determinant of the enhanced CO during swimming (see for example Axelsson et al. 1994; Altimiras and Larsen 2000; Cooke et al. 2003), suggesting that the relative balance of HR vs. SV for adjusting CO in fish depends on species-related differences as well as on the kind of the experimental design employed.

A relevant autocrine NO role has emerged in both mammalian and non-mammalian vertebrates in relation to the stretch-induced increase of contractility. In mammals, autocrine nNOS promotes left ventricular relaxation by regulating the rate of Ca^{2+} reuptake by SR Ca^{2+} ATPase (SERCA2a) (Zhang et al. 2008). On the other hand, myocardial eNOS mediates the sustained stretch-induced inotropism through a mechanism involving S-nitrosylation of thiol residues in the RYR Ca^{2+} release channels (Massion et al. 2005). In the *A. anguilla* heart, the high sensitivity to preload increases is significantly enhanced by a basal release of endogenous NO (Imbrogno et al. 2001), generated through a protein kinase B (Akt)-mediated activation of eNOS-like isoform. In particular, Garofalo et al. (2009b) demonstrated that this autocrine NO-mediated regulation of the Frank–Starling mechanism in the eel occurs through a non-classical cGMP-independent pathway, which involves a beat-to-beat regulation of phospholamban (PLN) S-nitrosylation-dependent calcium reuptake by SERCA2a and thus of myocardial relaxation (Garofalo et al. 2009a; Cerra and Imbrogno 2012) (Fig. 3). Further research is needed to evaluate the species-specific compartmentation of NOS isoforms and their fine temporal activation in the stretch-induced cardiac response in fish. However, the eel study of Garofalo et al. (2009b) indicates that the NO ability of fine-tuning myocardial function through NOS isoforms compartmentation, differences in their mode of stimulation and recruitment of distinct downstream pathways (S-nitrosylation or cGMP production) within various micro-domains of the same cell represent a fundamental and early event during the evolution of the vertebrate heart.

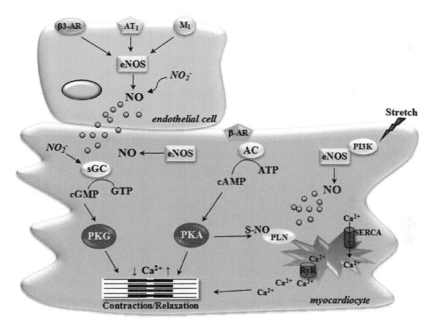

Fig. 3. Schematic diagram showing the role of NO as a spatial paracrine/autocrine integrator. Chemical stimuli such as ACh, Ang II, and β3-AR agonists converge on the eNOS-like-NO signaling which requires the obligatory involvement of the EE (Imbrogno et al. 2001, 2003, 2006). Stretch conditions activate the release of autocrine NO which directly modulates the SR Ca²⁺ reuptake through PLN S-nitrosylation. Modified from Imbrogno et al. 2011.

Extrinsic Regulation

Innervation

Being aneural in the hagfish or only innervated by cholinergic fibers in lampreys and elasmobranchs, the heart of other fish groups, including teleosts, possesses both a sympathetic and parasympathetic innervation through a 'vagosympathetic' trunk (Laurent et al. 1983; Nilsson 1983; Taylor 1992). The importance of the adrenergic and cholinergic innervation significantly varies according to species-specific traits and circumstances. The innervation patterns are also complicated by the variable contribution of the NANC (non-adrenergic non-cholinergic) nerve supply. This is represented by autonomic fibers that contain neurotransmitters such as substance P, serotonin, neuropeptide Y, vasointestinal peptide, etc., and 'unconventional' neurotransmitters such as ATP. Depending on the properties of the postsynaptic receptors, a certain neurotransmitter at any given synapse will be either excitatory or inhibitory, but never both. In addition, a nitrergic modulation related to nNOS localization has been identified in NANC fibers. Neurotransmitters and nNOS-derived NO appear to exert modulatory influence, altering signal transmission across synapses in many ways, either by increasing or preventing the neurotransmitter release from the presynaptic neuron or by altering the properties of the postsynaptic cell. Both these chemical agents can also act more diffusely, affecting neurons some distance from their site of release. Most of the relevant information deals with mammals, while research in fish is progressing at a slower rate and, therefore, the cardiac information on this issue is scarce (Iversen et al. 2010; Zaccone et al. 2012).

In many fish species, the peripheral nerves are accompanied by aggregates of CAs-containing chromaffin cells: they can provide a zonal CAs production, thereby contributing to the humoral cardiovascular regulation (Gannon and Burnstock 1969; Abrahamsson et al. 1979; Nilsson and Holmgren 1992; Tota et al. 2010). As mentioned below, this neurohumoral control can be fine tuned by a nitrergic tone.

Cholinergic Control

Released from the nerve terminals, Acetylcholine (ACh) hyperpolarizes the pacemaker cells, slowing the pacemaker rate (Saito 1973). In several fish species, the ACh-dependent control of HR is sensitive to temperature. For example, the level of cholinergic inhibition of HR is lower in cold-acclimated *A. anguilla* (Seibert 1979), while is greater in cold-acclimated rainbow trout with respect to warm-acclimated counterpart (Wood et al. 1979). Moreover, an ACh-dependent reduction of contractility has been reported in various species (Randall 1970; Holmgren 1977; Cameron and Brown 1981). In *A. japonica* (Chan and Chow 1976) and in *A. anguilla* (Imbrogno et al. 2001, Imbrogno 2013), exogenous ACh induces a biphasic contractile (inotropic) effect. In particular, in *A. anguilla*, the positive inotropism is mediated by M1 receptor subtype, while the negative one involves M2 receptors (Imbrogno et al. 2001). The M1-mediated positive inotropism involves an EE-NO-cGMP signal-transduction mechanism (Fig. 3), since the EE impairment by Triton X-100 treatment or by the pre-treatment with blockers of various steps of the NO-cGMP signaling abolished the positive effects of ACh without influencing the negative one (Imbrogno et al. 2001).

Adrenergic Control

Derived from both the circulation (adrenaline and noradrenaline) and the sympathetic nervous system terminals (noradrenaline), CAs target the cardiac adrenoceptors (ARs). In fish, the nervous activity plays a major role (Holmgren and Nilsson 1982; Smith et al. 1985; Randall and Perry 1992), while plasma CAs levels are low under basal conditions, and increase in response to various physical and environmental stimuli (e.g., exhaustive exercise, hypoxia, hypercapnia, etc.). However, the members of the genus *Anguilla* do not increase plasma CAs or cortisol, i.e., an indicator of chronic stress (Wendelar Bonga 1997), in response to stressful stimuli, thus representing stress-tolerant animals (McKenzie et al. 2003). In the eel, the cardiac ARs function has been correlated to seasonal changes in temperature (Peyraud-Waitzenegger et al. 1980).

Beside the classical β1- and β2-ARs, the recently discovered type of cardiac β-ARs, the β3-AR, provided new insights into the adrenergic control of the fish heart (Imbrogno et al. 2006). In particular, physio-pharmacological studies in the eel demonstrated for the first time in non-mammalian vertebrates that the β3-AR activation by the selective BRL37344 agonist depresses contractility through pertussis toxin-sensitive inhibitory Gi proteins and the NO-cGMP-PKG signal transduction pathway (Imbrogno et al. 2006) (Fig. 3). In addition, isoproterenol stimulation not only exerts its classic positive inotropism, but also induces in 30% of the cardiac preparations a negative inotropism which is probably mediated by β3-AR (Imbrogno et al. 2006). Therefore, the evidence of β3-ARs and their involvement in counteracting isoproterenol stimulation in the eel can contribute to clarify the adrenergic versatility that controls its heart function, shedding new light on the role of the balanced β1/β2 and β3-AR cardiac regulation in fish. Likewise, by counteracting the typical β1/β2-adrenergic-induced enhancement of contractility, the β3-adrenergic eNOS pathway might act as an integrated protection mechanism against excessive CAs stimulation. In addition to short-term cardio-depressant effects of β3-AR stimulation, Moens et al. (2009) proposed that chronic β3-AR activation may protect the heart from the long-term adverse effects of adrenergic over-stimulation by preserving eNOS in its coupled state.

Angiotensin II-dependent Regulation

The octapeptide AngII, principal effector of the renin-angiotensin system, induces its cardiac effects via binding to plasma membrane type 1 (AT_1) and type 2 (AT_2) receptors (see references in Cerra et al. 2001; De Gasparo 2002). AT_1 mediates most of the AngII-dependent cardiac effects (i.e., chronotropism and inotropism) and rate of protein synthesis in isolated myocyte preparations (see Imbrogno et al. 2003, 2013 for references). In contrast, the AT_2 receptor antagonizes cardiac AT_1 growth promoting effects via activation of a number of phosphatases. This receptor is also coupled with the NO-cGMP signaling, either directly, or indirectly, through enhanced bradykinin or eNOS expression (Dostal 2000). Several fish AngII receptors have been cloned (Marsigliante et al. 1996; Tran van Chuoi et al. 1999). For example, in the eel an angiotensin receptor cDNA sequence (see GenBank accession numbers AJ05132; Tran Van Chuoi et al. 1999) shows 60% homology with the mammalian AT_1 receptor (Russell et al. 2001).

Most of the AngII-mediated cardiovascular effects documented in fish appear to be species-specific. For example, in trout AngII injection produces a hypertension-dependent reflex bradycardia, which in turn reduces the CO (Russell et al. 2001 for references). In the eel *Anguilla rostrata* the AngII-induced hypertension depends on the increased CO rather than on changes in systemic resistances (Butler and Oudit 1995; Oudit and Butler 1995). As well as in the trout *Oncorhynchus mykiss*, in the American eel *A. rostrata*, AngII exerts both direct and indirect (i.e., via cardiac ARs) cardiac stimulatory effects (Oudit and Butler 1995; Bernier et al. 1999). Conversely, in the European *A. anguilla*, AngII exerts a direct mechanical cardio-suppression attained via the EE interaction which activates G protein-coupled AT1-like receptors that, in turn, trigger a NO-cGMP-PKG signal transduction pathway (Imbrogno et al. 2003; Imbrogno 2013) (Fig. 3). AT1 receptors are structurally and functionally localized in the endothelial cell caveolae (Ishizaka et al. 1998) together with many proteins involved in signal-transduction cascades such as G-proteins, G-protein-linked receptors (e.g., mAChR) and several signaling molecules (e.g., eNOS, protein kinase C, Ca^{2+} channels and plasmalemmal Ca^{2+}-ATPase). This intracardiac cross-talk between EE-NO-cGMP and chemical stimuli (Imbrogno et al. 2003) suggests that the co-localization of AT1 receptors and NOS in the restricted space of the EE caveolae represents a temporally and spatially delimited pathway where the AngII signal is generated and transduced.

NOS/NO-mediated Autocrine-paracrine Regulation

The role played by NO as a major autocrine/paracrine organizer of complex connection-integration signals has been highlighted by Imbrogno et al. 2011. Generated by NOS isoenzymes (i.e., eNOS, nNOS and iNOS) in one cell, NO can act on the adjacent cells (paracrine modulation) or on the cell itself (autocrine modulation) (Moncada et al. 1991). The EE-induced contractile modulation of myocardial performance, reported in the avascular, or poorly vascularized, fish hearts (Gattuso et al. 2002; Imbrogno et al. 2011), represents a hallmark of the paracrine NO signal. The basal EE-generated NO exerts a cGMP-dependent tonic negative inotropism, while, in contrast, the functional damage of the EE, elicits positive contractile effects (Imbrogno et al. 2011). As reported above in previous examples, this EE-eNOS signaling, located at the cross-road of many extrinsic and intrinsic neuro-endocrine pathways, coordinates many chemically activated cascades. Chemical stimuli such as ACh, AngII, Vasostatin-1, Catestatin, as well as β3-AR activation, all converge in their contractile effects via NO signaling, which requires the obligatory involvement of the EE (Imbrogno et al. 2001, 2003, 2004, 2006, 2010). Therefore, the EE not only represents a quantitatively important source of NO, but it is also necessary for coupling the intracavitary stimuli to the eNOS-like/NO system.

More recently, a NO-dependent autocrine pathway, related to the specific subcellular localization and regulation of the different NOS isoforms, as well as to NO target proteins, has been described (for references, see Seddon et al. 2007). Importantly, such subtle spatial compartmentation and signaling networks coordination between cardiac NO and specific intracellular effectors (Hare 2003) allows NO to achieve local control of different cellular functions (Iwakiri et al. 2006). At the same time, a relatively high local concentration of NO equivalents may be generated, which in turns provides a favorable environment for protein S-nitrosylation (Lima et al. 2010). Of note, a NO-induced modulation of the heterometric (the Frank-Starling response) regulation has been shown in the eel. In fact, Garofalo and coworkers (2009b) reported that the 'beat-to-beat' regulation of the *in vitro* working heart of *A. anguilla* is directly influenced by myocardial autocrine NO through a regulation of PLN S-nitrosylation-dependent calcium reuptake by SERCA2a pumps, thereby modulating myocardial relaxation. This evidence supports the notion that the NO-induced modulation of ventricular performance through NOS isoforms compartmentation, as well as differences in their mode of stimulation and recruitment of distinct downstream pathways was indeed a crucial and early event during vertebrate evolution.

Nitrite as a Bioactive NO Source

Nitrite (NO_2^-) is an important physiological reservoir of NO in various cells and tissues (Bryan et al. 2005) and thus acts as a key intrinsic signaling molecule in many biological processes. Its conversion into NO is achieved through both non-enzymatic and enzymatic pathways (Mayer et al. 1998; Modin et al. 2001; Cosby et al. 2003; Rassaf et al. 2007), including acidic disproportionation, conversion via xanthine-oxidoreductase (XO), deoxyhemoglobin and deoxymyoglobin, and the action of mitochondrial enzymes (see Angelone et al. 2012; for references). The biological responses mediated by nitrite include regulation of protein and gene expression (Bryan et al. 2005), hypoxic vasodilation (Cosby et al. 2003; Crawford et al. 2006), inhibition of mitochondrial respiration (Shiva et al. 2007) and cytoprotection following ischemia/reperfusion (Webb et al. 2004; Duranski et al. 2005). These actions can be dependent on either the reduction of nitrite to NO or the direct S-nitrosylation of thiol-containing proteins (Bryan et al. 2005; Perlman et al. 2009). Fish and other water-breathing organisms, compared to terrestrial animals, have an additional direct uptake of exogenous nitrite from the environmental water across the respiratory surfaces (Jensen 2009). Therefore, fish need to balance the advantageous access to an ambient pool of nitrite for internal NO production with the potentially dangerous effects of nitrite-polluted habitats (see Jensen and Hansen 2011).

In the eel, nitrite represents a significant source of bioactive NO with consequent influences on the heart function. It negatively affects the basal cardiac mechanical performance through a NOS-dependent mechanism, which involves a cGMP/PKG transduction signaling (Cerra et al. 2009). The effects of the nitrite-dependent NO production on heart contractility were confirmed also in antarctic teleosts, in which, recently, the involvement of two nitrite reductase, XO and cytochrome p450, has been demonstrated (Cerra et al. 2009; Garofalo et al. 2015b). Moreover, nitrite strikingly affects the Frank-Starling response through a NO/cGMP/PKG pathway and S-nitrosylation of both membrane and cytosolic proteins (Angelone et al. 2012).

Taken together, these data further support the importance of bioactive NO on fish biology and ecophysiology.

The NOS/NO System in the Lungfish: Implication in the Stress Response

The extant African freshwater lungfishes, which include *Protopterus dolloi*, *Protopterus aethiopicus*, *Protopterus annectens* and *Protopterus amphibious*, are obligate air breathers possessing true lungs and reduced gills (Burggrenn and Johansen 1986; Graham 1997). When water is lacking, they can survive for a long period (up to six years in the case of *P. amphibious*) by entering a dormancy state of corporal torpor (aestivation) (Lomholt 1993). To withstand protracted periods of water and food deprivation in a hot environment, the aestivating fish in air or in mud encases itself in a completely dried mucus cocoon (see for review Ip and Chew 2010; Fishman et al. 1986; Greenwood 1986). The aestivation period can be prolonged until the moment when water becomes once again available in the environment. Dominated by the complete dependence on aerial gas exchange and dehydration stress, the aestivating condition is characterized by metabolic depression, down-regulation of respiratory (reduced O_2 consumption and frequency of pulmonary breathing) and cardiovascular activities (low HR, blood pressure and flow, cardiac work) and mobilization of the internal stores (proteins and amino acids) together with ammonia detoxification to urea (Fishman et al. 1986; Storey and Storey 1990; Perry et al. 2008). The coordination of these processes allows the animal to sustain a low rate of waste production to minimize pollution of the internal milieu, and to prevent tissue degradation, cell death and to maintain morpho-functional integrity, especially in those organs that are crucial for survival (see for review Ip and Chew 2010). In particular, during the prolonged starvation and immobilization, the skeletal muscle is challenged by disuse atrophy. On the other hand, the cardiovascular system must be ready to sustain the sudden activity, when, upon contact with water, the lungfish immediately arouses from the maintenance phase of aestivation, comes out from the cocoon and starts to swim up to the surface to gulp air (Loong et al. 2005, 2007).

In the heart of *P. dolloi*, in which the eNOS-like isoform is located in the epicardium, EE and myocardiocytes of both freshwater (FW) and aestivating fish, the aestivation appears to increase the expression of the molecule, changing its localization patterns (Amelio et al. 2008). Moreover, using biochemical and immuno-fluorescence methods, Amelio et al. (2013a) evaluated the expression and the localization pattern of the eNOS-like isoform and related molecular activators, i.e., Akt and Hsp-90, in both the heart and skeletal muscles of *P. annectens* exposed to three different conditions: FW, six months of experimentally induced aestivation in air (6mAe), and six months of aestivation in air followed by six days after arousal (6mAe6d). Western blotting of cardiac homogenates indicated that, like in *P. dolloi* (Amelio et al. 2008), the heart of *P. annectens* expressed remarkable eNOS levels under both FW and aestivating conditions. Both aestivation and arousal were accompanied by increments in eNOS-like and Akt activities, paralleled by an increased Hsp-90 expression, suggesting adaptive response to the stress related to the prolonged aestivation. Within the first days after arousal, this molecular adaptation is still operative, presumably allowing the heart to face the organism requirements for recovery. Conceivably, a continuous NO release may contribute to sustain cardiac bioenergetics (metabolic depression and reduced myocardial O_2 consumption), cell survival (anti-apoptotic action) and mechanical performance of the heart exposed to the environmental stress (aestivation and arousal). In addition, during aestivation the unchanged expression of the gap junction protein Connexin 43, allow the heart of *P. annectens* to beat, although at lower rate (Delaney et al. 1974). That the morpho-functional integrity of the aestivating *P. annectens* heart can be well preserved despite the adverse conditions, is further supported by the apoptosis data (TUNEL assay) showing that the number of apoptotic nuclei as well as the expression of the apoptosis repressor ARC remained overall unchanged in the aestivating heart as compared with the FW control. This strategy appears highly conserved among all the species that experience long periods of inactivity during aestivation or hibernation and have accordingly evolved a striking capability to maintain cell survival to preserve organ structure and function (van Breukelen et al. 2010 and references therein). In contrast, in the skeletal muscle during aestivation phospho-eNOS/eNOS and phospho-Akt/Akt ratios decreased while the rate of apoptotic cell death increased and ARC expression was strongly reduced. Recently, the involvement of NOS/NO system in the morpho-functional readjustment in FW, 6mAe and 6mAe6d *P. annectens* has been described also in gills and lungs (Garofalo et al. 2015a).

On the whole, this evidence suggests that the NOS/Akt/Hsp-90 is implicated in the survival strategy orchestrated by the lungfish during prolonged aestivation, which also includes a selective regulation of its tissues and organs finely tuned to their hierarchically different functionality.

Conclusions

In this chapter we have illustrated the presence and function of the intracardiac NOS/NO system in two ancient fish, the teleost Anguilla and the dipnoan Protopterus. The available evidence shows that, despite the differences in phylogeny, morpho-functional traits and ecophysiological characteristics, the cardiac NOS system is equally present in these animals and NO exerts an important modulation on their hearts. Some paradigms have been discussed to show that NO modulates short-, medium- and long-term cardiac responses, all crucial for homeostasis. This recent information discloses new mechanisms of cardiac regulation and, at the same time, is of evolutionary interest since it indicates the deep phylogenetic root cardiac NO in vertebrates.

References

Aardal, S. and K.B. Helle. 1991. Comparative aspects of the endocrine myocardium. Acta Physiol. Scand. Suppl. 599: 31–46.

Abrahamsson, T., S. Holmgren, S. Nilsson and K. Pettersson. 1979. On the chromaffin system of the African lungfish, *Protopterus aethiopicus*. Acta Physiol. Scand. 107: 135–139.

Altimiras, J. and E. Larsen. 2000. Non-invasive recording of heart rate and ventilation rate in rainbow trout during rest and swimming. Fish go wireless! J. Fish Biol. 57: 197–209.

Amelio, D., F. Garofalo, D. Pellegrino, F. Giordano, B. Tota and M.C. Cerra. 2006. Cardiac expression and distribution of nitric oxide synthases in the ventricle of the cold-adapted Antarctic teleosts, the hemoglobinless *Chionodraco hamatus* and the red-blooded *Trematomus bernacchii*. Nitric Oxide 15: 190–198.

Amelio, D., F. Garofalo, E. Brunelli, A.M. Loong, W.P. Wong, Y.K. Ip, B. Tota and M.C. Cerra. 2008. Differential NOS expression in freshwater and aestivating *Protopterus dolloi* (lungfish): heart vs. kidney readjustments. Nitric Oxide 18: 1–10.

Amelio, D., F. Garofalo, W.P. Wong, S.F. Chew, Y.K. Ip, M.C. Cerra and B. Tota. 2013a. Nitric oxide synthase-dependent "on/off" switch and apoptosis in freshwater and aestivating lungfish, *Protopterus annectens*: skeletal muscle versus cardiac muscle. Nitric Oxide 32: 1–12.

Amelio, D., F. Garofalo, C. Capria, B. Tota, S. Imbrogno. 2013b. Effects of temperature on the nitric oxide-dependent modulation of the Frank-Starling mechanism: the fish heart as a case study. Comp. Biochem. Physiol. A Mol. Integr. Physiol. 164: 356–362.

Andreakis, S., R. D'Aniello, F.P. Albalat, J. Patti, G. Garcia-Fernàndez, P. Procaccini, A. Sordino and A. Palumbo. 2011. Evolution of the nitric oxide synthase family in metazoans. Mol. Biol. Evol. 28: 163–179.

Angelone, T., A. Gattuso, S. Imbrogno, R. Mazza and B. Tota. 2012. Nitrite is a positive modulator of the Frank-Starling response in the vertebrate heart. Am. J. Physiol. Regul. Integr. Comp. Physiol. 302: R1271–R1281.

Axelsson, M., B. Davison, M. Forster and S. Nilsson. 1994. Blood pressure control in the Antarctic fish Pagothenia Borchgrevinki. J. Exp. Biol. 190: 265–279.

Bernier, N.J., J.E. McKendry and S.F. Perry. 1999. Blood pressure regulation during hypotension in two teleost species: differential involvement of the rennin-angiotensin and adrenergic systems. J. Exp. Biol. 202: 1677–1690.

Bryan, N.S., B.O. Fernandez, S.M. Bauer, M.F. Garcia-Saura, A.B. Milsom, T. Rassaf, R.E. Maloney, A. Bharti, J. Rodriguez and M. Feelisch. 2005. Nitrite is a signaling molecule and regulator of gene expression in mammalian tissues. Nat. Chem. Biol. 1: 290–297.

Burggrenn, W.W. and K. Johansen. 1986. Circulation and respiration in lungfishes (Dipnoi). pp. 217–236. *In*: W.E. Bemis, W.W. Burggren and N.E. Kemp (eds.). The Biology and Evolution of Lungfishes. Alan Liss, New York (Supplement J. Morphol.).

Butler, D.G. and G.Y. Oudit. 1995. Angiotensin I and III mediated cardiovascular responses in the freshwater North American eel, *Anguilla rostrata*: effect of Phe deletion. Gen. Comp. Endocrinol. 97: 259–269.

Cameron, J.S. and S.E. Brown. 1981. Adrenergic and cholinergic responses of the isolated heart of the goldfish *Carassius auratus*. Comp. Biochem. Physiol. C. 70: 109–115.

Cerra, M.C. and S. Imbrogno. 2012. Phospholamban and cardiac function: a comparative perspective in vertebrates. Acta Physiol. (Oxf). 205: 9–25.

Cerra, M.C., M. Canonaco, R. Acierno and B. Tota. 1997. Different binding activity of A and B-type natriuretic hormones in the heart of two antarctic teleosts the red blooded *Trematomus bernacchii* and the hemoglobinless *Chionodraco hamatus*. Comp. Biochem. Physiol. 118A: 993–999.

Cerra, M.C., M.L. Tierney, Y. Takei, N. Hazon and B. Tota. 2001. Angiotensin II binding sites in the heart of *Scyliorhinus canicula*: an autoradiographic study. Gen. Comp. Endocrinol. 121: 126–134.

Cerra, M.C., T. Angelone, M.L. Parisella, D. Pellegrino and B. Tota. 2009. Nitrite modulates contractility of teleost (*Anguilla anguilla* and *Chionodraco hamatus*, i.e., the Antarctic hemoglobinless icefish) and frog (*Rana esculenta*) hearts. Biochim. Biophys. Acta 1787: 849–855.

Chan, D.K. and P.H. Chow. 1976. The effects of acetylcholine, biogenic amines and other vasoactive agents on the cardiovascular functions of the Eel, *Anguilla japonica*. J. Exp. Zool. 196: 13–26.

Cooke, S.J., E.C. Grant, J.F. Schreer, D.P. Philipp and A.L. Devries. 2003. Low temperature cardiac response to exhaustive exercise in fish with different levels of winter quiescence. Comp. Biochem. Physiol. A Mol. Integr. Physiol. 134: 157–165.

Cosby, K., K.S. Partovi, J.H. Crawford, R.P. Patel, C.D. Reiter, S. Martyr, B.K. Yang, M.A. Waclawiw, G. Zalos, X. Xu, K.T. Huang, H. Shields, D.B. Kim-Shapiro, A.N. Schechter, R.O. Cannon and M.T. Gladwin. 2003. Nitrite reduction to nitric oxide by deoxyhemoglobin vasodilates the human circulation. Nat. Med. 9: 1498–1505.

Crawford, J.H., T.S. Isbell, Z. Huang, S. Shiva, B.K. Chacko, A.N. Schechter, V.M. Darley-Usmar, J.D. Kerby, J.D. Lang, Jr., D. Kraus, C. Ho, M.T. Gladwin and R.P. Patel. 2006. Hypoxia, red blood cells, and nitrite regulate NO-dependent hypoxic vasodilation. Blood 107: 566–574.

De Gasparo, M. 2002. Angiotensin II and Nitric Oxide interaction. Heart Failure Rev. 7: 347–358.

Delaney, R.G., S. Lahiri and A.P. Fishman. 1974. Aestivation of the African lungfish *Protopterus aethiopicus*: cardiovascular and respiratory functions. J. Exp. Biol. 61: 111–128.

Dostal, D. 2000. The cardiac renin–angiotensin system: novel signaling mechanisms related to cardiac growth and function. Reg. Pept. 91: 1–11.

Dunmall, K.M. and J.F. Schreer. 2003. A comparison of the swimming and cardiac performance of farmed and wild Atlantic salmon, *Salmo salar*, before and after gamete stripping. Aquaculture 220: 869–882.

Duranski, M.R., J.J. Greer, A. Dejam, S. Jaganmohan, N. Hogg, W. Langston, R.P. Patel, S.F. Yet, X. Wang, C.G. Kevil, M.T. Gladwin and D.J. Lefer. 2005. Cytoprotective effects of nitrite during *in vivo* ischemia-reperfusion of the heart and liver. J. Clin. Invest. 115: 1232–1240.

Elrod, J.W., J.W. Calvert, S. Gundewar, N.S. Bryan and D.J. Lefer. 2008. Nitric oxide promotes distant organ protection: evidence for an endocrine role of nitric oxide. Proc. Natl. Acad. Sci. USA 105: 11430–11435.

Farrell, A.P. and D.R. Jones. 1992. The Heart. pp. 1–88. *In*: W.S. Hoar, D.J. Randall and A.P. Farrell (eds.). The Cardiovascular System. Academic Press, San Diego, USA.

Feelisch, M. and J.F. Martin. 1995. The early role of nitric oxide in evolution. Trends Ecol. Evol. 10: 496–499.

Fishman, A.P., A. Pack, R.G. Delaney and R.J. Galante. 1986. Aestivation in Protopterus. J. Morphol. 1: 237–248.

Gannon, J.B. and G. Burnstock. 1969. Excitatory adrenergic innervation of the fish heart. Comp. Biochem. Physiol. 29: 765–773.

Garofalo, F., D. Amelio, M.C. Cerra, B. Tota, B.D. Sidell and D. Pellegrino. 2009a. Morphological and physiological study of the cardiac NOS/NO system in the Antarctic (Hb⁻/Mb⁻) icefish *Chaenocephalus aceratus* and in the red-blooded *Trematomus bernacchii*. Nitric Oxide 20: 69–78.

Garofalo, F., M.L. Parisella, D. Amelio, B. Tota and S. Imbrogno. 2009b. Phospholamban S-nitrosylation modulates Starling response in fish heart. Proc. R. Soc. B 276: 4043–4052.

Garofalo, F., D. Amelio, J.M. Icardo, S.F. Chew, B. Tota, M.C. Cerra and Y.K. Ip. 2015a. Signal molecule changes in the gills and lungs of the African lungfish Protopterus annectens, during the maintenance and arousal phases of aestivation. Nitric Oxide. 44: 71–80.

Garofalo, F., D. Amelio, A. Gattuso, M.C. Cerra and D. Pellegrino. 2015b. Cardiac contractility in Antarctic teleost is modulated by nitrite through xanthine oxidase and cytochrome p-450 nitrite reductase. Nitric Oxide. In press.

Gattuso, A., R. Mazza, S. Imbrogno, A. Sverdrup, B. Tota and A. Nylund. 2002. Cardiac performance in *Salmo salar* with infectious salmon anaemia (ISA): putative role of nitric oxide. Dis. Aquat. Organ. 52: 11–20.

Graham, M.S. 1997. Air-Breathing Fishes: Evolution, Diversity and Adaptation. Academic Press, S. Diego.

Greenwood, P.H. 1986. The natural history of African lungfishes. J. Morph. (Suppl. 1): 163–179.

Hare, J.M. 2003. Nitric oxide and excitation-contraction coupling, J. Mol. Cell. Cardiol. 35: 719–729.

Holmgren, S. 1977. Regulation of the heart of a teleost, *Gadus morhua*, by autonomic nerves and circulating catecholamines. Acta Physiol. Scand. 99: 62–74.

Holmgren, S. and S. Nilsson. 1982. Neuropharmacology of adrenergic neurons in teleostfish. Comp. Biochem. Physiol. Part C: Comp. Pharmacol. 72: 289–302.

Icardo, J.M., S. Imbrogno, A. Gattuso, E. Colvee and B. Tota. 2005. The heart of *Sparus auratus*: a reappraisal of cardiac functional morphology in teleosts. J. Exp. Zoolog. A Comp. Exp. Biol. 303: 665–675.

Imbrogno, S. 2013. The eel heart: multilevel insights into functional organ plasticity. J. Exp. Biol. 216: 3575–3586.

Imbrogno, S., L. De Iuri, R. Mazza and B. Tota. 2001. Nitric oxide modulates cardiac performance in the heart of *Anguilla anguilla*. J. Exp. Biol. 204: 1719–1727.

Imbrogno, S., M.C. Cerra and B. Tota. 2003. Angiotensin II-induced inotropism requires an endocardial endothelium-nitric oxide mechanism in the *in vitro* heart of *Anguilla anguilla*. J. Exp. Biol. 206: 2675–2684.

Imbrogno, S., T. Angelone, A. Corti, C. Adamo, K.B. Helle and B. Tota. 2004. Influence of vasostatins, the chromogranin A-derived peptides, on the working heart of the eel (*Anguilla anguilla*): negative inotropy and mechanism of action. Gen. Comp. Endocrinol. 139: 20–28.

Imbrogno, S., T. Angelone, C. Adamo, E. Pulerà, B. Tota and M.C. Cerra. 2006. Beta3-Adrenoceptor in the eel (*Anguilla anguilla*) heart: negative inotropy and NO-cGMP-dependent mechanism. J. Exp. Biol. 209: 4966–4973.

Imbrogno, S., F. Garofalo, M.C. Cerra, S.K. Mahata and B. Tota. 2010. The catecholamine release-inhibitory peptide catestatin (Chromogranin A344-364) modulates myocardial function in fish. J. Exp. Biol. 213: 3636–3643.

Imbrogno, S., B. Tota and A. Gattuso. 2011. The evolutionary functions of cardiac NOS/NO in vertebrates tracked by fish and amphibian paradigms. Nitric Oxide 25: 1–10.

Imbrogno, S., F. Garofalo, D. Amelio, C. Capria and M.C. Cerra. 2013. Humoral control of cardiac remodelling in fish: Role of Angiotensin II. Gen. Comp. Endocrinol. 194: 189–197.

Ip, Y.K. and S.F. Chew. 2010. Nitrogen metabolism and excretion during aestivation. Prog. Mol. Subcell. Biol. 49: 63–94.

Ishizaka, N., K.K. Griendling, B. Lassègue and R.W. Alexander. 1998. Angiotensin II type 1 receptor: relationship with caveolae and caveolin after initial agonist stimulation. Hypertension 32: 459–466.

Iversen, N.K., A. Dupont-Prinet, I. Findorf, D.J. McKenzie and T. Wang. 2010. Autonomic regulation of the heart during digestion and aerobic swimming in the European sea bass (*Dicentrarchus labrax*). Comp. Biochem. Physiol. A Mol. Integr. Physiol. 156: 463–468.

Iwakiri, Y., A. Satoh, S. Chatterjee, D.K. Toomre, C.M. Chalouni, D. Fulton, R.J. Groszmann, V.H. Shah and W.C. Sessa. 2006. Nitric oxide synthase generates nitric oxide locally to regulate compartmentalized protein S-nitrosylation and protein trafficking. Proc. Natl. Acad. Sci. USA 103: 19777–19782.

Jensen, F.B. 2009. The role of nitrite in nitric oxide homeostasis: a comparative perspective. Biochim. Biophys. Acta 1787: 841–848.

Jensen, F.B. and M.N. Hansen. 2011. Differential uptake and metabolism of nitrite in normoxic and hypoxic goldfish. Aquat. Toxicol. 101: 318–325.

Katz, A.M. 1988. Molecular biology in cardiology: a paradigmatic shift J. Mol. Cell Cardiol. 20: 355–366.

Laurent, P., S. Holmgren and S. Nilsson. 1983. Nervous and humoral control of the fish heart: structure and function. Comp. Biochem. Physiol. 76A: 525–542.

Lepiller, S., N. Franche, E. Solary, J. Chluba and V. Laurens. 2009. Comparative analysis of zebrafish nos2a and nos2b genes. Gene 445: 58–65.

Lima, B., M.T. Forrester, D.T. Hess and J.S. Stamler. 2010. S-nitrosylation in cardiovascular signaling. Circ. Res. 106: 633–646.

Lomholt, J.P. 1993. Breathing in the aestivating African lungfish, *Protopterus amphibious*. Adv. Fish Res. 1: 17–34.

Loong, A.M., K.C. Hiong, S.M. Lee, W.P. Wong, S.F. Chew and Y.K. Ip. 2005. Ornithine-urea cycle and urea synthesis in African lungfishes, *Protopterus aethiopicus* and *Protopterus annectens*, exposed to terrestrial conditions for six days, J. Exp. Zool. A Comp. Exp. Biol. 303: 354–365.

Loong, A.M., J.Y. Tan, W.P. Wong, S.F. Chew and Y.K. Ip. 2007. Defense against environmental ammonia toxicity in the African lungfish, *Protopterus aethiopicus*: Bimodal breathing, skin ammonia permeability and urea synthesis. Aquat. Toxicol. 85: 76–86

Marsigliante, S., A. Muscella, S. Vilella, G. Nicolardi, L. Ingrosso, V. Ciardo, V. Zonno, G.P. Vinson, M.M. Ho and C. Storelli. 1996. A monoclonal antibody to mammalian angiotensin II AT1 receptor recognises one of the angiotensin II receptor isoforms expressed by the eel (*Anguilla anguilla*). J. Mol. Endocrinol. 16: 45–56.

Massion, P.B., M. Pelat, C. Belge and J.L. Balligand. 2005. Regulation of the mammalian heart function by nitric oxide. Comp. Biochem. Physiol. 142A: 144–150.

Mayer, B., S. Pfeiffer, A. Schrammel, D. Koesling, K. Schmidt and F. Brunner. 1998. A new pathway of nitric oxide/cyclic GMP signaling involving S-nitrosoglutathione. J. Biol. Chem. 273: 3264–3270.

McKenzie, D.J., M. Piccolella, A.Z. Dalla Valle, E.W. Taylor, C.L. Bolis and J.F. Steffensen. 2003. Tolerance of chronic hypercapnia by the European eel *Anguilla anguilla*. J. Exp. Biol. 206: 1717–1726.

Modin, A., H. Bjorne, M. Herulf, K. Alving, E. Weitzberg and J.O. Lundberg. 2001. Nitrite-derived nitric oxide: a possible mediator of "acidic-metabolic" vasodilation. Acta Physiol. Scand. 171: 9–16.

Moens, A.L., J.S. Leyton-Mange, X. Niu, R. Yang, O. Cingolani, E.K. Arkenbout, H.C. Champion, D. Bedja, K.L. Gabrielson, J. Chen, Y. Xia, A.B. Hale, K.M. Channon, M.K. Halushka, N. Barker, F.L. Wuyts, P.M. Kaminski, M.S. Wolin, D.A. Kass and L.A. Barouch. 2009. Adverse ventricular remodeling and exacerbated NOS uncoupling from pressure-overload in mice lacking the beta3-adrenoreceptor. J. Mol. Cell. Cardiol. 47: 576–585.

Moncada, S., R.M. Palmer and E.A. Higgs. 1991. Nitric oxide: physiology, pathophysiology, and pharmacology. Pharmacol. Rev. 43. 109–142.

Nilsson, S. 1983. Automatic Nerve Function in the Vertebrates. Springer-Verlag, Heidelberg, Germany and New York, USA.

Nilsson, S. and S. Holmgren. 1992. Cardiovascular control by purines, 5-hydroxytryptamine and neuropeptides. pp. 301–341. *In*: D. Randall and A.P. Farrell (eds.). Fish Physiology, Vol. XII. Academic Press, New York, USA.

Olson, R.K. 1998. The cardiovascular system. pp. 129–154. *In*: H.D. Evans (ed.). The Physiology of Fishes. CRC Press, Boca Raton, New York, USA.

Oudit, G.Y. and D.G. Butler. 1995. Angiotensin II and cardiovascular regulation in a freshwater teleost *Anguilla rostrata* Le Sueur, Am. J. Physiol. 269: R726–R735.

Palumbo, A. 2005. Nitric oxide in marine invertebrates: a comparative perspective. Comp. Biochem. Physiol. A Mol. Integr. Physiol. 142: 241–248.

Perlman, D.H., S.M. Bauer, H. Ashrafian, N.S. Bryan, M.F. Garcia-Saura, C.C. Lim, B.O. Fernandez, G. Infusini, M.E. McComb, C.E. Costello and M. Feelisch. 2009. Mechanistic insights into nitrite-induced cardio-protection using an integrated metabolomic/proteomic approach. Circ. Res. 104: 796–804.

Perry, S.F., R. Euverman, T. Wang, A.M. Loong, S.F. Chew, Y.K. Ip and K.M. Gilmour. 2008. Control of breathing in African lungfish (*Protopterus dolloi*): a comparison of aquatic and cocooned (terrestrialized) animals. Respir. Physiol. Neurobiol. 160: 8–17.

Petroff, M.G.V., S.H. Kim, S. Pepe, C. Dessy, E. Marban, J.L. Balligand and S.J. Sollott. 2001. Endogenous nitric oxide mechanisms mediate the stretch dependence of Ca^{2+} release in cardiomyocytes. Nat. Cell. Biol. 3: 867–873.

Peyraud-Waitzenegger, M., L. Barthelemy and C. Peyraud. 1980. Cardiovascular and ventilatory effects of catecholamines in unrestrained eels (*Anguilla anguilla* L.). J. Comp. Physiol. 138: 367–375.

Pinsky, D.J., S. Patton, S. Mesaros, V. Brovkovych, E. Kubaszewski, S. Grunfeld and T. Malinski. 1997. Mechanical transduction of nitric oxide synthesis in the beating heart. Circ. Res. 81: 372–379.

Randall, D.J. 1970. The circulatory system. pp. 133–172. *In*: W.S. Hoar and D.J. Randall (eds.). Fish Physiology, 4. Academic Press, New York (USA) and London (UK).

Randall, D.J. and S.F. Perry. 1992. Catecholamines. pp. 255–300. *In*: W.S. Hoar, D.J. Randall and A.P. Farrell (eds.). Fish Physiology, Vol. 12b. Academic Press, San Diego, USA.

Rassaf, T., U. Flögel, C. Drexhage, U. Hendgen-Cotta, M. Kelm and J. Schrader. 2007. Nitrite reductase function of deoxymyoglobin: oxygen sensor and regulator of cardiac energetics and function. Circ. Res. 100: 1749–1754.

Russell, M.J., A.M. Klemmer and K.R. Olson. 2000. Angiotensin signalling and receptor types in teleost fish. Comp. Biochem. Physiol. 128: 41–51.

Saito, T. 1973. Effects of vagal stimulation on the pacemaker action potentials of carp heart. Comp. Biochem. Physiol. 44A: 191–199.

Seddon, M., A.M. Shah and B. Casadei. 2007. Cardiomyocytes as effectors of nitric oxide signalling. Cardiovasc. Res. 75: 315–326.

Seibert, H. 1979. Thermal adaptation of heart rate and its parasympathetic control in the European eel *Anguilla anguilla* (L.) Comp. Biochem. Physiol. C 64: 275–278.

Shiels, H.A. and E. White. 2008. The Frank–Starling mechanism in vertebrate cardiac myocytes. J. Exp. Biol. 211: 2005–2013.

Shiva, S., Z. Huang, R. Grubina, J. Sun, L.A. Ringwood, P.H. MacArthur, X. Xu, E. Murphy, V.M. Darley-Usmar and mM.T. Gladwin. 2007. Deoxymyoglobin is a nitrite reductase that generates nitric oxide and regulates mitochondrial respiration. Circ. Res. 100: 654–661.

Smith, D.G., S. Nilsson, I. Wahlqvist and B.M. Eriksson. 1985. Nervous control of the blood pressure in the Atlantic cod *Gadus morhua*. J. Exp. Biol. 117: 335–347.

Storey, K.B. and J.M. Storey. 1990. Metabolic rate depression and biochemical adaptation in anaerobiosis, hibernation and estivation. Q. Rev. Biol. 65: 145–174.

Taylor, E.W. 1992. Nervous control of the heart and cardiorespiratory interactions. pp. 343–387. *In*: W.S. Hoar, D.J. Randall and A.P. Farrell (eds.). Fish Physiology XIIB. Academic Press, New York, USA.

Tibbits, G.F., K.D. Philipson and H. Kashihara. 1992. Characterization of myocardial Na(+)-Ca2+ exchange in rainbow trout. Am. J. Physiol. 262: C411–C417.

Tota, B., V. Cimini, G. Salvatore and G. Zummo. 1983. Comparative study of the arterial and lacunary systems of the ventricular myocardium of the elasmobranch and teleost fishes. Am. J. Anat. 167: 15–32.

Tota, B., R. Acierno and C. Agnisola. 1991. Mechanical performance of the isolated and perfused heart of the hemoglobinless Antarctic icefish *Chionodraco hamatus* (Lonnberg), effects of loading conditions and temperature. Phil. Trans. R. Soc. B 332: 191–198.

Tota, B., D. Amelio, D. Pellegrino, Y.K. Ip and M.C. Cerra. 2005. NO modulation of myocardial performance in fish hearts. Comp. Biochem. Physiol. A 142: 164–177.

Tota, B., M.C. Cerra and A. Gattuso. 2010. Catecholamines, cardiac natriuretic peptides and chromogranin A: evolution and physiopathology of a 'whip-brake' system of the endocrine heart. J. Exp. Biol. 213: 3081–3103.

Tran van Chuoi, M., C.T. Dolphin, S. Barker, A.J. Clark and G.P. Vinson. 1999. Molecular cloning and characterization of the cDNA encoding the angiotensin II receptor of European eel (*Anguilla anguilla*). GenBank Database AJ005132.

van Breukelen, F., G. Krumschnabel and J.E. Podrabsky. 2010. Vertebrate cell death in energy-limited conditions and how to avoid it: what we might learn from mammalian hibernators and other stress tolerant vertebrates. Apoptosis 15: 386–399.

Webb, A., R. Bond, P. McLean, R. Uppal, N. Benjamin and A. Ahluwalia. 2004. Reduction of nitrite to nitric oxide during ischemia protects against myocardial ischemia-reperfusion damage. Proc. Natl. Acad. Sci. USA 101: 13683–13688.

Wendelar Bonga, S.E. 1997. The stress response in fish. Physiol. Rev. 77: 591–625.

Wood, C.M., P. Pieprzak and J.N. Trott. 1979. The influence of temperature and anaemia on the adrenergic and cholinergic mechanisms controlling heart rate in the rainbow trout. Can. J. Zool. 57: 2440–2447.

Zaccone, D., M. Sengar, E.R. Lauriano, S. Pergolizzi, F. Macri', L. Salpietro, A. Favaloro, L. Satora, K. Dabrowski and G. Zaccone. 2012. Morphology and innervation of the teleost physostome swim bladders and their functional evolution in non-teleostean lineages. Acta Histochem. 114: 763–772.

Zhang, Y.H., M.H. Zhang, C.E. Sears, K. Emanuel, C. Redwood, A. El-Armouche, E.G. Kranias and B. Casadei. 2008. Reduced Phospholamban phosphorylation is associated with impaired relaxation in left ventricular myocytes from neuronal NO synthase deficient mice. Circ. Res. 102: 242–249.

3

Evolutionary Aspects on the Comparative Biology of Lungfishes: Emphasis on South-American Lungfish, *Lepidosiren paradoxa*

*Vera Maria Fonseca de Almeida-Val,**
Luciana Mara Lopes Fé and *Derek Felipe de Campos*

Introduction

Lungfishes are represented by three genera in the world: *Protopterus, Lepidosiren,* and *Neoceratodus*. These three genera are endemic of Africa, South America, and Australia, respectively, and evolved separately since the beginning of the continental drift that promoted the splitting of big continents. *Neoceratodus* is the most ancient genus, and the most derived are *Protopterus* and *Lepidosiren*. They are named ancient (or extant) fishes because they currently retain their primitive characteristics. Within the Osteichthyes (bony fishes), they belong to the subclass Dipnoi since they have lungs and the ability to breathe air. However, they resemble Sarcopterygii since they present, as coelacanth, lobed fins with an internal skeleton. Although they are not a diversified group of

Laboratory for Ecophysiology and Molecular Evolution, National Institute of Research in the Amazon (LEEM-INPA). Av. André Araujo, 2936, Aleixo. CEP 69060-000, Manaus, AM, Brazil.
* Corresponding author: veraval@inpa.gov.br

living fishes, they have a very diversified group of fossils dated from late Devonian period. Lungfishes have in common a number of features in their skull roof and dentition, making the recognition of this group easy among fossil records. However, modern lungfishes show reductions and fusions of the skull roof bones, and these specific bones show no homology with the skull roof bones of ray-fined fishes or tetrapods. Thus, the separation of Dipnoi as a discrete group is largely based on the structure and arrangement of the skull bones, the endoskeleton of the paired fins, and the teeth. The two living orders of Dipnoi are distinguishable mainly by the number of lungs they possess. There are two well established orders: Ceratodontiformes, to which *Neoceratodus* belongs and Lepidosireniformes that includes both *Lepidosiren* and *Protopterus* genera. It is well known that Ceratodontiformes have only one lung and the species belonging to Lepidosireniformes present two lungs.

Although the current distribution of the three living genera suggest a vicariance divergence while still in supercontinent Gondwana at Mesozoic era, fossil records show a spread freshwater distribution with subsequent extinction of different lineages after Gondwana breakup.

Numerous studies have been done on vertebrate evolutionary processes and have shown that lungfish constitute a step towards acquisition of the main characteristics of the first land vertebrates, the tetrapods. Particularly, South American and African lungfish, in contrast with the Australian lungfish, developed several metabolic and physiological abilities that resemble amphibians (Janssens and Cohen 1966; Hochachka and Somero 1973). However, given that *Lepidosiren* and *Protopterus* are more recently derived lungfish than *Neoceratodus,* these advanced adaptations shared with amphibians could also be viewed as examples of parallel evolution.

Lepidosiren paradoxa (locally called pirambóia) is the single living representative of lungfish in South America. Along with its other living relatives around the world: four species of *Protopterus* (*dolloi, aethiopicus, annectens* and *amphibius*) in Africa and the only extant member of genus *Neoceratodus* in Australia make this group one of the most important living vertebrates for studies of the evolutionary transition from water to land. In this chapter, we review some of these characteristics, from morphology to metabolism, trying to put together all characteristics that allow these interesting species to have a specialized way of life.

The presence of a true lung and the ability to aestivate make these fish (Lepidosireniformes) an important biological model for studies of several metabolic adaptations to conditions such as desiccation, oxygen depletion, temperature increase, storage and excretion of toxic nitrogenous waste, and metabolic depression. Besides these important physiological issues, this group is also interesting from another perspective—that of genome size (largest vertebrate genome) (Pedersen 1971).

The Divergence of Living Lungfish Genera

Lungfishes have an extensive fossil record in Australia, dating from the Devonian, about 380 million years ago (Long 1995). Fossil records come from almost every period of geological time and the remains of this group indicate that they lived in both freshwater and shallow sea environments in the Devonian. However, since the end of the Carboniferous period lungfishes inhabited only freshwater environments, and some species acquired the ability to aestivate during the Permian. They diversified into many species during the Devonian period, followed by a slow steady change during the later part of the Palaeozoic and Mesozoic. Based on fossil records of the Queensland (Australian) lungfish, Long (1995) has suggested that since the Mesozoic they have remained almost unchanged.

The Lepidosireniformes lungfishes (*Protopterus* and *Lepidosiren*) are longer and slimmer than *Neoceratodus* and have reduced pectoral and pelvic fins for sensory functioning. Both *Protopterus* and *Lepidosiren* can survive periods of dry season burrowing into the ground, an ability not shared by *Neoceratodus*. All three genera of living lungfish may be considered as 'living fossils'. Living fossils are known by the retention of particular anatomical forms and ways of life. According to Stanley (1984) most living fossil groups are:

  i) 'Depauperate in species';
 ii) Retain a large number of plesiomorphies;
iii) Ecologically eurytopic in many physiological and behavioral attributes;
 iv) Show great individual species longevity with;
  v) Broad area and habitat distribution.

Living lungfishes have all these attributes. Evolutionary rates have been claimed to be a consequence of mutation rates and, regarding living fossils, there has always been a trend to correlate the rates of genetic changes with the rates of anatomical and morphological change (Stanley 1984; Kumar and Hedges 1998). However, there is evidence in the literature that related groups with different rates of morphological evolution show similar rates of molecular changes (Avise et al. 1994).

Phylogeny

The Osteichthyes (bony fish) can be divided into two main groups, the ray-finned fishes (Actinopterygii) and the lobe-limbed vertebrates (Sarcopterygii). The Sarcopterygii comprise three groups: coelacanths (Actinistia), lungfishes (Dipnoi), and four-limbed vertebrates, or Tetrapoda, which are the amphibians, reptiles, birds, and mammals (Clack 2002). Ten years ago, Takezaki et al. (2004) discussed the many possible phylogenies between the Sarcopterygii (a term more commonly applied to designate Coelacanth and Lungfishes) and the tetrapods. Along the years, there have been many attempts to formulate a hypothesis about the origin of tetrapods, approaching the traditional sharing characters to infer ancestor-descendent relationships and the cladistics analysis (Romer 1966; Hennig 1966; Rosen et al. 1981). Several phylogenetic trees were proposed, and none of them came to an exact conclusion about who was the original group more closely related to the tetrapods. Takezaki et al. (2004) described, then, a hypothesis based on the sequencing of 44 nuclear genes, generating three phylogenetic trees based on Maximum-Parsimony (MP), Neighbor-Joining (NJ), and Maximum-Likelihood (ML). The results favored (NJ analysis) consistently and with high bootstrap probabilities a sister-group relationship between lungfishes and coelacanths. However, this resulted in an excess of Phylogenetically Informative Configurations (PICs), and this excess may be caused by paralogous relationships between loci, evidencing a high occurrence of gene duplications and differential gene loss in coelacanth, lungfish and tetrapod (see other sections in this chapter to more details). If this is true, the duplicated genes separated before the divergence of the lineages and the longer branches (obtained in phylogenetic trees) resulted from an excessive number of PICs in paralogous sequences. All in all, Takezaki et al. (2004) suggested that if the transition fish-to-tetrapod was estimated as an interval of 10 to 20 Myr long, and the phylogeny was dichotomous, more than 200 loci would have to be analyzed to resolve it. As the estimative of this transition based on morphological traits is less than 20 Myr (Carroll 1997), more genes should be added to this analysis. Thus, Takezaki and co-workers suggested the analysis of the whole genome sequences of a lungfish and a coelacanth to provide the answer to whether this trichotomy is resolvable.

As far as we understand, the combination of molecular and morphological studies, considering fossil data on distribution of these groups, will possibly give us a clue about the real ancestor of the tetrapods. So, important issues must be addressed about the biology of this group and, here, we provide some of the main genetic, physiological, and biological features lungfishes present, collecting the main literature contents about this group.

DNA content

The so-called 'C-value paradox' (named C-value enigma by T. Ryan Gregory) describes the profound variation in genome size among eukaryotes. Genome size (C-value) is positively related with cell size, and negatively related to cell division rates. Thus, developmental rates have been associated with genome size in numerous plant, vertebrate, and invertebrate groups. Gregory (2002) included developmental complexity as also being related to genome size. Studies of the C-value paradox seek to find answers to: (a) which mechanisms are responsible for genome size variation? (b) what are the relationships between genome size and nuclear, cellular and organism characteristics? and (c) which traits are responsible for the maintenance of non-coding DNA?

Considering the highly diverse fish group Actinopterygii, which has a considerable diversity in genome size, Hinegardner (1968) stated that "the advanced species of fishes have less DNA than the primitive ones" and, thus, that "evolution and specialization in the teleosts have been accompanied by loss of DNA". Hinegardner's ideas have led to some discussion since they were proposed, but the answers are not yet resolved (Ohno 1970; Cavalier-Smith 1985).

What is currently known is that the modern lungfishes have the largest genomes of all vertebrates. Vinogradov (2005) suggested that the largest genome size among animals is found in the South American lungfish, *Lepidosiren paradoxa* (80 pg/haploid nucleus) and that the former data for marbled lungfish, *Protopterus* (> 100 pg) was greatly overestimated (Pedersen 1971). However, Vinogradov measured *Protopterus dolloi*, instead of *P. aethiopicus* (tetraploid), which has the largest animal genome reported (133 pg/haploid nucleus).

Lungfish possess highly variable or duplicated isozyme systems and loss of gene duplication (Mesquita-Saad et al. 2002), sharing the same patterns with amphibians and other tetrapods (Allendorf et al. 1983; Whitt 1983; Buth 1983). Thus, the higher C-value in lungfish may be explained by an increase in non-coding DNA.

Reviewing the literature we can characterize lungfish genetic traits as follows:

  i) Medium to low diploid chromosome number;
 ii) Enormous size of individual chromosomes; and
iii) High contents of DNA (Oliveira et al. 1988).

It is not easy to identify what caused the large genome size in lungfish during the evolutionary process. Gregory (2002) suggested that the genome size is inversely related to developmental complexity, i.e., when there is an overall reduction in developmental complexity, larger genome sizes result, as occur among lungfish and salamanders. In amphibians (salamanders) and lungfish, genomes have become large in association with *paedomorphosis* (i.e., the maintenance of juvenile characters in the adult by the deletion of developmental steps) (Bemis 1984). In advanced teleost fish, the proposed mechanism is exactly the reverse—small genomes of specialized fish are associated with the terminal addition of more developmental steps. However, this correlation with developmental complexity may be a result of genome size rather than a cause of it, since the loss of DNA must be accompanied by the gain of more complex gene regulatory systems.

Cavalier-Smith (1991) proposed that the enormous size of lungfish genomes is due to unique DNA. Recent evidence for this has been given for *Neoceratodus* (Sirijovsky et al. 2005). A transposable element was described for this lungfish, *NfCR1*. It is the most common duplicated element in this lungfish genome but it is no longer functional and demonstrates high levels of mutation, which together with many other transposable elements, could account for most of the unique DNA making up the large genome of this species of lungfish.

Metabolic Adaptations

Glycogen Storage and Metabolism

Studies of the ultra structure of cardiac muscle, white muscle, and red muscle of South America lungfish show that this species has an overwhelming dependence on glycogen as storage of carbon and as an energy source. Fish from Lake Janauacá at Rio Solimões (Amazonas state, Brazil) were analyzed on board of the *River Vessel Alpha Helix*. Enzyme and ultra-microscopy analyses revealed that the musculature of this lungfish consists of both white and red muscle, and the heart, although presenting an overall structure similar to other fish species, is modified by one partial atrial septa and ventricular trabeculae, clearly foreshadowing the situation in terrestrial vertebrates (Johansen et al. 1968).

Lungfish white muscle differs from other fish species in the kind and amount of glycogen stored (Hochachka and Hulbert 1978; pers. observation). The ultra structure of this tissue is characterized by large amounts of glycogen, different from other fish species, although most of them do have their white muscle metabolism dependent on anaerobic glycolysis. The amount of stored glycogen granules is much lower in teleost fish than found in lungfish (Pritchard et al. 1971) with the exception of tuna white muscle that also stores glycogen in a monoparticulate granular form (Guppy et al. 1977). The depots of glycogen found in lungfish white muscle by Hochachka and Hulbert (1978) are large-diameter alpha particles or glycogen rosettes. Glycogen rosettes appear to represent a more efficient way of packaging large amounts of glycogen, as in liver and other tissues such as the kidney. This unique situation (presence of this form of glycogen storage in white muscle) is presumably of advantage to the animal during aestivation. On the other hand, lungfish red muscle shows glycogen-membrane complexes, which are myofibrilar and peripheral in location. They usually occur in close association with mitochondria, which are far more abundant than in the white muscle. Occasionally, sections of lungfish red muscle show excessive peripheral concentrations of glycogen granules, which are so numerous and so uniformly dispersed that Hochachka and Hulbert (1978) referred to them as glycogen 'seas'. In all observed cases, the glycogen 'seas' surround the immense nuclei that typify lungfish cells. Large glycogen reservoirs such as these glycogen 'seas' were considered clearly advantageous during hypoxic or anoxic episodes even in the red muscle, which is often considered to be a relatively aerobic, oxygen-dependent tissue. Furthermore, lungfish heart has a well-developed sarcoplasmic reticulum, unlike other fish hearts, where this structure is not so well developed. This unusually shaped sarcoplasmic reticulum is in close association with glycogen and more rarely with mitochondria. Thus, the ultrastructure of white and red muscles and the heart of *Lepidosiren paradoxa* were suggested to be adapted for diving, burrowing, and aestivating (Hochachka and Hulbert 1978).

Cardiorespiratory studies done by Costa and co-workers (2002) have indicated that *Lepidosiren*, in spite of having a potentially functional and anatomically well-developed Sarcoplasmic Reticulum (SR), does not present a high calcium-cycling capacity, which is not compatible with high heart frequencies as temperature increases. However, as temperature decreases, this organ has a fundamental role in calcium regulation. SR becomes essential to cardiac performance maintenance when temperature drops down to 15°C.

The activities of several selected enzymes used in energy metabolism of South American lungfish heart showed a high anaerobic and a relatively low aerobic capacity, in comparison with other fishes. Such an ability is guaranteed by the extremely high activities of LDH (lactate dehydrogenase) enzyme that is clearly used during prolonged submergence in the African lungfish (Dunn et al. 1983) and is most likely needed during periods of aestivation (Almeida-Val and Hochachka 1995).

Reviewing the main biochemical metabolic characteristics of air-breathing fish, we have suggested that lungfish heart is similar to the hearts of anoxia-resistant vertebrates in its substrate preferences (Almeida-Val and Hochachka 1995). Their enzyme profiles are in accordance with preferential use of carbohydrates instead of lipid (as in the aquatic turtle, Almeida-Val et al. 1994). In the same review, we suggested that glycogen storages in lungfish heart and skeletal muscles are organized to meet energy needs during submergence and recovery (Almeida-Val and Hochachka 1995).

The metabolic adaptations found in African lungfish by Dunn and co-workers (1983) during submergence and subsequent recovery are summarized as (i) glycogen depletion; (ii) lactate accumulation; (iii) change in lactate/pyruvate ratio; (iv) change in adenylate concentrations; and (v) significant creatine phosphate depletion in epaxial muscles during submergence.

The metabolic regulation in aestivation was studied by Frick et al. (2008a,b) which observe a down-regulation in oxidative metabolic of African lungfish (*P. dolloi*), and show induction of GDH and Asp-AT activities in liver, which suggest the catabolism of amino acid for ATP production. The glycogen stored is only used when necessary, probably to emerge from the cocoon. Although there is no data available for *L. paradoxa* during submergence, we have analyzed this species while aestivating in the mud. Some of these data will be further presented in the chapter.

Red Cell Features

Lungfish possess, among other water-soluble phosphates such as ATP and GTP that are commonly present in fish erythrocytes, the compound Inositol diphosphate (Bartlett 1978a,b). This is a novel compound and no explanation has been found for the presence of this phosphate along with the more commonly found erythrocytic phosphates in fish. Inositol pentaphosphate has been found in another Amazonian air-breathing fish, *Arapaima gigas* (Bartlett 1978a), in which a strict correlation has been shown between animal size and the turn-on of the air-breathing habit and the increase in the concentration of this phosphate (Val et al. 1992; Val 2000). It is possible that inositol polyphosphates play a role in the modulation of the hemoglobin oxygen affinity in animals that acquired air-breathing respiration patterns while still living in water, where they are obliged to dive and breath-hold periodically. The uptake of oxygen from lung and its release to the tissues is a difficult task when the animal must hold breath while maintaining normal energy requirements.

Another interesting study utilizing lungfish red cells investigated their membrane permeability. According to Kim and Isaacks (1978) lungfish red cells are permeable to glucose and urea, facilitating the transport of both metabolites. Glucose is the main metabolic fuel for both anaerobic and aerobic glycolysis in most tissues and urea is a less toxic end product of nitrogen metabolism in lungfish and plays a significant role in their adaptation to aestivation and osmoregulation of blood of several other vertebrate species.

Nitrogen Metabolism

Most fish excrete nitrogenous waste as ammonia via their gills, because of soft dissolution in the surrounding water. Although ammonia (NH_3) can be considered highly soluble in tissues and plasma in its gaseous form, it strongly attracts protons and forms the ammonium ion (NH_4^+) in cellular milieu and may be kept dilute prior to excretion, or converted into less toxic compounds such as urea or uric acid (Hochachka and Somero 1973; Walsh and Henry 1991).

Lungfish is found in regions with extreme environmental fluctuations, when the oxygen may decrease drastically in the dry season. In normal conditions they depend on both ammonia and urea as nitrogen wastes, ammonia is released at the gills as NH_4^+ serving as a counter-ion to Na^+, and the urea is released at the kidneys, but if environmental water availability is limited and the lungfish aestivates (African and South American), NH_4^+ excretion ceases, and the catabolically formed ammonia can be detoxified by conversion to urea (ureotelism) that may be voided via either the urine or gills (Mommsen and Walsh 1992; Chew et al. 2003). Ureotelism evolved prior to animals conquering the land, as a preadaptation such as air breathing.

In fish, urea is the predominant form of nitrogen excretion in marine elasmobranchs and coelacanth (*Latimeriac halumnae*). Some air-breathers such as the African and South American lungfish also form large amounts of urea (Smith 1931; Campbell 1973; Hochachka and Somero 1984; Mommsen and Walsh 1989; Chew et al. 2004). These latter species synthesize urea through the ornithine urea cycle (OUC) where arginine is catalyzed by arginase to form urea (Hochachka and Somero 1984). The OUC is the major pathway for nitrogen excretion in terrestrial vertebrates and the presence of this pathway in the African and South American lungfishes, as well as in tetrapods, provides the rationale for the evolutionary link between the appearance of the OUC and the invasion of land by vertebrates. The OUC avoids problems caused by hypercapnia, desiccation, and nitrogen excretion imposed by air breathing and terrestriality. In water, both African and South American lungfish are ammoniotelic and have urea plasma levels equivalent to other freshwater fish species. During aestivation, however, lungfish use the OUC to form and store urea and avoid toxicity caused by ammonia accumulation when there is no possibility of excretion to the surroundings (Hochachka and Somero 1973; Chew et al. 2003, 2004). In contrast, the Australian lungfish is ammotelic throughout its life.

Distribution, Habitat, and Biology of Lungfishes

The South American lungfish was the first living lungfish discovered during the expedition of the Viennese naturalist Johann Natterer to the Amazon River, Brazil in 1836 (Hyrtl 1845, *apud* Graham 1997). Besides that, this species is not as well studied as the single Australian species *Neoceratodus forsteri* or the four African species that belong to the genus *Protopterus*.

Lepidosiren paradoxa is found in the neotropics of South America, including Argentina, Bolivia, Brazil, Columbia, French Guiana, Paraguay, Peru, and Venezuela (Lowe-McConnell 1987; Planquette et al. 1996). It is preferentially found in the Amazon River basin, inhabiting stagnant or lentic water systems, such as swamps or lakes, which are associated with vegetation and low oxygen conditions. This species lives in a subtropical climate with temperatures ranging from 24 to 28°C. It is of particular interest that *Lepidosiren* has the most extensive distribution of all extant lungfishes, occurring as it does in the tributaries of the Amazon River and Parana-Paraguay River systems, being also found in Guiana.

The four *Protopterus* lungfish species have a broad geographical distribution in the African continent, living in shallow water areas or large rivers. The natural populations of *Protopterus aethiopicus* are extensively spread in eastern and central Africa, surrounding Congo and Nile Rivers and their tributaries and major rivers and lakes, such as Lakes Victoria, Tanganyika, Albert, Edward, George, and Kyoga. The other species, *Protopterus annectens*, is found in western Africa and in the Zambezi and Limpopo Rivers of southern Africa. *P. dolloi* primarily occurs in Congo basin, whereas *P. amphibius* is distributed in East Africa. Considering their natural distribution, sympatric populations of these four species are probably rare, but they may co-occur in some places (Mlewa et al. 2010). The morphological and ecological diversification of *Protopterus* species occurred during Cretaceous, when the climatic conditions of the rain tropical forests have changed dramatically, allowing the occupation of different habitats. Populations of Australian lungfish, *Neoceratodus forsteri*, are restricted to the river systems within Southeast Queensland, where the species inhabits river channels and tributary streams (Kind 2010).

South American and African lungfishes are morphologically similar. Fish of both genera exhibit elongated bodies, paired fins and one continuous diphycercal tail (Bemis et al. 1987). *Lepidosiren* and *Protopterus* have flexible wisps, which, while aiding minimally in locomotion, have tactile, and chemosensory capabilities that are useful in orientation and prey capture. On the other hand, Australian lungfishes are recognized by distinctive external features: robust elongated body and present pectoral and pelvic fins frequently called 'flippers' or 'paddles' and are used during foraging (Kind 2010).

The respiratory physiology of lungfish presents differences between *Neoceratodus* and *Protopterus* and *Lepidosiren*. Australian lungfishes are bimodal air-breathers. They present well-developed gills, which support the respiration in well-aerated waters, and have a dorsal lung, used when oxygen levels are low or in forced exercise. However, lepidosirenid lungfishes are obligate air-breathers, i.e., the lung supports their respiration. This evidence was observed in laboratory studies with records of lungfish breaking the surface to breathe (Lenfant and Johansen 1968; Johansen et al. 1976; Greenwood 1986). In wild lungfish species, Mlewa et al. (2007) observed decrease of aerial respiration in open waters, attributed to high oxygen levels. Oduor et al. (2003) suggested that in well-oxygenated waters *P. aethiopicus* may be capable of meeting their oxygen demands as aquatic breathers.

During high water season, *Lepidosiren paradoxa* swims slowly and emerge to breathe air, although not very actively. Comparison between juveniles and adults swimming movements results in different behavior (Almeida-Val et al. 2010). Adults take longer time between breaths (approximately 45 minutes) and remain longer in the bottom of acquaria, showing a higher capacity to suppress activities compared to young animals (20 minutes). Previous observations described smaller intervals between breaths for this species (4–5 or 2–10 minutes) (Sawaya 1946; Johansen and Lenfant 1967). However, our findings rely on measurements of five individuals of each size and controlled environmental conditions (dark/light regime; temperature; quiet room) that allowed the installation of a digital camera in front of the acquaria to register these behaviors (Almeida-Val et al. 2010).

Aestivation

As already mentioned, *Neoceratodus* does not aestivate, while *Lepidosiren* and *Protopterus* stay burrowed during part of the year to survive desiccation periods. The South American, along with the African lungfish, are distinguished from most of the other known air-breathing fishes because of their ability to burrow in the mud and aestivate (Johansen 1966; Johansen and Lenfant 1967). Furthermore, these species are profoundly tolerant of low oxygen (Johansen 1966; Johansen 1970; Lahiri et al. 1970).

Many of these swampy areas inhabited by *Lepidosiren* and *Protopterus* species dry out on an annual basis, but the dryness varies in intensity according to the weather cycle of each year. When most of the water has dried up and only mud remains, *Lepidosiren* burrows in the mud up to 50 cm to avoid drying out (Berra 2001), leaving two or three holes for breathing purposes.

Although these animals are air-breathers, their ability to survive in this dry environment also involves adaptive mechanisms against desiccation. In this case, *Lepidosiren* does not form a mucous 'seal' (cocoon) as does *Protopterus*. This latter genus is more resistant to desiccation. The degree of aestivation varies among species, and has been particularly well studied in *Protopterus*. For both species the excavation of the burrows is made through biting the soil and expelling mud through the gill openings. After finishing this procedure, the fish turns around and remains with its head facing the burrow opening, which allows it to obtain oxygen.

Metabolic changes have been reported during aestivation periods, both to save energy and to endure the lack of moisture (Smith 1931; Janssens 1964; Delaney et al. 1974; Fishman et al. 1986). During aestivation, lungfishes do not feed to sustain themselves, they metabolize fat initially and when these stores are depleted, they begin to metabolize their muscle mass. This latter activity is constrained by nitrogen excretion (Janssens and Cohen 1966). Urea accumulated in their tissues (Janssens 1964) has been described as one of the most important adaptations in the transition from water to air breathing animals (Hochachka and Somero 1973).

Measurements of heart rates during aquatic and immersion phases by Harder and co-workers (1999) showed interesting facts. *Lepidosiren* decreases its heart rate during the first hour after burrowing, entering into bradycardia (heart rates below 46% of the control) and resume its control levels after the second hour of mud immersion. During the following 24 hours, their heart rates remain as high as control (aquatic phase) levels. Interestingly, apnea periods decrease in the burrowed phase. This short-term change in cardiorespiratory pattern to aestivation is also characterized by a significant increase in the lung ventilation rate, what has also been observed in air-exposed individuals by Johansen and Lenfant (1967) and can partially contribute to the increased arterial carbon dioxide partial pressure in the animal. Inducing artificial aestivation has showed to decrease oxygen uptake by 13 to 36% of the initial rate in this species (Perez-Gonzalez and Ginkraut 1971; Abe and Steffensen 1996). We also analyzed the enzyme activities of South American lungfish during aestivation (Mesquita-Saad et al. 2002). Animals were collected when aestivating in Lago do Canteiro, Careiro Island in the Amazon River (Amazonas state, Brazil), during the descending water period in August, 1997 (Fig. 1). Tissues were immediately dissected in the field and promptly stored in liquid nitrogen for transport to the laboratory at Manaus within three hours of collection. At the laboratory they were transferred to a low temperature freezer (–80°C). Maximum enzyme activities were determined and compared with those available in the literature. The comparison between the two conditions (aestivated fish versus non-aestivating fish) showed that the heart undergoes an absolute shutdown in enzyme activity during aestivation periods. The rates of anaerobic enzyme LDH (U.gwt^{-1}) decreases more than 300 times, and

Fig. 1. One of the scarce register of Lepidosiren paradoxa aestivating in the field. Former PhD student, Lenise Mesquita took the photo during descending water period in August 1997 at Lago do Canteiro, Careiro Island, nearby Manaus (from Almeida-Val et al. 2010).

the oxidative enzyme, CS, decreases almost 10 fold. Also, the gluconeogenic enzyme MDH presents a 100-time decrease during aestivation period. In fact, the CS/LDH rate, which is very low in awakened fish, increases by two orders of magnitude in aestivating fish, probably to avoid lactate accumulation and optimize ATP production, even at lower rates (Almeida-Val et al. 2010).

Among the four *Protopterus* species, *P. annectens* is considered more dependent on aestivation (Smith 1931). During the dry season, *P. annectens* are present in shallow habitats that are hydrologically disconnected to the mean rivers and then this specie excavates a short burrow, where it accommodates and starts to secrete mucus that gradually hardens to form a cocoon (Johnels and Svensson 1954) to reduced water loss, following a down-regulation of O_2 pressure in blood and in heart rate. On the other hand, for the other three *Protopterus* species, the cocoon formation appears to be a rare event in their natural environment (Brien et al. 1959; Greenwood 1986). However, *P. aethiopicus*, *P. amphibius* and *P. dolloi* may be induced to enter into aestivation under laboratory conditions (Smith 1931; Brien et al. 1959; Janssens 1964; Greenwood 1986; Perry et al. 2008) and they can produce cocoons similar to those described for *P. annectens*.

For both genera, this state can go on for periods of seven to eight months and awakening occurs only when water becomes available again (Janssens 1964; reviewed by Val and Almeida-Val 1995; Abe and Steffensen 1996; Mlewa et al. 2010). However, the survival time ever recorded *Protopterus* cocooned was seven years (Lomholt 1993). Thus, more observations are needed before we can conclude what are the actual physiological changes that occur during the entire period of aestivation in African and South American lungfish.

Reproductive Behavior

Lepidosiren has external fertilization and spawns seasonally during the wet season. Adult males guard and aerate the hatchlings and young temporarily. The larvae (which resemble tadpoles) respire through external gills that look like feathers for one to two months after which they start to breathe air. It is interesting to note that the male is the one to guard eggs and developing larvae within nests, and during this period it develops vascular filaments on their paired fins (Kerr 1900; Krogh 1941; Sterba 1963). The role of the fin filaments in respiration of *Lepidosiren* eggs is suggested because they have the format of gills. However, the function of such filaments in respiration is still under debate; the older accepted view though, is that these organs are auxiliary to respiration of the adult and enables it to guard the nest without leaving to breathe air; while a different view states that the fin filaments permit the 'emission' of aerially obtained oxygen into the nest water, causing the oxygenation of the surrounding water of the eggs (Cunningham 1932; Cunningham and Reid 1933). Although Foxon (1933) has defended the idea of auxiliary gills, the function of these filaments still remains uncertain (Graham 1997).

Males of three *Protopterus* species construct spawning nets in the shallow mud areas and also provide parental care of eggs and larvae. However, the reproductive behavior of *P. amphibius* is not yet fully understood (Mlewa et al. 2010). This nest structure resembles tunnels, where eggs and larvae occupy the area at the base of the tunnel. The reproductive periods of *P. annectens* and *P. aethiopicus* occur in the beginning of the wet season, whereas *P. dolloi* reproduces in the dry season. South American lungfish such as, the *Protopterus* larvae also use aerial respiration. Then, to improve increased dissolved oxygen levels for eggs and larvae through the tunnel system, *Protopterus* spp. vigorously move their posterior portion of the body by 'tail lashing' behavior. For *P. dolloi*, it has been reported that this species releases inspired air bubbles or produce air pockets into tunnel nests.

Neoceratodus breeding season starts in spring and continues until the summer season. Similar to other related lungfish species, the spawning site is also restricted along shallow river margins, where the fertilized eggs develop adhered to aquatic vegetation (Kemp 1986; Brooks and Kind 2002). The characteristics of Australian lungfish spawning site include protection from predators, dissolved oxygen supply for developing eggs and larvae and food items, since *Neoceratodus* does not provide parental care (Mlewa et al. 2010).

Food and Feeding Behavior

Neoceratodus forsteri presents a different diet along its ontogenetic development; when recently hatched this species is essentially an ambush predator. This species possesses isolated conical tooth cusps used to catch and hold their prey. Their feeding includes crustaceans and worms, eventually supplemented by filamentous algae. Adult lungfish eat a diversity of food, and is considered opportunistic omnivorous, feeding on small fishes, snails, tadpoles, fruit, and other parts of terrestrial plants (Kemp 1986). Individual forage mainly at night amongst macrophytes; they detect the vibrations of prey using ampullary organs scattered in cluster around the head and are capable of perceiving the weak electric fields generated by animals (Watt et al. 1999).

Protopterus species are carnivorous and their food habitats include: mollusks, small fishes, crustaceous, aquatic insects, and annelids. Some works reported a great intake of plant material (Outugbai 2001; Oniye et al. 2006) but the presence of these materials in the diet seems to be accidental since the structure of the digestive apparatus is adapted for a carnivorous diet (Amongi et al. 2001).

Lepidosiren paradoxa is primarily carnivorous. Their food items include bony fish, shrimp, insects, clams, snails and some algae, weeds and terrestrial plants (stems) (reviewed by Berra 2001). Their larvae remain relatively inactive and attached to the nest until their yolk reserves have been depleted. After that, juveniles start foraging for food and eat insect larvae, crustaceans and snails since they are strictly aquatic. Adults primarily ingest their prey by sucking it into their mouths where they use their tooth plate and depressor mandibulae to crush food prior to swallowing.

There is no evidence of any natural predator of this species once they are adults and no features that may suggest some anti-predator characteristics. Lungfish share the characteristics of strong teeth and enlarged cranial ribs that serve as the site for the origin of the muscle that depresses the hyoid apparatus. This apparatus, together with the tongue, promotes the hydraulic transport and positioning of the prey inside the mouth (Bemis 1986). There is no evidence of the use of their strong teeth for anti-predator defense, although there are some reports that describe painful bites with their strong jaws and sharp teeth if provoked in captivity. Some authors presume that larger carnivorous fishes and other vertebrates prey on lungfishes, particularly when they are young but there is no direct evidence for this yet.

The morphology and function of the feeding apparatus of the South American lungfish, *Lepidosiren paradoxa*, has been studied and was described by Bemis and Lauder in 2005 as very specialized. They reported that this species retains many primitive teleost characteristics and that the process of initial prey capture shares functional features with other primitive vertebrates, while feeding cycles, including alternate adduction and transport phases, are specialized. In *Protopterus*, the capture of prey is by suction, it chews using sharp incisors capable of cutting flesh and highly modified lateral molariform tooth plates that can crush hard bodies prey (Greenwood 1986; Outugbai et al. 2004). This dentition is accompanied by powerful jaw adductor muscle; this mode of feeding is characteristic of all *Protoperus*, and is an important factor in their ability to inhabit harsh unproductive environments.

Concluding Remarks

Although many new developments have been made since we reviewed the biological features of *Lepidosiren paradoxa* (Val and Almeida-Val 1995; Almeida-Val et al. 2010) several questions remain to be answered regarding the impressive processes that allow these animals to remain burrowed during months of aestivation with low energy (ATP) turnover. Besides aestivation, it will be important to analyze the mechanisms involved in the degeneration of external gills, which occurs 45 days after hatching and has no noticeable effect on growth rate of the larvae. If apoptosis (cell programmed death) mechanisms are responsible for the 're-absorption' of these specialized 'gills', how do these processes manage to rely on such low metabolic costs? We know that these animals are remarkably efficient in maintaining tissue oxygenation using gills during the early yolk absorption process. Also remarkable is the specialized capacity observed in adult males that, while taking care of their eggs stirring the surrounding water of their nests, develop a considerable number of respiratory filaments on their pelvic fins. Such transformation mechanisms are very specialized and should be investigated from histological and molecular point of view.

These and all other questions raised in this chapter may provide a good background for future evolutionary ecology studies of this group. We have no doubt that the lungfishes are a living proof that morphological, genetic, physiological, biochemical, and molecular processes have resulted from evolutionary changes based on the selection of many adaptive and specialized traits.

Acknowledgments

This work was funded by INCT/ADAPTA (CNPq/FAPEAM). VMFAV is the recipients of a Research fellowship from CNPq.

References

Abe, A.S. and J.F. Steffensen. 1996. Respiração pulmonar e cutânea da Pirambóia, *Lepidosiren paradoxa*, durante a atividade e a estivação (Osteichthyes, Dipnoi). Rev. Bras. Biol. 56: 485–489.

Allendorf, F.W., R.F. Learry and K.L. Kundsen. 1983. Structural and regulatory variation of phosphoglucomutase in rainbow trout. pp. 123–142. *In*: M.C. Rattazzi, J.G. Scandalios and G.S. Whitt (eds.). Isozymes: Current Topics in Biological and Medical Research. Alan R. Liss, New York.

Almeida-Val, V.M.F. and P.W. Hochachka. 1995. Air-breathing fishes: metabolic biochemistry of the first diving vertebrates. pp. 44–45. *In*: P.W. Hochachka and T. Mommsen (eds.). Biochemistry and Molecular Biology of Fishes, Environmental and Ecological Biochemistry. Elsevier Science, Amsterdam.

Almeida-Val, V.M.F., L.T. Buck and P.W. Hochachka. 1994. Substrate and temperature effects on turtle heart and liver mitochondria. Am. J. Physiol. 266: R858–R862.

Almeida-Val, V.M.F., S.R. Nozawa, N.P. Lopes, P.H. Rocha Aride, L.S. Mesquita-Saad, M.N. Paula-Silva, R.T. Honda and M.S. Ferreira-Nozawa. 2010. pp. 128–147. *In*: J.M. Jorgensen and J. Joss (eds.). Biology of the South American Lungfish, *Lepidosiren paradoxa*. Science, Enfield.

Amongi, T., R.T. Muwazi and I. Adupa. 2001. The gastrointestinal tract of the African lungfish *Protopterus aethiopicus*: structure and function. J. Morphol. 248: 202.

Avise, J.C. 1994. Molecular Markers, Natural History and Evolution. Chapman & Hall, New York.

Bartlett, G.R. 1978a. Phosphates in red cells of two South American osteoglossids: *Arapaima gigas* and *Osteoglossum bicirrhosum*. Can. J. Zool. 56: 878–881.

Bartlett, G.R. 1978b. Water-soluble phosphates of fish red cells. Can. J. Zool. 56: 870–877.

Bemis, W.E. 1984. Paedomorphosis and the evolution of the Dipnoi. Paleobiology 10: 293–307.

Bemis, W.E. 1986. Vertebrate evolution: evolutionary biology of primitive fishes. Science 233: 114–115.

Bemis, W.E. and G.V. Lauder. 2005. Morphology and function of the feeding apparatus of the lungfish, *Lepidosiren paradoxa* (Dipnoan). J. Morphol. 187: 81–108.

Bemis, W.E., W.W. Burggren and N.E. Kemp. 1987. The Biology and Evolution of Lungfishes. Alan R. Liss, New York.

Berra, T.M. 2001. Freshwater Fish Distribution. Academic Press, San Diego.

Brien, P., M. Pool and J. Bouillon. 1959. Ethologie de les reproduction de *Protopterus dolloi*. Soc. R. Zool. Belg. 89: 9–48.

Brooks, S.G. and P.K. Kind. 2002. Ecology and Demography of the Queensland Lungfish (*Neoceratodus forsteri*) in the Burnett River, Queensland with Reference to the Impacts of Walla Weir and Future Water Infrastructure Development: Final Report. Agency for Food and Fibre Sciences, Queensland.

Buth, D.G. 1983. Duplicate isozyme loci in fishes: origins, distribution, phyletic consequences and locus nomenclature. pp. 381–400. *In*: M.C. Rattazzi, J.G. Scandalios and G.S. Whitt (eds.). Isozymes: Current Topics in Biological and Medical Research. Alan R. Liss, New York.

Campbell, J.W. 1973. Nitrogen excretion. pp. 279–316. *In*: C.L. Prosser (ed.). Comparative Animal Physiology. W.B. Saunders, Philadelphia.

Carrol, R.L. 1988. Vertebrate Paleontology and Evolution. W.H. Freeman, New York.

Cavalier-Smith, T. 1985. Selfish DNA and the origin of introns. Nature 315: 283–284.

Cavalier-Smith, T. 1991. Coevolution of vertebrate genome, cell, and nuclear sizes. pp. 51–86. *In*: G.G. Gianfranco (ed.). Symposia on the Evolution of Terrestrial Vertebrates Selected Symposia and Monographs V Zl. Mucchi, Modena, Italy.

Chew, S.F., T.F. Ong, L. Ho, W.L. Tam, A.M. Loong, K.C. Hiong, W.P. Wong and W.P. Ip. 2003. Urea synthesis in the African lungfish *Protopterus dolloi*—hepatic carbamoyl phosphate synthetase III and glutamine synthetase can be up-regulated by 6 days of aerial exposure. J. Exp. Biol. 206: 3615–3624.

Chew, S.F., N.K.Y. Chan, A.M. Loong, K.C. Hiong, W.L. Tam and Y.K. Ip. 2004. Nitrogen metabolism in the African lungfish (*Protopterus dolloi*) estivating in a mucus cocoon on land. J. Exp. Biol. 207: 777–786.

Clack, J.A. 2002. Gaining Ground: The Origin and Evolution of Tetrapods. Indiana University Press, Bloomington.

Costa, M.J., C.D. Olle, J.A. Ratto, L.C. Anelli, Jr., A.L. Kalinin and F.T. Rantin. 2002. Effect of acute temperature transitions on chronotropic and inotropic responses of the South American lungfish *Lepidosiren paradoxa*. J. Therm. Biol. 27: 39–45.

Cunningham, J.T. 1932. Experiments on the interchange of oxygen and carbon dioxide between the skin of *Lepidosiren* and the surrounding water, and the probable emission of oxygen by the male *Symbranchus*. Proc. Zool. Soc. Lond. 102: 875–887.

Cunningham, J.T. and D.M. Reid. 1933. Pelvic filaments of *Lepidosiren*. Nature 131: 913.

DeLaney, R.G., S. Lahiri and A.P. Fishman. 1974. Aestivation of the African lungfish *Protopterus aethiopicus*: Cardiovascular and respiratory functions. J. Exp. Physiol. 61: 111–128.

Dunn, J.F., P.W. Hochachka, W. Davison and M. Guppy. 1983. Metabolic adjustments to diving and recovery in the African lungfish. Am. J. Physiol. 245: R651–R657.

Fishman, A.P., A.I. Pack, R.G. Delaney and R.J. Galante. 1986. Estivation in *Protopterus*. pp. 237–248. *In*: W.E. Bemis, W.W. Burggren and N.E. Kemp (eds.). Biology and Evolution of Lungfishes. Alan R. Liss, New York.

Foxon, G.E.H. 1933. Pelvic filaments of *Lepidosiren*. Nature 131: 913–914.

Frick, N.T., J.S. Bystriansky, Y.K. Ip, S.F. Chew and J.S. Ballantyne. 2008a. Lipid, ketone body and oxidative metabolism in the African lungfish, *Protopterus dolloi* following 60 days of fasting and aestivation. Comp. Biochem. Physiol. Part A 151: 93–101.

Frick, N.T., J.S. Bystriansky, Y.K. Ip, S.F. Chew and J.S. Ballantyne. 2008b. Carbohydrate and amino acid metabolism in fasting and aestivating African lungfish (*Protopterus dolloi*). Comp. Biochem. Physiol. Part A 151: 85–92.

Graham, J.B. 1997. Air-breathing Fishes: Evolution, Diversity, and Adaptation. Academic Press, San Diego.

Greenwood, P.H. 1986. The natural history of African lungfishes. J. Morphol. 190 Suppl. 1: 163–179.

Gregory, T.R. 2002. Genome size and developmental complexity. Genet. 115: 131–146.

Guppy, M., W.C. Hulbert and P.W. Hochachka. 1977. The tuna power plant and furnace. pp. 153–181. *In*: G. Sharp and A. Dizon (eds.). Physiological Ecology of Tuna. Academic Press, New York.

Harder, V., R.H.S. Souza, W. Severi, F.T. Rantin and C.R. Bridges. 1999. The South American lungfish—adaptations to an extreme habitat. pp. 87–98. *In*: A.L. Val and V.M.F. Almeida-Val (eds.). Biology of Tropical Fishes. Editora do INPA, Manaus.

Hennig, W. 1966. Phylogenetics Systematics. University of Illinois, Urbana.

Hinegardner, R. 1968. Evolution of cellular DNA content in teleost fishes. Am. Nat. 102: 517–523.

Hochachka, P.W. and G.N. Somero. 1973. Strategies of Biochemical adaptation. W.B. Saunders Company, Philadelphia.

Hochachka, P.W. and W.C. Hulbert. 1978. Glycogen seas, glycogen bodies, and glycogen granules in heart and skeletal muscle of two air-breathing, burrowing fishes. Can. J. Zool. 56: 774–786.

Hochachka, P.W. and G.N. Somero. 1984. Biochemical Adaptations. Princeton University Press, Princeton.

Hyrtl, J. 1845. *Lepidosiren paradoxa*. Bohm. Gesell. Abh. 3: 605–668.

Janssens, P.A. 1964. The metabolism of the aestivating African lungfish. Comp. Biochem. Physiol. 11: 105–117.

Janssens, P.A. and P.P. Cohen. 1966. Ornithine-urea cycle enzymes in the African lungfish *Protopterus aethiopicus*. Science 152: 358–359.

Johansen, K. 1966. Air breathing in the teleost *Symbranchus marmoratus*. Comp. Biochem. Physiol. 18: 383–395.

Johansen, K. 1970. Air-breathing in fishes. pp. 361–411. *In*: W.S. Hoar and D.J. Randall (eds.). Fish Physiology. Academic Press, New York.

Johansen, K. and C. Lenfant. 1967. Respiratory functions in the South American lungfish, *Lepidosiren paradoxa*. J. Exper. Biol. 46: 205–218.

Johansen, K., J.P. Lomholt and G.M. Maloiy. 1976. Importance of air and water breathing in relation to size of the African lungfish *Protopterus amphibius* Peters. J. Exp. Biol. 65: 395–399.

Johnels, A.G. and G.S.O. Svensson. 1954. On the biology of *Protopterus annectens* (Owen). Arkiv. Zool. 7: 131–164.

Kemp, A. 1986. The biology of the Australian lungfish, *Neoceratodus forsteri* (Krefft 1870). J. Morph. Suppl. 1: 181–198.

Kerr, J.G. 1900. The external features in the development of *Lepidosiren paradoxa*, Fitz. Phil. Trans. of the R. Soci. of London. Series B, Containing Papers of a Biological Character 192: 299–330.

Kim, H.D. and R.E. Isaaks. 1978. The membrane permeability of non electrolytes and carbohydrate metabolism of Amazon fish red cells. Can. J. Zool. 56: 863–869.

Kind, P.K. 2010. The natural history of the Australian lungfish *Neoceratodus forsteri* (Krefft 1870). pp. 61–95. *In*: J.M. Jorgensen and J. Joss (eds.). The Biology of Lungfishes. Science Publishers, Enfield.

Krogh, A. 1941. The Comparative Physiology of Respiratory Mechanisms. University of Pennsylvania, Philadelphia.

Kumar, S. and B. Hedges. 1998. A molecular timescale for vertebrate evolution. Nat. 392: 917–920.

Lahiri, S., J.P. Szidon and A.P. Fishman. 1970. Potential respiratory and circulatory adjustments to hypoxia in the African lungfish. Fed. Proc. 29: 1141–1148.

Lenfant, C. and K. Johansen. 1968. Respiration in the African lungfish *Protopterus aethiopicus* I. Respiratory properties of blood and normal patterns of breathing and gas exchange. J. Exp. Biol. 49: 437–452.

Lomholt, J.P. 1993. Breathing in the aestivating African lungfish, *Protopterus amphibius*. Adv. Fish. Res. 1: 17–34.

Long, J.A. 1995. The Rise of Fishes: 500 Million Years of Evolution. Johns Hopkins University Press, Baltimore.

Lowe-McConnell, R.H. 1987. Ecological Studies in Tropical Fish Communities. Cambridge University Press, Cambridge.

Mesquita-Saad, L.S.B., M.A.B. Leitão, M.N. Paula-Silva, A.L. Val and V.M.F. Almeida-Val. 2002. Specialized metabolism and biochemical suppression during aestivation of the extant South American Lungfish—*Lepidosiren paradoxa*. Braz. J. Biol. 62: 495–501.

Mlewa, C.M., J.M. Green and R. Dunbrack. 2007. Are wild African lungfish obligate air breathers? Some evidence from radio telemetry. Afri. Zool. 42: 131–134.

Mlewa, C.M., J.M. Green and R.L. Dunbrack. 2010. The general natural history of the African lungfishes. pp. 97–127. *In*: J.M. Jorgensen and J. Joss (eds.). The Biology of Lungfishes. Science Publishers, Enfield.

Mommsen, T.P. and P.J. Walsh. 1989. Evolution of urea synthesis in vertebrates: The piscine connection. Science 243: 72–75.

Mommsen, T.P. and P.J. Walsh. 1992. Biochemical and environmental perspectives on nitrogen metabolism in fishes. Experientia 48: 583–593.

Ohno, S. 1970. The enormous diversity in genome size of fish as a reflection of nature's extensive experiments with gene duplication. Trans. Am. Fish. Soc. 99: 120–130.

Oduor, S.O., M. Schagerl and J.M. Mathooko. 2003. On the limnology of Lake Baringo (Kenya): I. temporal physico-chemical dynamics. Hydrobiologia 506: 121–127.

Oliveira, C., L.F.A. Toledo, F. Foresti, H.A. Britski and S.A. Toledo-Filho. 1988. Chromosome formulate of neotropical freshwater fishes. Rev. Bra. Gen. 11: 577–624.

Oniye, S.J., D.A. Adebote, S.K. Usman and J.K. Makpo. 2006. Some aspects of the biology of *Protopterus annectens* (Owen) in Jachi dam near Katsina, Katsina state, Nigeria. J. Fish. Aqu. Sci. 1: 136–141.

Otuogbai, T.M., A. Ikhenoba and I. Elakhame. 2004. Food and feeding habits of the African lungfish, *Protopterus annectens* (Owens) (Pisces: Sarcopterygii) in the flood plains of River Niger in Etasako east of Edo State, Nigeria. Afr. J. Trop. Hydrobio. Fish. 10: 14–26.

Pedersen, R.A. 1971. DNA content, ribosomal gene multiplicity and cell size in fish. J Exp. Zool. 177: 65–78.

Perez-Gonzalez, M.D. and C.N. Grinkraut. 1971. Comportamento e metabolismo respiratório da *Lepidosiren paradoxa* durante a vida aquática e em estivação. Bol. Zool. Biol. Mar. 28: 137–164.

Perry, S.F., R. Euverman, T. Wang, A.M. Loong, S.F. Chew, Y.K. Ip and K.M. Gilmour. 2008. Control of breathing in African lungfish (*Protopterus dolloi*): a comparison of aquatic and cocooned (terrestrialized) animals. Resp. Physiol. Neurobiol. 160: 8–17.

Planquette, P., P. Keith and P.Y. Lebail. 1996. Atlas des Poissons d'Eau Douce de Guyane, Tome 1. Museum National d'Histoire Naturelle, Paris.

Pritchard, A.W., J.R. Hunter and R. Lasker. 1971. The relation between exercise and biochemical changes in red and white muscle and liver in the jack mackerel, *Trachurus symmetricus*. Fish. Bull. 69: 379–386.

Romer, A.S. 1966. Vertebrate Paleontology. University of Chicago Press, Chicago.

Rosen, D.E., P.L. Forey, B.G. Gardiner and C. Patterson. 1981. Lungfishes, tetrapods, paleontology, and plesiomorphy. Bull. Am. Mus. Nat. Hist. 167: 159–276.

Sawaya, P. 1946. Sobre a biologia de alguns peixes de respiração aérea (*Lepidosiren paradoxa* Fitzinger e *Arapaima gigas* Cuvier). Bol. Facul. Fil. Ciên. Let. Univ. São Paulo 11: 255–286.

Sirijovski, N., C. Woolnough, J. Rock and J.M. Joss. 2005. *Nf*CR1, the first non-LTR retrotransposon characterized in the australian lungfish genome, *Neoceratodus forsteri*, shows similarities to CR1-like elements. J. Exp. Zool. 304B: 40–49.

Smith, H.M. 1931. Metabolism of the lungfish *Protopterus aethiopicus*. J. Biol. Chem. 88: 97–130.

Stanley, S.M. 1984. Simpon's inverse: Bradytely and the phenomenon of living fossil. pp. 272–277. *In*: N. Eldredge and S.M. Staley (eds.). Living Fossil. Springer-Verlag, New York.

Sterba, G. 1963. Freshwater Fishes of the World. Viking Press, New York.

Takezaki, N., F. Figueroa, Z. Zaleska-Rutczynska, N. Takahata and J. Klein. 2004. The phylogenetic relationship of Tetrapod, Coelacanth, and Lungfish revealed by the sequences of forty-four nuclear genes. Mol. Biol. and Evol. 21: 1512–1524.

Val, A.L. 2000. Organic phosphates in the red blood cell of fish. Comp. Bioch. Physiol. A: Comp. Physiol. 125: 417–435.

Val, A.L. and V.M.F. Almeida-Val. 1995. Fishes of the Amazon and their Environments. Physiological and Biochemical Features. Springer Verlag, Heidelberg.

Val, A.L., E.G. Affonso and V.M.F. Almeida-Val. 1992. Adaptive features of amazon fishes: blood characteristics of Curimatā (Prochilodus cf. nigricans, Osteichthyes). Physiol. Zool. 65: 832–843.

Vinogradov, A.E. 2005. Genome size and chromatin condensation in vertebrates. Chrom. 113: 362–369.

Walsh, P.J. and R.P. Henry. 1991. Carbon dioxide and ammonia metabolism and exchange. pp. 181–207. *In*: P.W. Hochachka and T.P. Mommsen (eds.). Biochemistry and Molecular Biology of Fishes. Elsevier, New York.

Watt, M., C.S. Evans and J.M. Joss. 1999. Use of electroreception during foraging by the Australian lungfish. Ani. Beh. 58: 1039–1045.

Whitt, G.S. 1983. Isozymes as probe and participants in developmental and evolutionary genetics. pp. 1–40. *In*: M.C. Rattazzi, J.G. Scandalios and G.S. Whitt (eds.). Isozymes: Current Topics in Biological and Medical Research. Alan R. Liss, New York.

4

Developmental Physiology of the Australian Lungfish, *Neoceratodus forsteri*

Casey A. Mueller

Introduction

Ancient fishes have survived physiologically unchanged for millions of years in part because of successful reproduction and development. Yet, knowledge of the developmental physiology of ancient fishes is scarce. Physiological studies are a useful tool for revealing the developmental requirements of these unique fish species, and can allow us to predict how species will fare in the future, particularly in the face of a changing global climate. Fundamental questions concerning the developmental physiology of ancient fishes include: What mode of reproduction does an ancient species utilize? What are the egg characteristics, including clutch size, egg size and egg morphology? What are the physiological consequences of such egg characteristics? Under what environmental conditions does embryonic development occur? By addressing such questions, we can define the ecophysiological requirements for successful development of a particular species. Physiological data can also be compared between species to place the process of embryonic development in an evolutionary context. Such comparisons may shed light on why one species is more successful, or has persisted longer, than others.

The Australian lungfish, *Neoceratodus forsteri* (Krefft 1870), is one such ancient fish species for which morphological and physiological embryonic

Department of Biological Sciences, California State University San Marcos, 333 Twin Oaks Valley Road, San Marcos, CA 92096, USA.
E-mail: caseyamueller@gmail.com

development is well documented. The Australian lungfish is a sarcopterygian (lobed-finned) dipnoan (lungfish), and the only surviving species of the ancient Ceratodontidae family. Five other freshwater lungfish species survive today, including the South American lungfish (*Lepidosiren paradoxa*) of the family Lepidosirenidae and the African lungfishes, of the Protopteridae, including the marbled lungfish (*Protopterus aethiopicus*), the gilled African lungfish (*Protopterus amphibius*), the West African lungfish (*Protopterus annectens*), and the slender African lungfish (*Protopterus dolloi*). The lungfishes are species of interest due to the group now almost universally believed to be the closest living relatives to the tetrapods (land vertebrates, including amphibians) (Yokobori et al. 1994; Brinkmann et al. 2004; Amemiya et al. 2013). The Australian lungfish is considered the most primitive and displays a number of traits in common with the amphibians, and in particular the urodeles, including soft anatomical features, neoteny and a large genome size (Joss 2006; Gregory 2013).

Using the Australian lungfish as an example, this chapter illustrates how physiological studies can be used to understand how developmental metabolism of an ancient fish is related to and influenced by the environment. In this chapter, an overview of embryonic development of the Australian lungfish with particular focus on the effects of oxygen and temperature on (a) the success and length of development and (b) the energy use of embryos are provided. Where possible, data are placed in the context of other fishes and amphibians to better understand how and why the development of the Australian lungfish is unique and how it may be related to the species' success.

Ecology and Reproduction

At present, the Australian lungfish has a restricted distribution, located in a handful of rivers in Southeast Queensland, Australia. The species is endemic to at least two river systems, the Mary and Burnett Rivers. The species has also been successfully introduced to the Brisbane, Albert, Stanley and Coomera Rivers and the Enoggera Reservoir, though the Enoggera Reservoir population is now extinct (Kemp 2011). Within these river systems, lungfish live in small groups in water depths of 3–10 m under submerged logs or in dense macrophytes. The fish is usually quiescent, and considered quite sluggish, during the day, but more active during the evening. Growing to 1.5 m and 45 kg, the species averages around 1 m in length and 20 kg in mass (Allen et al. 2002), and feeds mainly on small water snails and clams (Kemp 2011). The species is a facultative air breather, with one single lung in contrast to the paired lungs of all other extant lungfish species, which are obligate air breathers (Johansen and Lenfant 1967; Johansen et al. 1967; Johansen and Lenfant 1968). The Australian lungfish cannot aestivate like the South American and African lungfishes.

The Australian lungfish is listed as a vulnerable species under the Commonwealth Environment Protection and Biodiversity Conservation Act 1999, and its capture in the wild is prohibited. Natural events, such as drought and flooding, and human disturbances have impacted the lungfish population (Kemp 2011). Construction of impoundments on the Burnett River system has reduced available breeding sites, which are characterized by shallow flowing water with dense beds of aquatic vegetation. As the lungfish does not readily seek out new breeding grounds, their populations may decline as recruitment decreases (Kemp 1986). Construction of a dam on the Mary River was vetoed in 2009 by the Australian Government due to the irreversible environmental changes that would have impacted a number of threatened species, including the lungfish. An understanding of the ecophysiology of embryonic development can provide insight into the effects of these habitat changes on the already limited Australian lungfish population.

Australian lungfish spawn from August to December in temperatures ranging from 16 to 26°C (Kemp 1984, 1986). Spawning sites are usually characterized by a slow or moderate water current to a depth of 1.5 m (Kemp 1984). Unlike the African and South American lungfishes, the Australian lungfish does not build a nest for the eggs, nor is there parental care of the eggs. Adults spawn in pairs during the day and night, laying one or two eggs at a site before moving to another site to repeat the process, thus it is difficult to determine how many eggs are produced by a single pair. An initially sticky outer jelly layer helps adhere the eggs to water plants, and eggs that fall to the river bed rarely survive (Kemp 1994).

Embryonic Growth and Development

At approximately 3.5 mm in diameter with a volume of 22 mm³ (Mueller et al. 2011a), the heavily yolked fresh Australian lungfish egg is closer in size to those of large-egged amphibians than other fishes. The South American and West African lungfish eggs are also large at approximately 7 mm and 3.5–4 mm in diameter, respectively. As with all fishes and amphibians, Australian lungfish eggs are anamniotic. The fresh ovum is surrounded by an egg capsule that consists of a strong vitelline membrane surrounding the perivitelline space and numerous jelly layers that increase the total diameter of the egg to approximately 6.7 mm, with a total volume of 160 mm³.

The morphological development of the Australian lungfish embryo is well documented (Kemp 1982). The developmental staging system is similar to that devised for the South American and West African lungfishes and illustrates the amphibian-like development of the species. The embryos pass through four cleavage stages (stages 2–6), five blastulation stages (stages 7–11), five gastrulation stages (stages 12–16) and 14 neurulation stages (stages 17–30). The tail bud, gills and eyes develop from stages 30–35, the dorsal and post anal fins appear from stage 36, the lateral line develops from stage 40 and the yolk begins to reduce and the mouth moves forward from stage 42. The embryos only take on a fish-like appearance with the development of fins, just prior to hatching. Hatching occurs around stages 42–43, but can be as early as stage 37 or as late as stage 45 (Kemp 1982; Mueller et al. 2011a,b). The Australian lungfish hatches late in development in comparison to the South African lungfish, which hatches at stage 33 (Kerr 1900), and the West African lungfish, which hatches at stage 34 (Budgett 1901). In contrast to some amphibians and the South American and African lungfishes, the Australian lungfish embryo lacks external gills.

Temperature

In their natural environment, Australian lungfish embryos experience a temperature range from 16 to 26°C (Kemp 1984). Embryos incubated outside the limits of this range, at 10 and 30°C, fail to develop (Kemp 1981). The relationship between age and developmental stage until hatching at 15, 20 and 25°C is presented in Fig. 1. At 15°C, the lower end of the natural temperature range, only 40% of embryos hatch after 72 days and those that fail to hatch

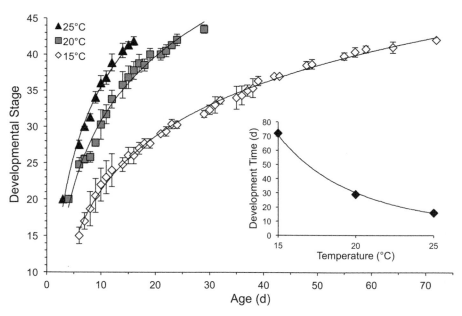

Fig. 1. Developmental stage as a function of age (d) at three incubation temperatures (°C). Stages derived from Kemp (1982). Data presented as mean ± 95% CI. The relationship between age and stage at each temperature is represented by the following equations: 15°C: stage = $0.00011age^3$ − $0.0185age^2$ + $1.2106age$ + 10.654, $r^2 = 0.98$, 20°C: stage = $0.0007age^3$ − $0.0689age^2$ + $2.5877age$ + 10.578, $r^2 = 0.98$ and 25°C: stage = $0.0012age^3$ − $0.1073age^2$ + $3.3257age$ + 11.456, $r^2 = 0.99$. Insert: Development time (d) against temperature (°C) described by the following equation: Development time = $211058T^{-2.955}$, $r^2 = 0.99$. Modified from Mueller et al. (2011b) with kind permission from Springer Science and Business Media.

show morphological abnormalities (Mueller et al. 2011b). All embryos hatch at 20 and 25°C (from a small sample size) and the time to reach hatching decreases from 20 to 25°C (Table 1). The effect of temperature on development time from 15 to 25°C is described by a temperature coefficient (Q_{10}), of 4.5. The relationship between development time to hatch and temperature is not linear, but best described by a power function (Fig. 1 insert). This is represented by a high Q_{10} of 6.17 from 15 to 20°C, compared to 3.28 for 20 to 25°C. While temperature strongly influences the development rate and hatching age, hatching occurs at the same morphological stage irrespective of temperature (Table 1). Hatchling wet mass, dry mass and water content are also constant across rearing temperatures from 15 to 25°C (Table 1). Likewise, residual yolk wet and dry mass, and yolk water content, do not change with temperature.

The incubation time of Australian lungfish embryos is relatively long despite developing at relatively warm temperatures. Two factors are likely to contribute to their slow development; a large egg size and a large genome. In fishes and amphibians, there is a strong positive correlation between egg size and time to hatching (Pauly and Pullin 1988; Bradford 1990). Furthermore, a large genome size, or high DNA content, slows cell division (Horner and MacGregor 1983; Gregory 2001) and there is evidence that this contributes to a longer development time (Goin et al. 1968; Bachmann 1972; Chipman et al. 2001). The Australian lungfish genome size has been estimated as 53–75 pg (Pedersen 1971; Rock et al. 1996; Gregory 2013), and this may also contribute to its slow rate of development. In fact, of all animals investigated, the extant lungfishes have the largest genomes (50–130 pg), followed somewhat closely by the urodele amphibians (13–120 pg) (Gregory 2002).

Table 1. Summary hatching data for Australian lungfish embryos incubated at three temperatures. Different superscript letters indicate a significant effect of temperature within each hatching parameter (P < 0.05, ANOVA, Tukey HSD). Data from Mueller et al. (2011a) with kind permission from Springer Science and Business Media.

Hatching Parameter	15°C	20°C	25°C
Age (d)	72[a]	28.7 ± 2.5[b]	15.8 ± 0.7[c]
Stage	41.0 ± 1.4	42.9 ± 1.0	41.9 ± 0.8
Wet Gut-free Body Mass (mg)	8.39 ± 1.10	8.93 ± 2.42	9.47 ± 0.48
Dry Gut-free Body Mass (mg)	1.74 ± 0.20	1.65 ± 0.50	1.71 ± 0.53
Gut-free Body Water Content (%)	78.56 ± 3.29	81.19 ± 4.90	81.84 ± 2.29
Wet Residual Yolk Mass (mg)	7.73 ± 0.48	8.83 ± 0.63	8.86 ± 0.72
Dry Residual Yolk Mass (mg)	4.22 ± 0.30	4.07 ± 0.46	4.23 ± 0.35
Residual Yolk Water Content (%)	47.11 ± 4.86	52.94 ± 6.72	51.81 ± 7.09
Total O_2 Consumed (ml)	0.358	0.233	0.160
Cost of Development (wet mass, ml mg^{-1})	0.043	0.026	0.017
Cost of Development (dry mass, ml mg^{-1})	0.206	0.141	0.094

Oxygen

The nature of Australian lungfish spawning, in which eggs are laid individually or in pairs in water plant beds, ensures a relatively high oxygen environment around the eggs. Most eggs are laid in spawning locations with oxygen partial pressures (Po_2) of 12–18 kPa (Kemp 1984). In laboratory raised eggs at 20°C, survival to 20 days of age is 60% under 20.9 kPa and progressively decreases to 45 at 15 kPa, 15 at 10 kPa and 5% at 5 kPa (Mueller et al. 2011a). Eggs have been found in natural spawning sites at a Po_2 as low as 2 kPa, but as eggs reared in the laboratory at this low oxygen level fail to develop beyond four days of development, these eggs are unlikely to survive.

Development time of Australian lungfish eggs is most strongly influenced by environmental oxygen during late embryonic and early larval development (Fig. 2). At day 13 of development at 20°C, embryos are at stage 34 when incubated in 5 and 10 kPa, a stage or two behind embryos incubated at 15 and 20.9 kPa (Mueller et al. 2011a). After this point development continues to slowly diverge at all Po_2 levels, a trend that continues after hatching. The slowing of development in response to decreased Po_2 is not unique to the lungfish and occurs due to the aerobic nature of fish embryos (Alderdice et al. 1958; Garside 1959; Hamor and Garside 1976). The time of hatching is also

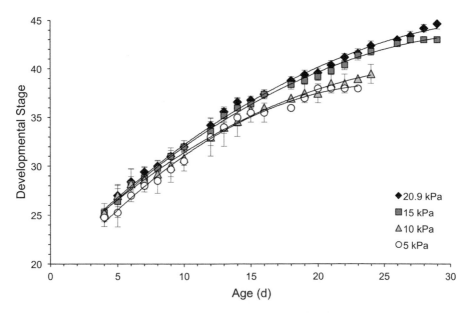

Fig. 2. Developmental stage as a function of age (d) for embryos incubated under four Po_2 conditions at 20°C. Data presented as mean ± 95% CI. The relationship between age and stage at each Po_2 is represented by the following equations: 20 kPa: stage = –0.019age^2 + 1.36age + 20.50, r^2 = 0.99, 15 kPa: stage = –0.020age^2 + 1.37age + 20.27, r^2 = 0.99, 10 kPa: stage = –0.021age^2 + 1.27age + 20.72, r^2 = 0.99, 5 kPa: stage = –0.031age^2 + 1.57age + 18.45, r^2 = 0.99. Modified from Mueller et al. (2011a) with kind permission from Springer Science and Business Media.

influenced by P_{O_2}, as both the age and stage of hatching decreases as oxygen availability declines (Table 2). Therefore, unlike temperature which does not change the time of hatching, hypoxia shifts the developmental event, causing premature hatching. A change in hatching age and stage occurs whether embryos are exposed to a particular P_{O_2} for all of embryonic development or just from day 15 (Mueller et al. 2011a). Therefore, exposure to hypoxia late in development acts as a hatching trigger, but whether this is occurs in the natural environment is unknown. Likewise, hatching occurs prematurely when chum salmon (*Oncorhynchus keta*) and rainbow trout (*Oncorhynchus mykiss*) are exposed to hypoxia just prior to hatching (Alderdice et al. 1958; Latham and Just 1989). Premature hatching occurs in response to low oxygen, as a free swimming larva can seek oxygen rich environments while an embryo cannot. Therefore, plasticity in hatching allows embryos to respond to and leave a poor oxygen environment. Despite the generally high oxygen environment in which the embryos develop, embryos have been found in hypoxic conditions (Kemp 1984), and therefore hatching plasticity may be important for increased embryo survival. Australian lungfish embryos display a unique behavior in that they will on occasion return to the mainly intact egg capsule through the hole made during hatching. Thus the embryos have the ability to respond to low P_{O_2}, but they still have the protective benefits of the egg capsule if required.

Table 2. Summary of hatching data for Australian lungfish embryos incubated at four P_{O_2} treatments at 20°C. Different superscript letters indicate a significant effect of P_{O_2} within each hatching parameter (P < 0.05, ANOVA, Tukey HSD). Data from Mueller et al. (2011b) with kind permission from Springer Science and Business Media.

Hatching Parameter	5 kPa	10 kPa	15 kPa	20.9 kPa
Age (d)	17.6 ± 2.5^a	20.0 ± 2.2^{ab}	23.9 ± 1.6^{bc}	26.0 ± 1.6^c
Stage	37.2 ± 0.4^a	38.8 ± 1.2^a	41.6 ± 0.7^b	42.8 ± 0.8^b
Wet Gut-free Body Mass (mg)	6.33 ± 1.13^a	7.35 ± 1.32^{ab}	10.78 ± 1.89^{bc}	11.29 ± 1.74^c
Dry Gut-free Body Mass (mg)	1.04 ± 0.09^a	1.18 ± 0.13^{ab}	1.63 ± 0.24^c	1.50 ± 0.12^{bc}
Gut-free Body Water Content (%)	82.8 ± 4.2	83.6 ± 2.1	84.5 ± 1.3	86.1 ± 1.7
Wet Residual Yolk Mass (mg)	8.77 ± 0.92^{ab}	9.44 ± 0.50^{ab}	8.82 ± 0.32^a	$9.63 + 0.63^b$
Dry Residual Yolk Mass (mg)	4.72 ± 0.48	4.77 ± 0.29	4.17 ± 0.28	4.28 ± 0.31
Residual Yolk Water Content (%)	45.9 ± 5.6^a	49.3 ± 3.6^{ab}	52.8 ± 2.5^{ab}	55.1 ± 4.4^b
Total O_2 Consumed to Hatch (ml)	0.031	0.052	0.102	0.211
Total O_2 Consumed up to stage 36 (ml)	0.031	0.023	0.028	0.026
Stage 36 Dry Gut-free Mass (mg)	1.04 ± 0.09^a	0.96 ± 0.12^a	1.15 ± 0.08^{ab}	1.34 ± 0.08^b
Cost of Development to stage 36 (dry mass, ml mg^{-1})	0.030	0.024	0.024	0.019

Incubation under varying Po_2 also alters mass at hatching, reflecting the shift in stage of hatching. Hatchling wet and dry masses are greater at higher Po_2, due to the later hatching stages (Table 2). However, water content of the hatchling body is constant. Despite differences in body mass, residual yolk wet and dry masses are not as strongly affected by Po_2. The large initial yolk content of the eggs ensures that the embryos always have residual yolk available at hatching, irrespective of environmental conditions.

Embryonic Oxygen Consumption

Fish embryos require oxygen from the environment for survival, development and growth, and oxygen consumption provides a measure of metabolic rate and energy use. The immediate oxygen environment of the embryo is the result of not only the environmental oxygen level surrounding the egg but also the resistances to gas exchange that lie between the environment and the embryo. The egg structure itself creates a number of resistances to the consumption of oxygen. A boundary layer of decreased oxygen, with little convective movement, may form in the region immediately surrounding the egg capsule of fish eggs (Rombough 1988b). Oxygen then moves through the capsule via diffusion and is therefore relatively slow (Seymour 1994). The diffusive nature of oxygen exchange across the egg capsule is described by the Fick equation:

$$\dot{M}o_2 = Go_2 \cdot (Po_{2(out)} - Po_{2(in)}) \tag{1}$$

in which the rate of oxygen flux ($\dot{M}o_2$, nmol h^{-1}) is the product of the capsule oxygen conductance (Go_2, nmol h^{-1} kPa^{-1}) and the difference in oxygen partial pressure (Po_2, kPa) between the outside and inside of the capsule.

Oxygen demand increases during fish development as embryos convert inert yolk into metabolically active tissue (Hayes et al. 1951; Alderdice et al. 1958; DiMichele and Powers 1984; Collins and Nelson 1993; Darken et al. 1998). Embryonic $\dot{M}o_2$ of Australian lungfish follows this trend with an uninterrupted increase during development (Fig. 3). As a result of faster development at higher temperatures, $\dot{M}o_2$ increases with temperature at the same embryonic age (Fig. 2 in Mueller et al. 2011b). However, higher $\dot{M}o_2$ at the same developmental stage at warmer temperatures reveals the direct effect of temperature on metabolic rate, particularly at later stages of development when oxygen demand is the highest (Fig. 3). Interestingly, maximum $\dot{M}o_2$ prior to hatching is similar at 20 and 25°C, which may represent a limitation to gas exchange, whether due to constraints of the egg or within the embryo itself. An increase in oxygen availability also augments metabolic rate at the same developmental stage, particularly later in embryonic development (Fig. 4A). Lower $\dot{M}o_2$ at Po_2 below 20.9 kPa indicates that embryos are unable to maintain their $\dot{M}o_2$ in hypoxic conditions, and are therefore oxyconformers, i.e., the critical Po_2 (P_C) at which $\dot{M}o_2$ begins to decline is at or above 20.9 kPa.

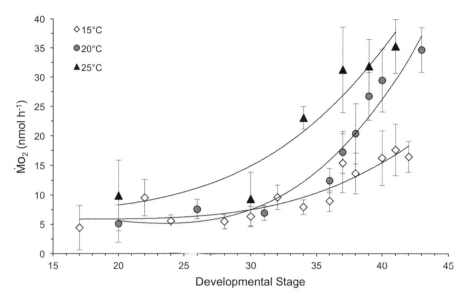

Fig. 3. Oxygen consumption rate ($\dot{M}O_2$, nmol h^{-1}) during embryonic development as a function of developmental stage at three incubation temperatures. Data presented as mean ± 95% CI. The relationship between stage and $\dot{M}O_2$ at each temperature is represented by the following equations: 15°C: $\dot{M}O_2 = 0.0009\text{stage}^3 - 0.0504\text{stage}^2 + 0.9367\text{stage}$, $r^2 = 0.84$, 20°C: $\dot{M}O_2 = 0.0022\text{stage}^3 - 0.1118\text{stage}^2 + 1.6453\text{stage}$, $r^2 = 0.95$ and 25°C: $\dot{M}O_2 = 0.0014\text{stage}^3 - 0.0637\text{stage}^2 + 1.1114\text{stage}$, $r^2 = 0.90$.

Other fish embryos demonstrate an increase in P_C as incubation proceeds, matching their increased metabolic demand (Rombough 1988a, 2007). Under hypoxia, the reduced $\dot{M}O_2$ may act as an 'oxygen signal' to trigger premature hatching. The inability of late-stage embryos to regulate $\dot{M}O_2$ is most likely due to the presence of the egg capsule and the resistances it imposes on gas exchange, as discussed in more detail later.

Egg Capsule Go$_2$

Most fish eggs take in water soon after fertilization and generally do not change in morphology throughout incubation (Alderdice et al. 1984). This is true of the South American lungfish (Kerr 1900), however, the Australian lungfish exhibits a unique change in egg morphology. At approximately half way through incubation the vitelline membrane and some jelly breaks away from the inner layer of the capsule. A further layer of jelly dissolves and the membrane and jelly remnants lie at the bottom of the egg (Kemp 1982). These morphological changes are important for ensuring the embryo has enough room in the egg to continue development. The rigid membrane does not have the ability to expand via the uptake of water, as occurs in amphibian eggs (Salthe 1963, 1965; Seymour 1994), and thus failure of the membrane to break

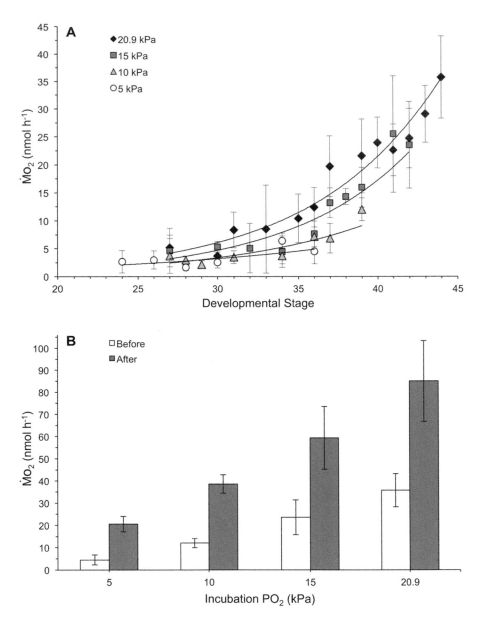

Fig. 4. (A) Oxygen consumption rate ($\dot{M}o_2$, nmol h^{-1}) during embryonic development as a function of environmental Po_2 at 20°C. Data presented as mean ± 95% CI. The relationship between embryonic stage and $\dot{M}o_2$ at each incubation Po_2 is shown by the fitted curve line and described by the following equations: 20.9 kPa: $\dot{M}o_2 = 0.14e^{0.126stage}$, $r^2 = 0.92$, 15 kPa: $\dot{M}o_2 = 0.10e^{0.129stage}$, $r^2 = 0.84$, 10 kPa: $\dot{M}o_2 = 0.13e^{0.110stage}$, $r^2 = 0.77$, 5 kPa: $\dot{M}o_2 = 0.39e^{0.070stage}$, $r^2 = 0.46$. Modified from Mueller et al. (2011a) with kind permission from Springer Science and Business Media. (B) $\dot{M}o_2$ before and immediately after hatching (24–48 hours) under selected Po_2 incubation treatments at 20°C. $\dot{M}o_2$ before and after hatching are separated by one or two developmental stages. Data presented as mean ± 95% CI.

down results in rapid embryo death. From a gas exchange perspective, the change in capsule morphology has consequences for the egg capsule oxygen conductance, GO_2, and therefore oxygen uptake of the embryo.

Using morphological measurements of the egg capsule, and Krogh's coefficient of oxygen diffusion (a measure of oxygen permeability) of the jelly (Ko_2; nmol mm^{-1} h^{-1} kPa^{-1}), egg capsule GO_2 can be estimated using the following equation:

$$GO_2 = \frac{4 \cdot \pi \cdot r_i \cdot r_o \cdot Ko_2}{r_o - r_i} \qquad (2)$$

where GO_2 depends on the effective surface area of a sphere, $ESA = 4 \cdot \pi \cdot r_i \cdot r_o$, in which the radius is the geometric mean of the inner (r_i: mm) and outer radii (r_o: mm) of the jelly layers, and the thickness of the jelly ($L = r_o - r_i$: mm) (Seymour and Bradford 1987).

Capsule morphology in Australian lungfish eggs remains unchanged until stage 32, the point of membrane breakdown. After stage 32, the remnants of the vitelline membrane and jelly create an uneven capsule. The top and sides of the egg capsule are thinner but the remnants of membrane and jelly create a thicker barrier at the bottom of the egg. By measuring capsule morphology at the top, bottom and sides of the egg an overall decrease in capsule thickness is demonstrated (Fig. 5A). The removal of the very rigid membrane within the egg allows for water uptake by the egg, which increases effective surface area and perivitelline volume (Fig. 5B,C). The decrease in thickness and increase in effective surface area, both important morphological variables in Equation 2, result in an increase in GO_2 (Fig. 5D).

The change in Australian lungfish egg morphology at stage 32 is rather abrupt, particularly in comparison to amphibian eggs which tend to show a gradual increase in perivitelline volume and GO_2 as water is absorbed into the egg (Salthe 1965; Seymour et al. 1991). After stage 32 the morphology of the egg tends to remain mostly unchanged, with only a small further increase in effective surface area and perivitelline volume. The breakdown of the membrane can be viewed as the beginning of a drawn out hatching process, and while water absorption occurs after stage 32, the loss of structural integrity of the capsule in the absence of the rigid membrane also contributes to morphological changes in the latter half of incubation.

Despite the importance of the membrane breakdown to embryonic survival and gas exchange, there appears to be no plasticity when the event occurs. While hatching age is plastic in response to altered Po_2, the change in capsule morphology always begins at stage 32, irrespective of Po_2 (Fig. 5) or temperature (see Fig. 6 in Mueller et al. 2011a). Furthermore, additional water absorption after stage 32 that may increase surface area of the capsule does not vary with incubation conditions. Therefore, there appears to be no selective pressure for GO_2 to be responsive to the environment, even when the capsule does impose a resistance to gas exchange under natural incubation conditions.

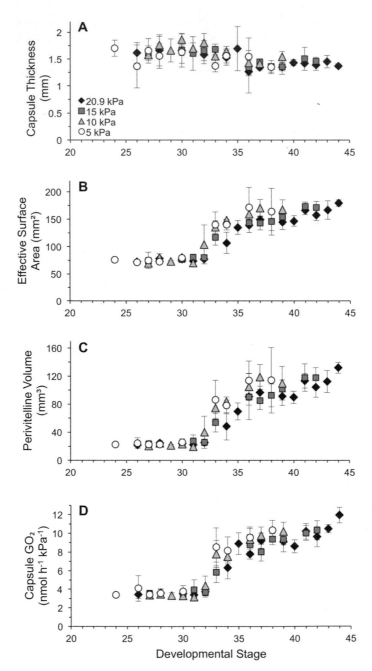

Fig. 5. (A) Egg capsule thickness (mm), (B) capsule effective surface area (mm^2), (C) perivitelline fluid volume (mm^3), and (D) capsule oxygen conductance (Go$_2$, nmol h^{-1} kPa^{-1}) of eggs in relation to developmental stage during incubation at four Po$_2$ treatments at 20°C. Data presented as mean ± 95% CI. Modified from Mueller et al. (2011a) with kind permission from Springer Science and Business Media.

The effect of egg capsule morphology and Go_2 on $\dot{M}o_2$ is evident when examining $\dot{M}o_2$ at different temperatures before and after the membrane breakdown at stage 32. Before this time $\dot{M}o_2$ is similar at different temperatures, yet after this stage $\dot{M}o_2$ diverges more noticeably as the restriction of the membrane is removed and the temperature effect on metabolism is uninhibited (Fig. 3). Furthermore, the removal of the resistances to gas exchange imposed by the capsule allows for a significant increase in $\dot{M}o_2$ upon hatching at all Po_2 levels (Fig. 4B). $\dot{M}o_2$ before and after hatching was measured on different days, and therefore $\dot{M}o_2$ after hatching represents the next developmental stage, but the increase after hatching is clearly greater than the normal day-to-day and stage-to-stage increase in $\dot{M}o_2$ that occurs during development within the egg (Fig. 4A). An increase in metabolism upon hatching occurs in other fishes (Hayes et al. 1951; Eldridge et al. 1977; Davenport 1983) and is attributed to the removal of the egg capsule and not to greater activity of the larva. This is particularly true for the Australian lungfish, as the larvae are quiescent and lie on their side unless disturbed, including when they are within a respirometry chamber.

Perivitelline Po_2

In addition to capsule morphology and Go_2, Eq. 1 also indicates the important effect of the Po_2 difference between the outside and inside of the egg capsule. At low environmental Po_2 the Po_2 gradient across the egg capsule is small, and thus $\dot{M}o_2$ is reduced. The large eggs of the Australian lungfish permit direct measurement of perivitelline Po_2 inside the egg using a needle-mounted fiber-optic oxygen microprobe. At warmer temperatures, when $\dot{M}o_2$ is higher, perivitelline Po_2 initially declines prior to stage 32, when the restrictive vitelline membrane is still in place and Go_2 has not increased (Fig. 6). After stage 32 the Po_2 plateaus at 20°C and 25°C, as the increase in Go_2 contributes to the greater movement of oxygen into the egg.

Using the Fick equation (Eq. 1), measured $\dot{M}o_2$ and Go_2 at each stage, and assuming an external Po_2 of 20.9 kPa, perivitelline Po_2 ($Po_{2(in)}$ in Eq. 1) can also be calculated. A comparison of the directly measured to calculated perivitelline Po_2 at each stage allows assessment of, firstly, how accurate the Go_2 estimation is and, secondly, if there are other resistances to gas exchange beyond the egg capsule. Prior to stage 32 and membrane breakdown, measured and calculated perivitelline Po_2 are very similar at all incubation temperatures (Fig. 7). This indicates that the egg morphological measurements used to calculate Go_2 are accurate in estimating the movement of oxygen across the capsule. However, after membrane breakdown, measured perivitelline Po_2 is lower than that predicted from $\dot{M}o_2$, Go_2 and the Fick equation. The difference between measured and calculated perivitelline Po_2 increases with temperature. At stage 40 at 15°C, measured perivitelline Po_2 is 1 kPa (or 5%) below that predicted by $\dot{M}o_2$ and Go_2. This difference increases to nearly 4 kPa (22%) at

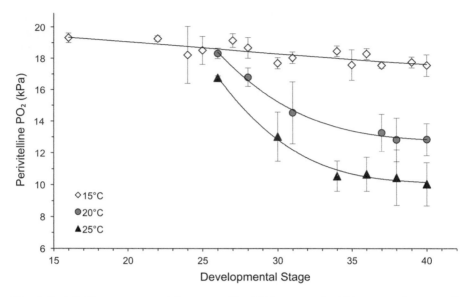

Fig. 6. Perivitelline oxygen partial pressure (P_{O_2}, kPa) measured within eggs, using a needle mounted fiber-optic oxygen microprobe, in relation to developmental stage at three incubation temperatures. Data presented as mean ± 95% CI. Modified from Mueller et al. (2011a) with kind permission from Springer Science and Business Media.

20°C and over 6 kPa (39%) at 25°C. In fact, measured perivitelline P_{O_2} is closer to that predicted by the Fick equation if capsule G_{O_2} remains constant, i.e., G_{O_2} prior to membrane breakdown (Fig. 7). On initial inspection this suggests that the estimation of G_{O_2} from capsule morphology is not very accurate at later developmental stages. However, perivitelline P_{O_2} was measured in the middle of the egg and at later stages this is located some distance (~ 3 mm) from the inner surface of the egg capsule. Therefore, the measurement encompasses not only the egg capsule but also some perivitelline fluid, which also imposes a resistance to gas exchange. In essence, the increase in capsule G_{O_2}, which improves the movement of oxygen across the egg capsule, creates a new resistance in the perivitelline fluid which mitigates the effect of increased G_{O_2}, and reduces the oxygen immediately surrounding the embryo.

The movement of oxygen through the perivitelline fluid can be examined by measuring P_{O_2} at various depths within the egg, as well as assessing fluid convection. In Australian lungfish eggs at 20°C just prior to hatching (stages 38–39), P_{O_2} within the perivitelline fluid decreases from approximately 17 kPa just inside the capsule to under 14 kPa next to the embryo (Fig. 8 in Mueller et al. 2011a). Australian lungfish embryos are ciliated (Kemp 1996) and this helps to circulate the perivitelline fluid, preventing the formation of a boundary layer next to the embryo. However, at the center of the egg the convective current is 0.18 mm s^{-1}. This convection rate is much slower than predicted from temperature corrected convection rates in the perivitelline fluid of amphibian

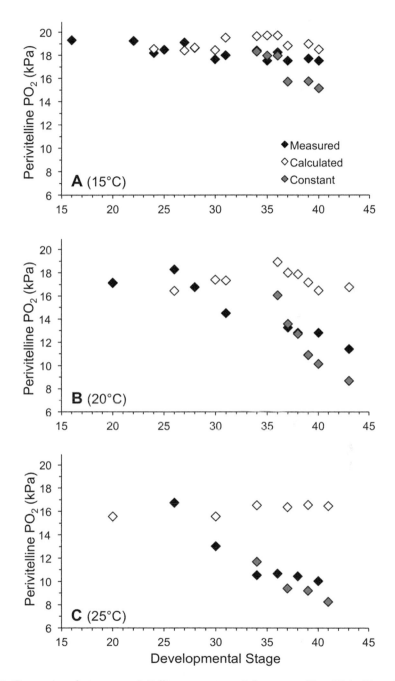

Fig. 7. Comparison between perivitelline oxygen partial pressure (P_{O_2}, kPa) either directly measured within the egg (black diamonds), calculated using the Fick equation (Eq. 1), measured $\dot{M}_{O_2}$ and G_{O_2} (white diamonds), or calculated using the Fick equation, measured $\dot{M}_{O_2}$ and a constant G_{O_2} value (3.5 nmol h^{-1} kPa^{-1}) (gray diamonds) at (A) 15°C, (B) 20°C or (C) 25°C.

eggs at 20°C (Burggren 1985; Mueller and Seymour 2011). The volume of the perivitelline fluid of lungfish eggs at this stage is approximately 120 mm^3 (Fig. 4C). This is much larger, for example, than the 30 mm^3 perivitelline volume in the egg of the amphibian, *Pseudophryne bibronii*, which has a similar embryo mass to the lungfish (Seymour and Bradford 1987; Mueller et al. 2012). In comparison to *P. bibronii* eggs, in which the embryo is curled around on itself and occupies most of the perivitelline space, the lungfish embryo rarely moves and instead lies at the bottom of the egg, further limiting potential mixing of the fluid.

Cost of Embryonic Development

The effects of temperature and environmental oxygen on metabolism has consequences for the 'cost of development', defined as the oxygen or energy required for all developmental processes, including growth (cell proliferation), differentiation (cell specification) and tissue maintenance. Developmental costs can be determined by respirometry, providing an 'oxygen cost of development', or by calorimetry, which measures the 'energy cost of development'. Oxygen and energy cost measures are interchangeable, if the amount of oxygen consumed per amount of energy is known.

The cost of development is a convenient index for comparisons of energy use between individuals under different environmental conditions or between species. The total amount of oxygen consumed during embryonic development (ml) divided by the resulting gut-free hatchling mass (mg) determines the oxygen cost of development (ml mg^{-1}) until hatching. Gut-free mass eliminates any remaining endogenous energy from the yolk after hatching. Either wet or dry gut-free mass can be used, however dry mass eliminates water content differences that may exist between treatments or between species. Thus, the cost of development is a useful measure for determining the conditions under which embryos utilize the least amount of energy, i.e., those conditions that result in 'efficient' energy use.

The cost of development is a balance between how environmental conditions alter the total oxygen consumed, by influencing $\dot{M}O_2$ and incubation time, and the mass produced. In the case of the Australian lungfish, temperature influences $\dot{M}O_2$ with a Q_{10} of 1.9 from 15 to 25°C. This is less than the Q_{10} of 4.5 for development rate over the same temperature range. Therefore, development rate is more temperature sensitive than $\dot{M}O_2$, hence the total oxygen consumed decreases as temperature increases (Table 1). As temperature does not affect hatching mass, the effect of temperature on total oxygen consumed is directly reflected in the cost of development at different temperatures (Table 1). Therefore, the amount of oxygen required to build 1 mg of embryonic tissue declines with increasing temperature, and therefore incubation at warmer temperatures can be considered more efficient.

The balance between $\dot{M}O_2$, development rate and mass produced under different environmental PO_2's can also be examined in the context of the cost of development. As the time to reach hatching is greater at higher PO_2's, and $\dot{M}O_2$ also increases with PO_2, it is not surprising that the total oxygen consumed to hatch increases with greater oxygen availability (Table 2). However, the total oxygen consumed is confounded by the premature hatching that occurs under hypoxia. To achieve an accurate measure of developmental costs under different PO_2's the total oxygen consumed and embryo mass can be examined at the same developmental stage, prior to hatching at any PO_2. Until stage 36, the total oxygen consumed by an embryo decreases with increasing PO_2 (Table 2). Dry mass at stage 36 increases with PO_2, and therefore the cost of development to stage 36 is lowest at 20.9 kPa and highest at 5 kPa, with intermediate developmental costs at 10 and 15 kPa. Therefore, the cost of development to a comparable stage decreases as PO_2 increases and stage-specific development under higher PO_2's is less costly.

Interspecific Energetic Comparisons

Compared to embryos of other fishes and amphibians, Australian lungfish embryos have a low cost of development for their hatchling mass (Fig. 8). The regression describing the relationship between cost of development and hatchling gut-free dry mass ($C = 0.3M^{0.22}$) for fishes and amphibians predicts that the cost of development should be around 0.338 ml mg^{-1} for the Australian lungfish. In comparison, the actual cost at each temperature (Table 1) is only 60, 42 and 28% of the predicted value at 15, 20 and 25°C, respectively.

When examining the cost of development between species two main factors can influence differences in the cost of development; temperature and incubation time. In the fishes and amphibians examined, cost of development shows a mild decreasing trend with increasing temperature (Fig. 9). The weak relationship between cost of development and temperature is most likely due to differences in egg sizes, which confounds the effect of temperature. However Fig. 9 indicates that, while Australian lungfish eggs are large, the relatively warm temperature range at which the species develops partially accounts for its low cost of development. There is also a relationship between incubation time and cost of development, with a longer incubation time incurring higher costs (Fig. 10). As cost of development encompasses growth, differentiation and maintenance, longer incubation times can increase the energy expended on maintenance, and thus overall cost (Vleck et al. 1980). It is clear that given its incubation time the cost of development of the Australian lungfish is low compared to other species with similar development times, being just 35, 37 and 33% of the predicted regression value at 15, 20 and 25°C, respectively.

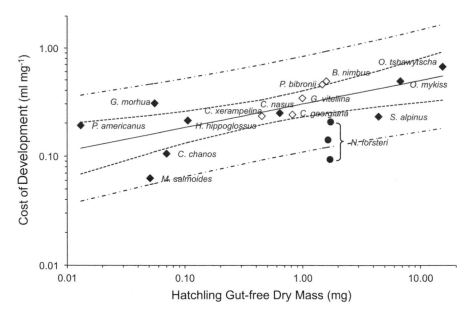

Fig. 8. Cost of development (C, total oxygen consumed per mg of dry mass, ml mg^{-1}) as a function of hatchling gut-free dry mass (M, mg) for the Australian lungfish (closed circles) compared to a regression of fishes (closed diamonds) and amphibians (open diamonds) described by the equation: C = 0.30M$^{0.22 \pm 0.13}$, r^2 = 0.52. Inner dashed lines represent 95% confidence bands for the regression means, outer dashed lines represent 95% confidence bands for the data. Modified from Mueller et al. (2011b) with kind permission from Springer Science and Business Media. Details for the total oxygen consumed and hatchling gut-free dry mass used to calculate the cost of development for each species, and the data sources, can be found in Table 2 of Mueller et al. (2011b).

Therefore, the Australian lungfish has a long incubation time but is able to maintain a low energetic cost. One potential explanation for the unusually low energetic cost in lungfish embryos may be found by examining the species at later stages. Just prior to hatching, the $\dot{M}o_2$ of the lungfish at 20°C is 0.78 µl h^{-1}, which is only 15% of the predicted value for its hatchling gut-free dry mass (M, mg) based on the regression:

$$\dot{M}o_2 = 2.85M^{1.13}, r^2 = 0.94 \qquad (3)$$

that describes temperature corrected pre-hatch $\dot{M}o_2$ (µl h^{-1}) for other fishes and amphibians (see Fig. 7 in Mueller et al. 2011b). The low metabolic demand of lungfish embryos is quite apparent, and is a characteristic carried through to juvenile stages. Juvenile (~ 51 g) Australian lungfish stimulated to swim have an active $\dot{M}o_2$ of up to 3.57 ml h^{-1} (Grigg 1965). Using the allometric relationship between standard metabolic rate (SMR, ml h^{-1}) and mass (M, g) for fishes at 20°C (White et al. 2006);

$$SMR = 0.091M^{0.879}, r^2 = 0.92 \qquad (4)$$

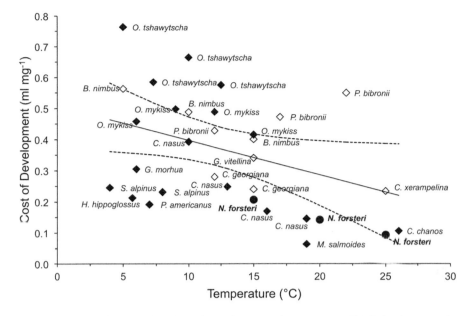

Fig. 9. Cost of development (C, ml mg^{-1}) as a function of temperature (T, °C) for the Australian lungfish (closed circles) compared to a regression of fishes (closed diamonds) and amphibians (open diamonds) described by the equation: C = –0.01T + 0.51, r^2 = 0.14. Dashed lines represent 95% confidence bands for the regression mean. Modified from Mueller et al. (2011b) with kind permission from Springer Science and Business Media. Details for data sources as in Fig. 8.

and a Q_{10} of 1.65, the predicted SMR at 25°C for a 51 g fish is 3.70 ml h^{-1}. Therefore, the active $\dot{M}o_2$ of juvenile lungfish is lower than the SMR of fishes in general, and consequently the SMR of the lungfish juveniles must be even lower, which is perhaps reflected in the sluggish nature of the adults (Kemp 1986).

The low cost of development of the Australian lungfish, and the apparent pattern of low metabolic demand throughout its life history, suggests that there may be a relationship between the cost of development and the metabolism of the resultant adult. In mammals and insects, SMR tends to be positively related to maximum metabolic rate (Keister and Buck 1974; Bozinovic 1992; Reinhold 1999; White and Seymour 2004), and mammals show a correlation between mitochondrial density and aerobic capacity (Taylor et al. 1989). Therefore, it can be postulated that a sluggish animal may only need to produce the physiological and biochemical machinery needed for its low metabolic demand, i.e., fewer mitochondria, less muscle, decreased gas transport and a reduced aerobic capacity compared to an active animal. The energetic cost of building such reduced physiological and biochemical needs of a sluggish fish may be less than an active fish, and hence this may result in the lower energetic costs during development. This suggestion is based on the results of the Australian lungfish alone, but it demonstrates how developmental physiology

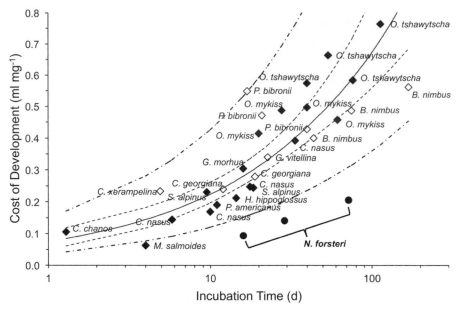

Fig. 10. Cost of development (C, ml mg^{-1}) as a function of incubation time (t, d) for the Australian lungfish (closed circles) compared to a regression of fishes (closed diamonds) and amphibians (open diamonds) described by the equation: $C = 0.08t^{0.48}$, $r^2 = 0.75$. Inner dashed lines represent 95% confidence bands for the regression means, outer dashed lines represent 95% confidence bands for the data. Modified from Mueller et al. (2011b) with kind permission from Springer Science and Business Media. Details for data sources as in Fig. 8.

can be related to adult physiology and be used as a potential predictive tool for adult performance. To examine this concept further, metabolic parameters at the organismal, tissue and cellular levels are required in the Australian lungfish, and also other species. Comparisons of mitochondrial density and gas transport pathways, as well as aerobic scope, across a continuum of sluggish to active fishes, will help to uncover if there is a link between developmental energy use and the energetic capacities of adults.

Conclusions

Our knowledge of the embryonic physiology of the Australian lungfish has significantly advanced in recent years and now far surpasses what we know about the developmental physiology of other ancient fishes. For example, we know the temperature and oxygen conditions under which the embryos will develop successfully, and how environmental conditions influence their growth, metabolism and hatch timing. In addition, we now understand the gas exchange consequences of the unique changes in egg morphology that occur during embryonic incubation, knowledge that advances our basic understanding of embryonic physiology of anamniotes. Using egg morphology and direct measurements of internal Po_2, it is clear that the large egg size of the Australian lungfish, including an initial thick jelly coat and late stage increase in perivitelline volume, creates significant barriers to gas exchange. A large egg size is ideal for such gas exchange studies, and thus the large eggs of other lungfish species makes them ideal candidates for examining embryonic respiratory physiology also.

The recent work on Australian lungfish embryos has raised many questions, and there is certainly still much to be discovered about the developmental physiology of the species. Examining respiratory physiology of the embryos at the molecular and cellular level will assist in addressing the differences in whole animal metabolism found between lungfish embryos, other fishes and amphibians. For example, examination of gas exchange tissue morphology and physiology will help elucidate what gas exchange characteristics contribute to the low oxygen demand of the embryos. The development of other regulatory systems, such as the cardiovascular, osmoregulatory and ionoregulatory systems, remain to be examined, with such studies likely to further our understanding of development in both an ontogenetic and evolutionary context.

Physiological studies, including developmental physiology, can be used to formulate and address evolutionary questions. One central question concerning ancient fishes is "Why or how have they survived for so long?" Ancient fishes tend to be found in those environments where the feeding and locomotive abilities of the modern teleosts do not provide a significant advantage (Ilves and Randall 2007). Therefore, the adaptations of ancient fishes to their environment, including that of their developmental stages, will help answer how such species have survived. For example, regardless of if there is a relationship between embryonic energetic costs and adult performance, the low cost of development in Australian lungfish embryos may be advantageous for developing in adverse environments. The ability to successfully develop in poor conditions has surely contributed to the long evolutionary history of ancient fishes. The low metabolic demand of the Australian lungfish may result from a time in the species' history when embryos experienced a limited oxygen supply, rather than being a present day adaptation. The lungfish's air breathing capacity results in relatively high tolerance of aquatic hypoxia, and having embryos that are matched in this ability is important if the species is to exploit certain environments. Interspecific studies will address if ancient fishes share metabolic attributes, both during development and as adults, and if such attributes are a result of their history or a factor of their long term success. Studies on the embryos of ancient fishes will help to place developmental physiology in an evolutionary context, with shared physiological traits being utilized to understand the evolution of embryonic development from ancient fishes to modern teleosts and beyond.

Acknowledgements

I would like to thank Roger Seymour for his guidance during the undertaking of the work reported in this chapter, Jean Joss for providing lungfish eggs and the Australian Geographic Society and University of Adelaide for funding the research. Thank you John Eme and an anonymous reviewer for their comments on earlier versions of this chapter.

References

Alderdice, D.F., W.P. Wickett and J.R. Brett. 1958. Some effects of temporary exposure to low dissolved oxygen levels on pacific salmon eggs. J. Fish. Res. Board Can. 15: 229–249.

Alderdice, D.F., J.O.T. Jensen and F.P.J. Velsen. 1984. Measurement of hydrostatic pressure in salmonid eggs. Can. J. Zool. 62: 1977–1987.

Allen, G.R., S.H. Midgley and M. Allen. 2002. Field Guide to the Freshwater Fishes of Australia. Western Australian Museum, Perth.

Amemiya, C.T., J. Alfoldi, A.P. Lee, S. Fan, H. Philippe, I. MacCallum, I. Braasch, T. Manousaki et al. 2013. The African coelacanth genome provides insights into tetrapod evolution. Nature 496: 311–316.

Bachmann, K. 1972. Nuclear DNA and developmental rates in frogs. Q. J. Florida Acad. Sci. 35: 225–231.

Bozinovic, F. 1992. Scaling of basal and maximum metabolic rate in rodents and the aerobic capacity model for the evolution of endothermy. Physiol. Zool. 65: 921–932.

Bradford, D.F. 1990. Incubation time and rate of embryonic development in amphibians: the influence of ovum size, temperature, and reproductive mode. Physiol. Zool. 63: 1157–1180.

Brinkmann, H., A. Denk, J. Zitzler, J.J. Joss and A. Meyer. 2004. Complete mitochondrial genome sequences of the south american and the australian lungfish: testing of the phylogenetic performance of mitochondrial data sets for phylogenetic problems in tetrapod relationships. J. Mol. Evol. 59: 834–848.

Budgett, J.S. 1901. On the breeding habits of some west-african fishes, with an account of the external features in development of *Protopterus annectens*, and a description of the larva of *Polypterus lapradei*. T. Zool. Soc. Lond. 16: 115–136.

Burggren, W. 1985. Gas exchange, metabolism, and "ventilation" in gelatinous frog egg masses. Physiol. Zool. 58: 503–514.

Chipman, A.D., O. Khaner, A. Haas and E. Tchernov. 2001. The evolution of genome size: what can be learned from anuran development? J. Exp. Zool. B 291: 365–374.

Collins, L.A. and S.G. Nelson. 1993. Effects of temperature on oxygen consumption, growth, and development of embryos and yolk-sac larvae of *Siganus randalli* (pisces: Siganidae). Mar. Biol. 117: 195–204.

Darken, R.S., K.L.M. Martin and M.C. Fisher. 1998. Metabolism during delayed hatching in terrestrial eggs of a marine fish, the grunion *Leuresthes tenuis*. Physiol. Zool. 71(4): 400–406.

Davenport, J. 1983. Oxygen and the developing eggs and larvae of the lumpfish, *Cyclopterus lumpus*. J. Mar. Biol. Assoc. UK 63: 633–640.

DiMichele, L. and D.A. Powers. 1984. The relationship between oxygen consumption rate and hatching in *Fundulus heteroclitus*. Physiol. Zool. 57: 46–51.

Eldridge, M.B., T. Echeverria and J.A. Whipple. 1977. Energetics of pacific herring (*Clupea harengus pallasi*) embryos and larvae exposed to low concentrations of benzene, a monoaromatic component of crude oil. T. Am. Fish. Soc. 106: 452–461.

Garside, E.T. 1959. Some effects of oxygen in relation to temperature on the development of lake trout embryos. Can. J. Zool. 37: 689–698.

Goin, O.B., C.J. Goin and K. Bachmann. 1968. DNA and amphibian life history. Copeia 3: 532–540.

Gregory, T.R. 2001. Coincidence, coevolution, or causation? DNA content, cell size, and the c-value enigma. Biol. Rev. 76: 65–101.

Gregory, T.R. 2002. Genome size and developmental complexity. Genetica 115: 131–146.

Gregory, T.R. 2013. Animal genome size database. http://www.genomesize.com.

Grigg, G.C. 1965. Studies on the Queensland lungfish, *Neoceratodus forsteri* (krefft). III. Aerial respiration in relation to habits. Aust. J. Zool. 13: 413–421.

Hamor, T. and E.T. Garside. 1976. Developmental rates of embryos of Atlantic salmon, *Salmon salar* l., in responses to various levels of temperature, dissolved oxygen, and water exchange. Can. J. Zool. 54: 1912–1917.

Hayes, F.R., I.R. Wilmot and D.A. Livingstone. 1951. The oxygen consumption of the salmon egg in relation to development and activity. J. Exp. Zool. 116: 377–395.

Horner, H.A. and H.C. MacGregor. 1983. C value and cell volume: their significance in the evolution and development of amphibians. J. Cell Sci. 63: 135–146.

Ilves, K.L. and D.J. Randall. 2007. Why have primitive fishes survived? pp. 515–536. *In*: D.J. McKenzie, A.P. Farrell and C.J. Brauner (eds.). Primitive Fishes. Fish Physiology, Vol. 26. Elsevier, Amsterdam.

Johansen, K. and C. Lenfant. 1967. Respiratory function in the South American lungfish, *Lepidosiren paradoxa* (Fitz). J. Exp. Biol. 46: 205–218.

Johansen, K. and C. Lenfant. 1968. Respiration in the African lungfish *Protopterus aethiopicus* ii. Control of breathing. J. Exp. Biol. 49: 453–468.

Johansen, K., C. Lenfant and G.C. Grigg. 1967. Respiratory control in the lungfish, *Neoceratodus forsteri* (Krefft). Comp. Biochem. Physiol. 20: 835–854.

Joss, J.M.P. 2006. Lungfish evolution and development. Gen. Comp. Endocrinol. 148: 285–289.

Keister, M. and J. Buck. 1974. Respiration: some exogenous and endogenous effects on rate of respiration. pp. 469–509. *In*: M. Rockstein (ed.). The Physiology of Insecta. Academic Press, New York.

Kemp, A. 1981. Rearing of embryos and larvae of the Australian lungfish, *Neoceratodus forsteri*, under laboratory conditions. Copeia 4: 776–784.

Kemp, A. 1982. The embryological development of the Queensland lungfish, *Neoceratodus forsteri* (krefft). Mem. Queenl. Mus. 20: 553–597.

Kemp, A. 1984. Spawning of the Australian lungfish, *Neoceratodus forsteri* (Krefft) in the Brisbane River and in Enoggera Reservoir, Queensland. Mem. Queenl. Mus. 21: 391–399.

Kemp, A. 1986. The biology of the Australian lungfish, *Neoceratodus forsteri* (Krefft 1870). J. Morphol. Suppl. 1: 181–198.

Kemp, A. 1994. Pathology in eggs, embryos, and hatchlings of the Australian lungfish, *Neoceratodus forsteri* (Osteichthyes: Dipnoi). Copeia 1994: 935–943.

Kemp, A. 1996. Role of epidermal cilia in development of the Australian lungfish, *Neoceratodus forsteri* (Osteichthyes: Dipnoi). J. Morphol. 228: 203–221.

Kemp, A. 2011. Comparison of embryological development in the threatened Australian lungfish *Neoceratodus forsteri* from two sites in a Queensland river system. Endanger. Species Res. 15: 87–101.

Kerr, J.G. 1900. The external features in the development of *Lepidosiren paradoxa*, Fitz. Philos. T. Roy. Soc. Lond. B 192: 299–330.

Krefft, G. 1870. Description of a gigantic amphibian allied to the genus *Lepidosiren*, from the wide-bay district, Queensland. Proc. Zool. Soc. Lond. 1870: 221–224.

Latham, K.E. and J.J. Just. 1989. Oxygen availability provides a signal for hatching in the rainbow trout (*Salmo gairdneri*) embryo. Can. J. Fish. Aquat. Sci. 46: 55–58.

Mueller, C.A. and R.S. Seymour. 2011. The importance of perivitelline fluid convection on oxygen uptake of *Pseudophryne bibronii* eggs. Physiol. Biochem. Zool. 84: 299–305.

Mueller, C.A., J.M.P. Joss and R.S. Seymour. 2011a. Effects of environmental oxygen on development and respiration of Australian lungfish (*Neoceratodus forsteri*) embryos. J. Comp. Physiol. B 181: 941–952.

Mueller, C.A., J.M.P. Joss and R.S. Seymour. 2011b. The energy cost of embryonic development in fishes and amphibians, with emphasis on new data from the Australian lungfish, *Neoceratodus forsteri*. J. Comp. Physiol. B 181: 43–52.

Mueller, C.A., S. Augustine, S.A.L.M. Kooijman, M.R. Kearney and R.S. Seymour. 2012. The trade-off between maturation and growth during accelerated development in frogs. Comp. Biochem. Physiol. A 163: 95–102.

Pauly, D. and R.S.V. Pullin. 1988. Hatching time in spherical, pelagic, marine fish eggs in response to temperature and egg size. Environ. Biol. Fish. 22: 261–271.

Pedersen, R.A. 1971. DNA content, ribosomal gene multiplicity, and cell size in fish. J. Exp. Zool. 177: 65–78.

Reinhold, K. 1999. Energetically costly behaviour and the evolution of resting metabolic rate in insects. Funct. Ecol. 13: 217–224.

Rock, J., M. Eldridge, A. Champion, P. Johnston and J. Joss. 1996. Karyotype and nuclear DNA content of the Australian lungfish, *Neoceratodus forsteri* (Ceratodidae: Dipnoi). Cytogenet. Cell Genet. 73: 187–189.

Rombough, P.J. 1988a. Growth, aerobic metabolism, and dissolved oxygen requirements of embryos and alevins of steelhead, *Salmo gairdneri*. Can. J. Zool. 66: 651–660.

Rombough, P.J. 1988b. Respiratory gas exchange, aerobic metabolism, and effects of hypoxia during early life. pp. 59–161. *In*: W.S. Hoar and D.J. Randall (eds.). The Physiology of Developing Fish. Fish Physiology, Vol. 11. Academic Press, San Diego.

Rombough, P.J. 2007. Oxygen as a constraining factor in egg size evolution in salmonids. Can. J. Fish. Aquat. Sci. 64: 692–699.

Salthe, S.N. 1963. The egg capsules in the amphibia. J. Morphol. 113: 161–171.

Salthe, S.N. 1965. Increase in volume of the perivitelline chamber during development of *Rana pipiens* Schreber. Physiol. Zool. 38: 80–98.

Seymour, R.S. 1994. Oxygen diffusion through the jelly capsules of amphibian eggs. Israel J. Zool. 40: 493–506.

Seymour, R.S. and D.F. Bradford. 1987. Gas exchange through the jelly capsule of the terrestrial eggs of the frog, *Pseudophryne bibronii*. J. Comp. Physiol. B 157: 477–481.

Seymour, R.S., F. Geiser and D.F. Bradford. 1991. Gas conductance of the jelly capsule of terrestrial frog eggs correlates with embryonic stage, not metabolic demand or ambient Po_2. Physiol. Zool. 64: 673–687.

Taylor, C.R., E.R. Weibel, R.H. Karas and H. Hoppeler. 1989. Matching structures and functions in the respiratory system. pp. 27–65. *In*: S.C. Wood (ed.). Comparative Pulmonary Physiology: Current Concepts. Marcel Dekker, Inc., New York.

Vleck, C.M., D. Vleck and D.F. Hoyt. 1980. Patterns of metabolism and growth in avian embryos. Am. Zool. 20: 405–416.

White, C.R. and R.S. Seymour. 2004. Does basal metabolic rate contain a useful signal? Mammalian bmr allometry and correlations with a selection of physiological, ecological, and life-history variables. Physiol. Biochem. Zool. 77: 929–941.

White, C.R., N.F. Phillips and R.S. Seymour. 2006. The scaling and temperature dependence of vertebrate metabolism. Biol. Lett. 2: 125–127.

Yokobori, S., M. Hasegawa, T. Ueda, N. Okada, K. Nishikawa and K. Watanabe. 1994. Relationship among coelacanths, lungfishes and tetrapods: a phylogenetic analysis based on mitochondrial cytochrome oxidase i gene sequences. J. Mol. Evol. 38: 602–609.

5

Aestivation in African Lungfishes: Physiology, Biochemistry and Molecular Biology

Shit F. Chew,[1] Biyun Ching,[2] You R. Chng,[2] Jasmine L.Y. Ong,[2] Kum C. Hiong,[2] Xiu L. Chen[2] and Yuen K. Ip[2,3,]*

Introduction

Suspended animation has long fascinated scientists because of its great application potentials in fields ranging from medicine to space travel. Animals become inactive during suspended animation, with absolutely no intake of food and water, and hence producing minimal or no urine and faecal materials, for an extended period. They enter into a state of torpor, possibly slowing down the biological time in relation to the clock time. If suspended animation can be achieved in humans, surgeons would have more time to operate on patients during critical moments when the blood circulation stops, and the dream of long distance space travel can be realized. In nature, suspended animation

[1] Natural Sciences and Science Education, National Institute of Education, Nanyang Technological University, 1 Nanyang Walk, Singapore 637616, Republic of Singapore.
[2] Department of Biological Sciences, National University of Singapore, Kent Ridge, Singapore 117543, Republic of Singapore.
[3] The Tropical Marine Science Institute, National University of Singapore, Kent Ridge, Singapore 119227, Republic of Singapore.
* Corresponding author: dbsipyk@nus.edu.sg

is expressed in some adult animals undergoing hibernation or aestivation. 'Aestivation' is a loose term that signifies little more than an animal undergoing a state of torpor to survive arid conditions (except for aquatic aestivators like certain sponges and sea cucumbers) at high temperature, in many cases during summer. The term has been used to describe the listless state of endotherms, like ground squirrels and cactus mouse at the height of summer heat, and ectotherms, like amphibians and African lungfishes that make cocoons to encase themselves for weeks to more than a year during the hot dry season.

From the behavioural point of view, aestivation can be defined as inactivity at high environmental temperature, particularly during dry seasons for terrestrial animals (Gregory 1982; Peterson and Stone 2000), and is regarded by Ultsch (1989) as 'a non-mobile fossorialism'. From the physiological point of view, aestivation has often been associated with metabolic depression (Storey 2002), because conservation of metabolic fuels has been regarded as an important adaptation during long periods of fasting. While this association is clearly present in endothermic mammalian aestivators, it is debatable whether it can be universally applied to aestivating ectothermic animals. For instance, it has been proposed that metabolic depression (Storey and Storey 1990; Guppy and Withers 1999) would decrease both urea production and respiratory water loss, in addition to conserving metabolic fuels, in aestivating turtle (Hailey and Loveridge 1997). However, whether metabolic depression in turtles is an adaptation to aestivation per se or simply a response to fasting (Belkin 1965; Sievert et al. 1988) remains an open question. In fact, the decrease in O_2 consumption in laboratory-aestivating yellow mud turtle *Kinosternon flavescens* is identical to that of fully hydrated turtles that are fasting for an equivalent period (Seidel 1978; Hailey and Loveridge 1997).

Although aestivation normally occurs in the summer, it is not part of a chronobiological rhythm. Rather, it appears to be an episodic event requiring an initial stimulus. In comparison with hibernation, which appears in response to cold temperature, aestivation is more intriguing and fascinating because a state of corporal torpor is achieved at high environmental temperature. While hibernation has attracted wide attention, aestivation has eluded the attention of many scientists probably because of two reasons. Firstly, animals capable of aestivation are normally found in the tropics and therefore, unlike hibernators, they are not easily accessible to scientists in the temperate regions. Secondly, it is difficult to induce animals, especially vertebrates, to aestivate under laboratory conditions. For instance, the induction of African lungfishes to aestivate in a pure mucus cocoon (Chew et al. 2004; Ip et al. 2005a), facilitating the control of air composition (Loong et al. 2008a) and humidity, has been achieved only recently by the authors in Singapore. There are altogether six species of lungfishes in the world spreading over three continents (Australia, South America and Africa), but only African lungfishes can undergo aestivation in a cocoon for an extended period. Therefore, this chapter aims to examine past and recent literature on aestivation in African lungfishes, with special emphasis on the physiological, biochemical and molecular changes during the induction, maintenance and arousal phases of aestivation. Special efforts will be made to address some controversial and enigmatic issues currently confronting researchers studying aestivation in African lungfishes.

African Lungfishes and Aestivation

Lungfishes, or dipnoans, belong to a special group of fishes called Sarcopterygians. They hold an important position in the evolutionary tree with regard to water-land transition, during which many important physiological and biochemical adaptations occur (e.g., air-breathing, urea synthesis, redirection of blood flow, heart partitioning). These adaptations supposedly facilitate the migration of fishes to terrestrial environments, leading to the evolution of tetrapods. Indeed, some data on molecular analyses of sarcopterygian phylogeny reveal that lungfishes are related closer to tetrapods than the coelacanth, *Latimeria chalumnae* (Brinkmann et al. 2004; Amemiya et al. 2013), although others indicate an unresolved trichotomy between all three groups (Takezaki et al. 2004).

Lungfishes can breathe air using 'lungs' that develop as outgrowths of the oesophagus, and air-breathing facilitates some lungfish species to survive on land for extended periods during drought. There are altogether six extant lungfish species (*Neoceratodus forsteri*, *Lepidosiren paradoxa*, *Protopterus aethiopicus*, *Protopterus annectens*, *Protopterus dolloi* and *Protopterus amphibius*) in the world, four of which (*Protopterus* spp.) can be found in Africa. African lungfishes are obligate air-breathers, meaning that their gills do not allow them to breathe exclusively in water and that they have to gulp atmospheric air to supply their O_2 requirements. They are ureogenic and possess a full complement of hepatic ornithine-urea cycle enzymes (Janssens and Cohen 1966, 1968a; Mommsen and Walsh 1989) that comprises carbamoyl phosphate synthetase (Cps) III instead of Cps I (Chew et al. 2003; Loong et al. 2005, 2012a). However, they are ammonotelic in water, and would turn transiently ureotelic after feeding (Lim et al. 2004; Iftika et al. 2008). In general, African lungfishes inhabit shallow water, although in Lake Victoria, *P. aethiopicus* can be found some distance off-shore as well as in water over 20 m deep (Greenwood 1986). Without limbs to facilitate locomotion on land, lungfishes would have to passively tolerate desiccation, and aestivation could be the only means for survival during desiccation at high temperatures. One of the most remarkable features of African lungfishes is their ability to aestivate in a mucus or mud cocoon during desiccation. They can aestivate without food and water intake in a subterranean mud cocoon for ~ 4 years (Smith 1931), which could be the longest aestivation period known for vertebrates. Johansen et al. (1976b) reported that they had kept *P. amphibius* in their laboratory naturally aestivating in a mud cocoon collected from the field for nearly three years. Aestivation does not involve a decrease in temperature, yet the metabolism of *P. aethiopicus* aestivating in a mud cocoon (probably in hypoxia) may drop to 10% of the awake rate, with a complete suspension of metabolic waste excretion except for CO_2 (Smith 1930).

Under natural conditions, aestivation covers the time between two wet seasons, which is normally only a fraction of a year (Johnels and Svensson 1954). However, there could be great variation in the duration of the dry season and thus of aestivation each year. In certain localities, like Lake Victoria (Smith 1931) and Lake Edward (Poll and Damas 1939), *Protopterus* spp. may live for years without being forced into aestivation by drought (Johnels and Svensson 1954). Although aestivation in a cocoon has become the textbook example among aestivating African lungfishes (Greenwood 1986), Brien et al. (1959) reported that the burrows of *P. dolloi*, as observed in Stanley Pool (Congo River), remained wet in their lower part without cocoon formation. According to Greenwood (1986), the burrows of *P. aethiopicus* described by Wasawo (1959) resembled the combined dry-season burrow and breeding nest of *P. dolloi*. However, when the lakes dry out, *P. aethiopicus* secretes mucus that gradually hardens to form a cocoon, which may reduce water loss. Thus, Otero (2011) concluded that there were variations in aestivation behaviour and in burrow conditions (notably with or without a cocoon) among African lungfishes, depending probably on the characteristics of the freshwater environment and the intensity of the drought.

Traditionally, aestivation experiments on African lungfishes were performed either in mud or in cloth bags in the laboratory (Janssens and Cohen 1968a,b; DeLaney et al. 1977; Fishman et al. 1986). The only difference between the two methods is that aestivation in cloth bags promotes more rapid evaporation of surface water, thereby leading to more rapid and intense dehydration, than does aestivation in mud (DeLaney et al. 1977). Actually, Fishman et al. (1986) had tried to induce *P. aethiopicus* to aestivate in air, but they stopped the experiment too early (i.e., three days) before the formation of the dried mucus cocoon, which would take six–eight days in 80–90% humidity as demonstrated by Chew (2004) and Ip et al. (2005a). Therefore, Fishman et al. (1986) designed cloth bags to achieve more effective dehydration and aestivation in *P. aethiopicus*. Chew et al. (2004) were the first to achieve induction of aestivation in *P. dolloi* in pure mucus cocoons in air inside plastic boxes. Subsequently, it has been confirmed that *P. annectens*, *P. aethiopicus* (Loong et al. 2005, 2007, 2008a,b), and *P. amphibius* (Y.K.I. and S.F.C., unpubl. observation) can also be induced to aestivate in pure mucus cocoons in air. During the induction phase of aestivation in air, the fish hyperventilates and secretes a substantial amount of mucus which turns into a cocoon in six–eight days. The maintenance phase of aestivation begins when the fish is completely encased in a dried mucus cocoon, and there is a complete cessation of feeding and locomotor activities. The fish can perpetuate to aestivate under such conditions for more than a year. The aestivating lungfish can be aroused from aestivation by the addition of water. Upon arousal, the fish struggles out of the cocoon and swims, albeit sluggishly, to the water surface to gulp air. Feeding begins approximately 7–10 days after arousal, and the fish grows and develops as normal thereafter. It is apparent that adaptive (physiological, biochemical and molecular) changes in various organs of the aestivating African lungfish would vary during the three phases of aestivation. With new insights into the ability of African lungfishes to aestivate in air, some controversial issues have arisen and they must be addressed.

Current Issues Concerning Aestivation in African Lungfishes

Aestivation or Terrestrialization?

Since the discovery of *P. dolloi* to aestivate in pure mucus cocoon in air by Chew et al. (2004), there has been some confusion over the actual definition of aestivation in African lungfishes. Several studies have described the lungfish aestivating in air as undergoing 'terrestrialization' instead of 'aestivation', and there was a lack of consistency on how the word 'terrestrialization' was adopted (Wilkie et al. 2007; Perry et al. 2008). 'Terrestrialization' was a term used originally by Wood et al. (2005; experiment series 2 in that study) and subsequently by Wilkie et al. (2007) to describe simply a prolonged period of aerial exposure in *P. dolloi* with the occasional addition of water to prevent the complete formation of a dried cocoon. This was actually an extension of the method used by Chew et al. (2003) to study aerial exposure on the same species earlier.

Wilkie et al. (2007) and Staples et al. (2008) studied *P. dolloi* in water (control) or exposed to air for an extended period (five months) without a complete cocoon formation. In essence, the fish was sustained in a prolonged induction phase of aestivation which they described as terrestrialization (Wood et al. 2005; Wilkie et al. 2007). Terrestrialization was achieved by spraying water onto the bottom of the container in which *P. dolloi* was induced to aestivate. Since the bottom of the container was wet, an incomplete cocoon was formed only along the dorsal-lateral cutaneous surface, with the ventral surface in direct and constant contact with water throughout the five-month period. Unlike aestivating fish, those undergoing terrestrialization exhibited occasional movement and were not confronted with desiccation which should theoretically lead to tissue dehydration. Thus, it is not surprising that Wilkie et al. (2007) and Staples et al. (2008) reported that water was absorbed through the ventral cutaneous surface of fish exposed to air for five months, resulting in a substantial increase, instead of a decrease, in muscle water content. However, aestivation occurs with desiccation and it would not be possible for an aestivating African lungfish to gain water from the environment. Therefore, observations made by Wilkie et al. (2007) and Staples et al. (2008) could not be manifested by fish during the maintenance phase of aestivation, whereby the ventral surface is encased completely in a dry mucus cocoon. In addition, if not because of the artificial extension of the induction phase to five months, tissue urea content would not have built up to high concentrations (13-fold). Without high levels of tissue urea, the magnitude of water retention in the muscle would have been dismal during an induction period of six–eight days before the formation of a complete cocoon. Furthermore, Staples et al. (2008) reported that five months of terrestrialization had no effects on activities of citrate synthase, glycogen phosphorylase, phosphofructokinase and pyruvate kinase in the liver, but led to decreases in citrate synthase and pyruvate kinase activities in muscle. For isolated muscle mitochondria, state 3 and state 4 respiration were reduced by 74 and 89%, respectively following air-exposure, although liver mitochondria were not affected. However, it is noteworthy that these results provide information on prolonged air-exposure, instead of the maintenance phase of aestivation.

In a separate study, Perry et al. (2008) determined respiratory rates in *P. dolloi* aestivating in pure mucus cocoon air in Singapore. They discovered that the respiratory rate of the aestivating fish was comparable to that of the control in water. Because of the apparent contradiction with the general notion that aestivation should be associated with metabolic rate reduction, Perry et al. (2008) decided to adopt the term 'terrestrialization' from Wood et al. (2005) to describe the fish used in their study without considering the differences in experimental conditions and in cocoon formation. The fish studied by Perry et al. (2008) were in fact aestivating in air, because the experimental conditions under which aestivation occurred were identical to those regarded as aestivation in air by Wood et al. (2005; experiment series 1) and Chew et al. (2004) (i.e., all under the same conditions in Singapore). African lungfishes (*P. annectens* or *P. dolloi*) aestivating in air or in a cloth sac are both characterized by (1) the complete encasement in a totally dried mucus cocoon, (2) complete torpor with irritabilities to non-tactile stimuli (Fishman et al. 1986; Perry et al. 2008), and (3) decreases in heart beat rate and blood pressure (Fishman et al. 1986; Y.K.I. and S.F.C., unpubl. results).

Induction, Maintenance and/or Arousal?

Aestivation comprises three phases: induction, maintenance and arousal. During the induction phase, African lungfishes detect environmental cues and turn them into some sort of internal signals that would instil the necessary changes at the behavioural, structural, physiological and biochemical levels in preparation for aestivation. After entering the maintenance phase, they have to preserve the biological structures and sustain a slow rate of waste production to avoid pollution of the internal environment. Upon the return of favourable environmental conditions, they must arouse from aestivation, excrete the accumulated waste products, and feed for repair and growth. Completion of aestivation occurs only if arousal is successful; if not, the animal would have apparently succumbed to certain factors during the maintenance phase. As there was a lack of effort in the past to identify and examine phenomena associated specifically with a certain phase of aestivation in African lungfishes, it becomes difficult to evaluate the physiological implications of the observed phenomena. In fact, many studies in the past focused only on the maintenance phase.

Based on classical information in the literature, the maintenance phase of aestivation in African lungfishes in mud or cloth bags is generally characterized by: (1) encasement in a mucus cocoon (Smith 1930), (2) immobility (DuBois 1892; Smith 1930), (3) a decrease in O_2 consumption to less than 50% of its resting rate in water, corresponding to a striking decrease in metabolic activity (Smith 1930; Swan et al. 1968; Lahiri et al. 1970), (4) changes in endocrine activity and possible production of a neurohumoral antimetabolite (Leloup 1958; Godet 1959, 1962; Godet et al. 1964; Swan et al. 1969), (5) virtual cessation of erythrocytopoiesis (Jordan and Speidel 1931), (6) an increase in muscle glycogen and the cessation of ammonia production indicating a shift in metabolic pathways (Janssens 1964), (7) an accumulation of urea and other nitrogenous metabolic end-products in relation to marked oliguria (Smith 1930; Forster and Goldstein 1966; Janssens and Cohen 1968a,b), and (8) remarkable changes in the circulation and respiration as typified by slowing of the heart rate (from 25 beats per minute while in water to three beats per minute in the cocoon) and a slowing of the respiratory frequency (from six per hour to one-two per hour; Smith 1930).

To understand aestivation, it is important to distinguish the processes and mechanisms involved in the three different phases: induction, maintenance and arousal. Although many of the distinctive features of the maintenance phase of aestivation in African lungfishes can be readily defined, neither induction factors/mechanisms nor maintenance mechanisms are well understood, and there is a dearth of information on the process/mechanism of arousal from aestivation.

Aestivation in Mud or in Air?

There is no doubt that African lungfishes can aestivate in a pure mucus cocoon in air in the laboratory for a year or more (Chew et al. 2004; Loong et al. 2005, 2008a,b), but it is questionable whether that would occur in the wild. It has been confirmed through field work that African lungfishes aestivate in subterranean mud cocoons, but so far, there is no citing of fish aestivating in air. If a muddy substratum is available, the advantage of aestivating in mud is that the African lungfish can reduce dehydration and avoid predation. However, the fish would be exposed to hypoxia/anoxia, and it may have difficulty sensing the return of water. Furthermore, it would have to struggle out of the cocoon in hardened mud upon arousal, which can be physically demanding after long periods of muscle disuse. In contrast, aestivation in air would mean a high rate of dehydration and exposure to predators, but the aestivating fish would avoid hypoxic exposure, and be able to arouse immediately upon raining and return to water at the earliest. If an African lungfish is stranded in a hard substratum, it is logical to deduce that it would secrete mucus to form a cocoon and to undergo aestivation in air, simply because it is capable of doing so. However, it would be difficult to discover an African lungfish aestivating in air because of two reasons. Firstly, it is possible that a fish aestivating in air can be easily consumed by predators. Secondly, a fish aestivating in air would be aroused from aestivation once water becomes available, and it would not remain in dormancy for long periods waiting to be discovered, as in the case of fish aestivating deep in mud.

Aestivation in Normoxia or Hypoxia?

Aestivation in a cloth bag, mud or an artificial substratum may prescribe exposure to hypoxia. Therefore, interpretation of classical information on physiology and biochemistry of aestivating African lungfishes is problematic due to the uncertainty over whether the aestivating fish has been exposed to hypoxic conditions and, if so, the degree of hypoxia involved. Because of that, the identification of phenomena incidental to aestivation independent of hypoxia becomes elusive. For instance, it has long been accepted that a profound decrease in metabolic rate occurs in African lungfishes aestivating in a mud cocoon or an artificial substratum (Smith 1935; Janssens and Cohen 1968a,b), although there was no knowledge on whether aestivation takes place in hypoxia or normoxia. Yet, it has been reported recently that *P. dolloi* aestivating in a completely dried mucus cocoon in air (normoxia) has a respiratory rate comparable to that of control fish immersed in water (Perry et al. 2008, the application of the term 'terrestrialization' to these fish was inappropriate; see comments by Loong et al. 2008a). By contrast, the respiratory rate of fish immersed in water was greatly reduced by aerial hypoxia (Perry et al. 2005a). Hence, Loong et al. (2008a) reasoned that there could be a greater reduction in metabolic rate in fish aestivating in hypoxia than in normoxia, resulting in a greater suppression in nitrogen metabolism in the former than in the latter. They (Loong et al. 2008a) determined ATP and creatine phosphate in three different regions of *P. annectens* based on ^{31}P NMR spectroscopy. Indeed, their results supported the proposition that metabolic depression in aestivating African lungfishes was triggered by hypoxia and might not be an integral part of aestivation. Therefore, aestivation in African lungfishes should be regarded as a state of summer corporal torpor with or without metabolic rate reduction, depending on the environmental conditions involved.

Preservation of Biological Structures or Conservation of Metabolic Fuels?

During long-term fasting, animals generally enter into a protein catabolic state, mobilizing amino acids as metabolic fuels and releasing ammonia of endogenous origins. However, unlike carbohydrates and lipids, there is no known protein store in animals, and proteins have to be mobilized from biological structures that have specific functions. Skeletal, smooth and cardiac muscles are protein structures with contractile properties but cardiac muscles must be spared from the catabolic process until very critical moments. Although skeletal muscle is the most prominent protein source, aestivating African lungfishes have to preserve muscle structure and strength in preparation for arousal. This has to be achieved in spite of the aestivating animal being in a state of corporal torpor which is associated with skeletal muscle disuse. In mammals, muscle disuse can lead to a decrease in protein synthesis and an increase in protein degradation, resulting in muscle atrophy (Childs 2003). However, in aestivating African lungfishes, a drastic increase in proteolysis, as in the case of fasting alone, does not occur, and they can effectively preserve muscle structure and strength through suppression of protein degradation and amino acid catabolism. Therefore, as suggested by Ip and Chew (2010), suppression of protein degradation during the maintenance phase of aestivation should be regarded primarily as an adaptation to preserve proteinaceous structures and functions, and conservation of metabolic fuel stores can at best be regarded as a secondary phenomenon.

Modifications of Structures/Functions or Static Preservation of Structures?

In the past, the occurrence of organic structural modifications in aestivating animals has been largely neglected, but to date, aestivation in African lungfishes are known to be associated with structural and functional modifications in at least the heart (Icardo et al. 2008), the kidney (Ojeda et al. 2008; Amelio et al. 2008) and the intestine (Icardo et al. 2012). Icardo et al. (2008) reported that the myocytes in the trabeculae associated with the free ventricular wall of *P. dolloi* showed structural signs of low transcriptional and metabolic activity (heterochromatin, mitochondria of the dense type) while in water (Icardo et al. 2008). These signs were partially reversed in aestivating fish (euchromatin, mitochondria with a light matrix), and paradoxically, aestivation appeared to trigger an increase in transcriptional and synthetic myocardial activities, especially at the level of the ventricular septum (Icardo et al. 2008). In addition, Ojeda et al. (2008) demonstrated structural modifications in all the components of the renal corpuscle of aestivating *P. dolloi*. These changes could be reversed after arousal, indicating that the renal corpuscle was a highly dynamic structure capable of modifying its architecture in response to different phases of aestivation. Morphological down-regulation during the maintenance phase and quick restoration of morphology during the arousal phase were observed in the intestine of *P. annectens* (Icardo et al. 2012). Thus, aestivation should not only be regarded as the result of a general depression of metabolism, as it involves the complex interplay between up-regulation and down-regulation of diverse cellular activities. Unlike fasting in non-aestivating animals, aestivation would logically involve variations in rates of protein degradation and protein synthesis because reconstructing and regenerating cells and tissues during the induction and arousal phases, respectively, would involve a rapid protein turnover with perhaps little production of nitrogenous wastes.

Increased Urea Synthesis or Decreased Ammonia Production?

Due to the lack of water to facilitate nitrogenous waste excretion, ammonia must be turned into less toxic products for retention. In the past, increased urea synthesis took centre stage in nitrogen metabolism in aestivating African lungfishes (Smith 1935; Janssens and Cohen 1968a,b), but the conversion of ammonia to urea is energy intensive. Furthermore, modification and preservation of biological structures during the induction and maintenance phases of aestivation, respectively, prescribe a low rate of ammonia production which would ameliorate the demand for ammonia to be detoxified to urea. Thus, the focus should be on decreased ammonia production instead of increased detoxification of ammonia, especially during the maintenance phase of aestivation (Chew et al. 2004; Ip et al. 2005a; Loong et al. 2007).

Hepatic Cps III or Cps I?

Based on colourimetric enzymatic analyses, previous works on *P. aethiopicus* and *P. annectens*, suggested the presence of a mitochondrial Cps I and a cytosolic arginase in the liver (Janssens and Cohen 1966, 1968a; Mommsen and Walsh 1989). In contrast, ureosmotic coelacanths and elasmobranchs possess hepatic Cps III, which utilizes glutamine as a substrate (Mommsen and Walsh 1989, 1991). The previous assertion of the presence of Cps I in lungfishes lent support to a lungfish/tetrapod clade, and suggested that the replacement of Cps III with Cps I, and mitochondrial arginase with cytosolic arginase, occurred before the evolution of extant lungfishes (Mommsen and Walsh 1989, 1991). On the contrary, Chew et al. (2003) using a radiometric enzyme assay system demonstrated that, like marine elasmobranchs, the slender African lungfish *P. dolloi* possessed Cps III in the liver, and not Cps I, as has been shown previously in *P. aethiopicus* and *P. annectens*. Subsequently, Loong et al. (2005) reported that contrary to previous reports (Janssens and Cohen 1966, 1968a; Mommsen and Walsh 1989) and similar to *P. dolloi* (Chew et al. 2003), *P. aethiopicus* and *P. annectens* also possessed Cps III in their liver. Liver glutamine levels are very low in *P. aethiopicus* (< 0.001 mmol l^{-1}) (Ip et al. 2005a) perhaps indicating its high demand for urea synthesis. Taken together, these results suggest that the evolution of Cps III to Cps I could not have occurred before the evolution of the extant lungfishes as previously suggested (Mommsen and Walsh 1989, 1991).

Recently, based on very short partial sequences of *cps*, which did not include the glutamine binding sites, from five lungfishes, Laberge and Walsh

(2011) suggested that they formed a monophyletic clade within the CPS I clade, despite the fact that phylogenetic analysis provided no direct evidence on the preferred substrate of the lungfish Cps. Furthermore, Laberge and Walsh (2011) concluded that the mitochondrial Cps in lungfish was 'derived' from, and likely to have a physiological function similar to, CPS I. However, it is highly unlikely that lungfish mitochondrial Cps could have 'derived' from CPS I because no CPS I is known to be present in invertebrates and elasmobranchs. Thus, Loong et al. (2012a) made an effort to obtain the full sequence of *cps III* from the liver of *P. annectens*, and affirmed that the deduced amino acid sequence comprised the critical Cys-His-Glu catalytic triad (cysteine 301, histidine 385 and glutamate 387) together with methionine 302 and glutamine 305, which are characteristic signatures of Cps III. A comparison of the translated amino acid sequence of Cps III from *P. annectens* with Cps/CPS sequences from other animals revealed that it shared the highest similarity with elasmobranch Cps III. A phylogenetic analysis indicated that *P. annectens* Cps III was evolved from Cps III of elasmobranchs. Furthermore, Loong et al. (2012a) confirmed once again that the Cps III from *P. annectens* used mainly glutamine as the substrate, and its activity decreased significantly when glutamine and ammonia were included together in the assay system, eliminating the argument of the presence of two Cps isoforms (Cps III and Cps I). Subsequently, based on the molecular and biochemical characterization of CPS I from the liver of the hylid tree frog, *Litoria caerulea*, Ip et al. (2012) confirmed that the evolution of CPS I from Cps III occurred during the evolution of amphibians instead of during the evolution of lungfishes as suggested by Mommsen and Walsh (1989) and Laberge and Walsh (2011).

Induction Phase of Aestivation

Processing of External Stimuli and Internal Cues for Aestivation

In the past, several induction factors have been proposed for aestivation in African lungfishes (Fishman et al. 1986). These include (1) dehydration, leading to oliguria/anuria and metabolic acidosis, (2) air-breathing on land, leading to CO_2 retention and respiratory acidosis, (3) starvation, effecting the metabolic, circulatory and respiratory changes, and (4) stress, leading to release of neurohumoral mediators and/or affecting thyroid function. It is highly probable that multiple factors are involved in initiating aestivation, and there are synergistic effects between factors. However, the possible involvement of salinity changes and ionic composition of the ambient water have been neglected until recently.

Naturally, aestivation occurs when an African lungfish is stranded in a puddle of water or in semi-solid mud during the dry season. The continual excretion of ammonia into a small volume of external medium would lead to high concentrations of environmental ammonia. The situation would be aggravated by the constant evaporation of the external medium under high temperature, further concentrating ammonia and other ions and resulting in high ambient salinity. Indeed, Ip et al. (2005b) demonstrated that *P. dolloi* exposed to water at salinity 3 for six days exhibited consistently lower daily urea excretion rate as compared with the freshwater control. Simultaneously, there were decreases in urea contents in various tissues and organs. Ip et al. (2005b) therefore concluded that *P. dolloi* could respond to salinity changes in the external medium as it dried up, suppressing ammonia production in preparation for aestivation. As the experimental animals were kept in water, water shortage did not occur. At salinity 3, the osmolality of the external medium (90 mosmol kg^{-1}) was still lower than that of the blood (260–280 mosmol kg^{-1}). In fact, there was no significant change in the body mass of the experimental fish after being exposed to salinity 3 water for six days. In addition, the blood osmolality of fish exposed to salinity 3 was comparable to that of fish exposed to fresh water. Hence, dehydration as the initiating factor of a reduction in ammonia production could be eliminated. Since both control (in fresh water) and experimental fish (in salinity 3) were fasted for the same period (six days), the observed effects on endogenous ammonia production were unrelated to fasting. In addition, both groups had free access to air, and had comparable blood pH, P_{O_2} and P_{CO_2} at the end of the six-day period, and thus the results obtained by Ip et al. (2005b) could not be a consequence of metabolic/respiratory acidosis or CO_2 retention. Therefore, Ip et al. (2005b) concluded that changes in salinity and ionic composition of the external medium could act as important signals to initiate aestivation in *P. dolloi* during the induction phase as the external medium dried up.

If indeed changes in ambient salinity and ionic composition constitute parts of the external signal for inducing aestivation in African lungfishes, it is important to explore the location of the sensors involved. Fish gills interface directly with the external aqueous environment, and there is extensive literature devoted to all aspects of osmoregulation and osmotic stress adaptation in fish gills across multiple levels of biological organization (Evans et al. 2005; Fiol and Kültz 2007; Hwang and Lee 2007; Kültz et al. 2007; Hwang et al. 2011). Multiple osmosensors, including calcium sensing receptor, likely act in concert to convey information about osmolality changes to downstream signalling and effector mechanisms (Fiol and Kültz 2007). Recently, translationally controlled tumour protein has been implicated to be involved in the osmosensing process in fish gills (Evans and Somero 2008). However, at present, there is a dearth of information on the possible relationship between osmosensing in the gills and the induction of aestivation in African lungfishes. Fish skin differs from other exposed vertebrate skin, most notably at the surface where living epidermal cells are in direct contact with the environment. Masini et al. (1999) reported that the skin of *P. annectens* was capable of producing proopiomelanocortin-derived peptides and proposed that peptides produced in the skin could be utilized by intracutaneous circuits or they could be delivered through circulation centrally to control body homeostasis, provoking a response of the entire organism to environmental changes. In general, adaptive and acclimatory responses of fishes to salinity stress involve not only effective osmosensing but also efficient osmotic stress signalling. The osmosensory signal transduction network in fishes is complex and includes calcium, mitogen-activated protein kinase, 14-3-3 and macromolecular damage activated signalling pathways (Fiol and Kültz 2007). This network controls, among other targets, osmosensitive transcription factors such as tonicity response element binding protein and osmotic stress transcription factor 1, which, in turn, regulate the expression of genes involved in osmotic stress acclimation. Furthermore, the systemic response to osmotic stress in euryhaline fishes is coordinated via hormone- and paracrine factor-mediated extracellular signalling (Fiol and Kültz 2007), which often involves the brain. However, no information is available at present concerning osmotic (dehydration) stress signalling in aestivating African lungfishes.

The internal cues involved in the initiation of aestivation in African lungfishes are uncertain at present. Chew et al. (2005) discovered that *P. dolloi* was capable of maintaining low concentrations of ammonia in its body by up-regulating the rate of urea synthesis to detoxify ammonia when exposed to environmental ammonia. Simultaneously, *P. dolloi* was able to increase its rate of urea excretion, but urea accumulated in the muscle, liver, and plasma of specimens exposed to environmental ammonia despite the significant increase in urea excretion rate (Chew et al. 2005). Similar observations were made on *P. dolloi* that fasted for 40 days, and urea contents in various tissues increased significantly in fasted *P. dolloi* (Chew et al. 2004), despite being immersed in water and having the capacity to up-regulate urea excretion under certain conditions (Lim et al. 2004; Wood et al. 2005). Since fasting is known to be one of the inducing factors of aestivation, urea accumulation can be an important part of the induction mechanism. Thus, Ip et al. (2005c) undertook a series of experiments to determine whether ammonia (as NH_4Cl) injected intra-peritoneally into *P. dolloi*, would be excreted directly instead of being detoxified to urea, and to examine whether injected urea would be retained in its body, leading to decreases in liver arginine and brain tryptophan levels as observed during aestivation in air. Despite being ureogenic, *P. dolloi* rapidly excreted the excess ammonia within the subsequent 12 hours after NH_4Cl was injected into its peritoneal cavity. By contrast, when urea was injected intra-peritoneally into *P. dolloi*, only a small percentage (34%) of it was excreted during the subsequent 24 hours. At hour 24, significant quantities of urea were retained in various tissues of *P. dolloi*. The intra-peritoneal injection of urea led to an apparent reduction in endogenous ammonia production, a significant decrease in the hepatic arginine content, and a significantly lower level of brain tryptophan in this lungfish. All these three phenomena had been observed in aestivating *P. dolloi* (Chew et al. 2004). Hence, Ip et al. (2005c) concluded that urea synthesis and accumulation could be one of the essential internal cues for initiating and perpetuating aestivation in *P. dolloi*, and urea might have a physiological role besides being an accumulating nitrogenous end-product. In addition, Muir et al. (2008) reported that urea depressed the metabolism of living organs of the wood frog, *Rana sylvatica*, *in vitro*, although its effect varied with temperature and seasonal acclimatization. Thus, the conception that urea is accumulated simply as an end-product of ammonia detoxification in aestivating African lungfishes, pending excretion during subsequent arousal, needs to be re-evaluated.

It has long been suspected that the brains of African lungfishes would process the internal cues and produce some hormones or bioactive compounds to coordinate the whole-body aestivation process (Fishman et al. 1992). Hiong et al. (2013) did a pioneer study on differential gene expression in the brain of *P. annectens* during the induction (six days) or maintenance (six months) phases of aestivation as compared with the freshwater control using suppression subtractive hybridization PCR. During the induction phase of aestivation, the mRNA expression levels of *prolactin* and *growth hormone* were up-regulated in the brain of *P. annectens*, which indicate for the first time the possible induction role of these two hormones in aestivation. Also, the up-regulation of mRNA expression levels of *tyrosine 3-monooxygenase/ tryptophan 5-monooxygenase activation protein γ polypeptide* and the down-regulation of *phosphatidylethanolamine binding protein*, suggested that there could be a reduction in biological and neuronal activities in the brain. The mRNA expression levels of *cold inducible RNA-binding protein* and *glucose regulated protein 58* were also up-regulated in the brain, probably to enhance their cytoprotective effects.

Prolactin, growth hormone and somatolactin are three pituitary hormones whose genes are considered to have evolved from a common ancestral gene (Rand-Weaver and Kawauchi 1993). Prolactin affects a number of physiological processes and among them are the control of mammary gland development, initiation and maintenance of lactation, immune modulation, osmoregulation, control of hypothalamic releasing-inhibiting factors, and behavioural modification (Bole-Feysot et al. 1998; Manzon 2002). At the cellular level, prolactin exerts mitogenic, morphogenic, and secretory activities. In fish, the major function of prolactin is osmoregulation in fresh water to prevent the loss of Na^+ (Specker et al. 1985; Ayson et al. 1993), and in terrestrial adaptation (Sakamoto et al. 2002). The up-regulation in the mRNA expression level of *prolactin* in the brain of *P. annectens* during the induction phase of aestivation indicates that prolactin might have an important role in the aestivation process, although the exact mechanism involved is unknown at present. Somatostatin is a cyclic tetradecapeptide that acts as a negative regulator for the secretion of growth hormone. It is synthesized as a large precursor molecule before it is processed into its active form. The novel findings reported by Hiong et al. (2013) on the down-regulation of *preprosomatostatin 2*, together with the up-regulation of *growth hormone*, in the brain of *P. annectens* during the induction phase of aestivation indicate their possible involvement in the induction process. In humans, growth hormone secretion was associated with sleep and the maximal growth hormone level occurs within minutes of the onset of slow wave sleep (Holl et al. 1991).

It is widely known that the γ-aminobutyric acid (GABA) type A receptor (GABAAR) complex is involved in many behavioural and neuroendocrine activities of vertebrates ranging from teleosts (Dolda'n et al. 1999) to mammals (Olsen and Sieghart 2009). This neuroreceptor belongs to an evolutionary superfamily of ligand-gated ion protein complex in which α1,4,5 subunits are largely responsible for inhibitory interneuronal activities (Olsen and Sieghart 2009). GABAergic actions are known to control reproductive (Fraser et al. 2002), motor (Facciolo et al. 2010), feeding (Martyniuk et al. 2005), and visual (Mora-Ferrer and Neumeyer 2009) activities. Moreover, anoxia induces increased GABAergic activities in brains of vertebrates that tolerate prolonged periods of oxygen depletion. It is probable that an increase in GABAergic activity may depress neuronal activity, thereby reducing ATP consumption during periods of oxygen deprivation and metabolic suppression (Ellefsen et al. 2009). Recently, Giusi et al. (2011) reported that the mRNA expression pattern of *gabaar* α5 was the highest in the brain of *P. annectens* after six days of aestivation in air. They suggested that Gabaar α5 might operate during the induction phase of aestivation, in which initial inhibitory GABAergic actions might be eliciting a phasic type of transmission in strategically positioned extrasynaptic neuronal clefts, as in mammals (Winsky-Sommerer 2009).

Mucus Secretion and Cocoon Formation

During the torrid season, African lungfishes escape desiccation in its natural habitat by burrowing into the mud, forming a cocoon in which it remains for months until the waters return. As the mud of the burrow hardens, the fish becomes covered in its entirety with a presumably waterproof cocoon that is open only at the mouth for breathing (Smith 1930; Johnels and Svensson 1954). However, being surrounded by mud is not a prerequisite for cocoon formation. In 2004, Chew et al. reported for the first time that *P. dolloi* could be induced to aestivate inside a pure mucus cocoon in air in the laboratory. The aestivating lungfish undergoes a series of physiological adjustments while encased in the cocoon. Hence, formation of the mucus cocoon is an important step of the aestivation process. Mucus is produced and secreted by the skin, but there is a dearth of knowledge concerning mucus secretion and cocoon formation in African lungfishes at present.

Because of its watery environment, fish skin is subjected to at least two types of stresses: (1) osmotic gradients between the cells and the water, and (2) physical forces from the water itself and from other environmental hazards. In addition, disease causing organisms such as fungi, bacteria, and water-borne parasites have easy access to the skin. In the case of amphibious fishes like African lungfishes, additional stresses confronting the skin during emersion include UV radiation, high atmospheric partial pressure of O_2, high environmental temperature and desiccation. One of the major cutaneous responses to air exposure is the production and secretion of mucus. Mucus plays a vital role in maintaining fish health and providing a physical and biochemical barrier between the fish and the environment, and is important for respiration, ionic and osmotic regulation, reproduction, excretion, and protection against microorganisms, toxins, pollutants and hydrolytic enzymes (Shephard 1994; MacPherson et al. 2005; Lillehoj et al. 2013). In general, mucus comprises glycoproteins secreted by goblet cells, though other cells including those in the submucosal glands can also produce it (Shephard 1994; Thornton and Sheehan 2004; Martinez-Anton et al. 2006). While all fish skins produce some sort of mucus, only the skins of African lungfishes can produce mucus that forms a protective cocoon upon desiccation. However, the composition of mucus of African lungfish has not been examined. It is apparent that the skins of African lungfishes can up-regulate the production of the mucus materials drastically and continuously for an extended period (three-four days) in the absence of food and water supply (Y.K.I. and S.F.C., unpubl. observation) to form a mucus cocoon that completely encases the body.

To date, there are only few reports on topics related to the skin of African lungfishes. The first report was published 80 years ago by Smith and Coates (1937), who worked on the histology of skins from fish immersed in fresh water and from fish that had aestivated for six months in mud. They (Smith and Coates 1937) reported that the skin of non-aestivating *P. annectens* in water comprised a stratified epithelium resting on a delicate basement membrane. Very conspicuous in the epidermis were the mucous cells, which presumably produced large amounts of mucus for cocoon formation, but unfortunately they did not examine skins of fish during cocoon formation in the induction phase of aestivation. After six months of aestivation, the entire epidermis was narrowed with an atrophic appearance due largely to disuse and the inactivity of mucus production (Smith and Coates 1937). Seventy years later, Masini et al. (1999) reported that the skin of *P. annectens* was capable of producing proopiomelanocortin-derived peptides. Then, Sturla et al. (2001) demonstrated the presence of a type of ionocytes, which resembled the α-chloride cells, in the skin of *P. annectens*, indicating that the skin was metabolically active and could be involved in iono- and osmo-regulation in water. While the dried cocoon material is impermeable to water, there was a thin layer of liquid between the cocoon and the skin to keep the skin moist (Y.K.I. and S.F.C., unpubl. observation). The origin and composition of this liquid is uncertain, but it is logical to hypothesize that the liquid is secreted by the skin after cocoon formation. Hence, it is not surprising to find ionocytes in the skin and efforts should be made in the future to explore the secretory function of the skin of African lungfishes during aestivation.

Permeability of the Skin

Since African lungfishes would have to defend against environmental ammonia toxicity as the ambient ammonia concentration builds up during the induction phase of aestivation, Loong et al. (2007) undertook a study to determine how the African lungfish, *P. aethiopicus*, defended against ammonia toxicity when confronted with high concentrations (30 or 100 mmol l^{-1}) of environmental ammonia. Using an Ussing-like apparatus, they (Loong et al. 2007) reported that the skin of *P. aethiopicus* had low permeability (1.26 x 10^{-4} μmol min^{-1} cm^{-1}) to NH$_3$ *in vitro*. Indeed, the influx of exogenous ammonia into fish exposed to 30 mmol l^{-1} NH$_4$Cl was low (0.117 μmol min^{-1} 100 g^{-1} fish). As a result, *P. aethiopicus* could afford to maintain relatively low ammonia contents in its plasma, muscle, liver and brain even after six days of exposure to 100 mmol l^{-1} NH$_4$Cl. In addition, Loong et al. (2007) obtained results which suggested that *P. aethiopicus* was capable of decreasing the NH$_3$ permeability of its body surface in response to ammonia exposure. After six days of exposure to 100 mmol l^{-1} NH$_4$Cl, the NH$_3$ permeability constant of the skin (0.55 x 10^{-4} μmol min^{-1} cm^{-1}) decreased to half of the control value. A decrease in the already low cutaneous NH$_3$ permeability and an increased urea synthesis, working in combination, allowed *P. aethiopicus* to effectively defend against environmental ammonia toxicity without elevating the plasma ammonia level. Hence, unlike other fishes (see Ip et al. 2001; Ip et al. 2004a,b; Chew et al. 2006; Ip and Chew 2010; Chew and Ip 2014 for reviews), glutamine and alanine contents did not increase in the muscle and liver, and there was no accumulation of glutamine in the brain, even when *P. aethiopicus* was immersed in water containing 100 mmol l^{-1} NH$_4$Cl (Loong et al. 2007). The mechanisms involved in the alteration of NH$_3$ permeability in the skin of *P. aethiopicus* are unclear at present, but results obtained by Loong et al. (2007) imply that the permeability of the skin to water and ions can be altered during the induction, and perhaps also the arousal phases of aestivation.

Hyperventilation and Metabolic Rate

During the induction phase of aestivation, there could be an increase in metabolic rate in African lungfishes. Hyperventilation occurs and the ventilation rate increases two- to five-fold during the first 30 days of aestivation and then returns to the control range (2–10 per hour) within 45 days (DeLancy et al. 1974). The arterial P$_{O_2}$ increases from the control range of 25–40 to 50–58 mmHg during the first 10 days and then returns to the control range (DeLaney et al. 1974). It could be that the increase in metabolic rate results from the structural and functional modifications of cells and tissues in preparation for the maintenance phase of aestivation.

Ammonia Production, Urea Synthesis and Energy Metabolism

Although decreased ammonia production was suspected to occur during aestivation (Janssens and Cohen 1968a; Carlisky and Barrio 1972), its importance during both the induction and maintenance phases has not been confirmed until recently (Chew et al. 2003, 2004; Loong et al. 2005; Ip et al. 2005a). This is because the traditional focus of nitrogen metabolism in aestivating African lungfishes was on increased urea synthesis (Smith 1930, 1935; Janssens 1964; Janssens and Cohen 1968a,b). Chew et al. (2003) demonstrated that urea concentrations increased significantly in muscle (eight-fold), liver (10.5-fold), and plasma (12.6-fold) of *P. dolloi* exposed to air for six days without entering into aestivation. There was also a significant increase in the rate of urea excretion in fish exposed to air for three days or more. Taken together, these results indicate that *P. dolloi* increased the rate of urea synthesis to detoxify ammonia during this period. Aerial exposure also led to an increase in the hepatic ornithine-urea cycle capacity, with significant increases in the activities of Cps III (3.8-fold), argininosuccinate synthetase + argininosuccinate lyase (1.8-fold) and glutamine synthetase (2.2-fold). In addition, the ammonia excretion rate in the experimental fish decreased significantly but without substantial increases in ammonia contents in the muscle, liver or plasma, indicating that endogenous ammonia production was drastically reduced. The apparent decrease in ammonia production in *P. dolloi* was associated with significant decreases in concentrations of glutamate, glutamine, lysine and total free amino acid in the liver (Chew et al. 2003). Therefore, Chew et al. (2003) interpreted that a decrease in proteolysis and amino acid catabolism could have occurred. However, in retrospect, the reduction in ammonia production during the induction phase of aestivation should not be viewed as an adaptation responding solely to ammonia toxicity and conservation of metabolic fuels (Chew et al. 2003). Actually, there could be an increase in protein synthesis, which would also result in decreases in ammonia production and in the total free amino acid content. African lungfishes secrete large quantities of mucus for cocoon formation during the induction phase of aestivation. The composition of the mucus cocoon is unclear at present, but mucus usually comprises nitrogenous compounds (Bayomy et al. 2002). Thus, there could be an increase in the synthesis of certain proteins for increased mucus production during the induction phase. Furthermore, structural changes are important facets of aestivation in African lungfishes (Icardo et al. 2008; Ojeda et al. 2008), and structural changes cannot occur without increased protein synthesis. Hence, the results obtained by Chew et al. (2003) could be interpreted as the occurrence of increased protein synthesis and turnover instead of decreased protein degradation during the induction phase of aestivation. Since African lungfishes hyperventilate during the initial period of the induction phase of aestivation, the reduction in ammonia production may not occur in association with metabolic depression, and there could indeed be an increase in metabolic rate instead.

In a separate study, Loong et al. (2005) reported that the rates of urea synthesis in *P. aethiopicus* and *P. annectens* exposed to air for six days increased only 1.2- and 1.5-fold, respectively, which were smaller than that in *P. dolloi*. However, unlike *P. dolloi*, aerial exposure had no significant effects on the hepatic Cps III activities of *P. aethiopicus* and *P. annectens*. Rather, aerial exposure induced relatively greater degrees of reductions in ammonia production in *P. aethiopicus* (34%) and *P. annectens* (37%) compared with *P. dolloi* (28%). Thus, Loong et al. (2005) concluded that there were subtle differences in responses by various species of African lungfishes to aerial exposure, and it would appear that *P. aethiopicus* and *P. annectens* depended more on a reduction in ammonia production than an increase in urea synthesis to ameliorate ammonia toxicity during the induction phase of aestivation. Subsequently, Loong et al. (2012a) reported that there were significant increases (9- to 12-fold) in the mRNA expression level of *cps III* in the liver of *P. annectens* during the induction phase (days 3 and 6) of aestivation in air. Aestivation in hypoxia or in mud had a delayed effect on the increase in the mRNA expression level of *cps III*, which extended beyond the induction phase of aestivation, reiterating the importance of differentiating effects that are intrinsic to aestivation from those intrinsic to hypoxia.

Loong et al. (2012b) identified aestivation-specific gene clusters through the determination of differential gene expressions in the liver of *P. annectens* after six days of aestivation in a mucus cocoon in air (normoxia) using suppression subtractive hybridization PCR. They (Loong et al. 2012b) reported that six days of aestivation in normoxia led to up-regulation of mRNA expression levels of several genes related to urea synthesis, including *cps, argininosuccinate synthetase* and *glutamine synthetase*, confirming that increased urea synthesis, despite being energy-intensive, was an important adaptive response of aestivation. Their results also offered indirect support to the proposition that urea synthesis in this lungfish involved a type of Cps that used glutamine as a substrate. In addition, the up-regulation of several mRNAs encoding protein involved in lipoprotein metabolism indicated that there could be an increase in fatty acid synthesis from carbon chains released from amino acid catabolism during the induction phase of aestivation. Some of the carbon chains from amino acids could also be channelled into glycogen, and there were down-regulation of some genes related to glycolysis.

Argininosuccinate synthetase and argininosuccinate lyase are involved in arginine synthesis for various purposes. In the liver, mRNA expression levels of *argininosuccinate synthetase* and *argininosuccinate lyase* increased significantly during the induction phase of aestivation, probably to increase arginine production to support increased urea synthesis (Chng et al. 2014).

Possible Changes in the Blood

Loong et al. (2012b) also reported that up- and down-regulation of several gene clusters occurred in the liver of *P. annectens* after six days of aestivation in normoxia. These aestivation-specific genes were involved in the prevention of clot formation, activation of the lectin pathway for complement activation, conservation of minerals (e.g., iron and copper) and increased production of haemoglobin beta. Specifically, there could be two reasons for the up-regulation of mRNA expression levels of two genes encoding proteins involved in iron metabolism, *transferrin* and *ferritin*, in the liver of *P. annectens*. Firstly, both genes could be up-regulated due to oxidative stress and inflammatory conditions during the induction and early maintenance phases of aestivation. Secondly, transferrin and ferritin could be induced due to a high turnover rate of free and bound iron as a result of the increase in synthesis of a certain type of haemoglobin and/or haemoglobins in general. Indeed, Loong et al. (2012b) observed an up-regulation of the gene expression of *haemoglobin beta* in the liver of *P. annectens* after six days of aestivation in normoxia. Furthermore, since there were up- and down-regulation of the mRNA expression levels of genes related to ribosomal proteins and translational elongation factors, there could be simultaneous increases in protein degradation and protein synthesis during the first six days (the induction phase) of aestivation, confirming the importance of reconstruction of protein structures in preparation for the maintenance phase of aestivation.

Oxidative Defence

With hyperventilation and possible increase in metabolic rate, it is highly probable that African lungfishes would be confronted with oxidative stress during the induction phase of aestivation. Surprisingly, little is known of the importance of antioxidant molecules and enzymatic oxidative defence in African lungfishes during the induction period. Ascorbate is well known for its anti-oxidative and anti-stress properties. The majority of tetrapods can synthesize ascorbate from glucose in their kidney and/or liver, but only certain fish species can do so in the kidney. The activity of L-gulonolactone oxidase, an enzyme involved in the terminal step of ascorbate synthesis, has been detected in the kidney of *P. aethiopicus* (Touhata et al. 1995) indicating the capacity for synthesis of ascorbate. Recently, Ching et al. (2014) cloned and sequenced *L-gulonolactone oxidase* from the kidney, and examined its mRNA expression levels in the liver and the brain, of *P. annectens*. Their results revealed for the first time that *L-gulonolactone oxidase* was not only expressed in the kidney but also the brain. Furthermore, the L-gulonolactone oxidase protein and enzyme activity were both detectable in the brain of *P. annectens* (Ching et al. 2014). The expression of *L-gulonolactone oxidase* in the brain could be important for the aestivating lungfish in ensuring a continual supply of ascorbate therein to counter oxidative stress when the main site (kidney) of ascorbate production shuts down. Indeed, transient increases in ascorbate and total ascorbate + dehydroascorbate concentrations occurred in the brain of *P. annectens* after six days of aestivation in air (Ching et al. 2014). This could be attributed to an increase in oxidative stress due to hyperventilation and a possible increase in metabolic rate during the induction phase of aestivation, leading to increases in ascorbate as an oxidative defence measure.

Reduction in Urine Production

With a total absence of water intake through the digestive tract or cutaneous surface, it is absolutely essential for aestivating African lungfishes to reduce or stop urine production to conserve water. Amelio et al. (2008) demonstrated that endothelial nitric oxide synthase (eNos) was present in vascular endothelial cells and podocytes of renal corpuscles, and restricted to the apical pole of tubular epithelial cells, in the kidney of *P. dolloi* kept in fresh water. During the induction phase of aestivation, eNos expression was slightly reduced in glomerular vessels without changes at the level of epithelial tubular cells. The trend observed in *P. dolloi* resembles the modifications of glomerular eNos expression encountered in the ground squirrel during hibernation (Sandovici et al. 2004). Therefore, Amelio et al. (2008) proposed that the reduced eNos expression and activity in *P. dolloi* during the induction phase of aestivation could have a role in preparing the kidney for the reduced renal activity and glomerular filtration rate.

Maintenance Phase of Aestivation

Changes in the Brain

In contrast to the induction phase, there was a drastic suppression of the mRNA expression level of *prolactin* in the brain during the maintenance phase, which was supported by quantitative real-time PCR results (Hiong et al. 2013). This corroborates the proposition that prolactin could be an important part of the signalling mechanisms to induce and maintain aestivation in *P. annectens*.

Kreider et al. (1990) examined the effects of aestivation on concentrations of thyrotropin-releasing hormone, norepinephrine, dopamine, and serotonin in the telencephalon, diencephalon, medulla and spinal cord of *P. annectens*. There was a significant decrease in the concentration of thyrotropin-releasing hormone in the diencephalon, with no alteration in other regions, of *P. annectens* after three months of aestivation but not after three months of fasting. However, aestivation for three months had no significant effect on norepinephrine, dopamine and serotonin in various regions of the brain and spinal cord of *P. annectens*. Therefore, they (Kreider et al. 1990) concluded that thyrotropin-releasing hormone could be involved in the maintenance of aestivation in African lungfishes.

Giusi et al. (2011) reported that an increase in the mRNA expression level of *gabaar α4* subunit occurred in the brain of *P. annectens* after six months of aestivation in air, indicating its participation in suppressing neuronal activities during the maintenance phase of aestivation. They suggested that Gabaar α4 subunit might activate the same neuronal pathway that the α4 agonist gaboxadol exploits to induce non-rapid eye movement sleep in mammals (Ebert et al. 2006). They further proposed that the brain of aestivating *P. annectens* could maintain a precise balance between cell death and neurogenesis, especially in the telencephalon and cerebellum, during the maintenance phase of aestivation, which could be an essential pre-requisite to subsequent arousal from aestivation.

Furthermore, Giusi et al. (2012) determined the gene expression of some key α-amino-3-hydroxy-5-methyl-4-isoxazole-propionic acid receptor channels subtypes (Glur1 and Glur2), as well as heat shock proteins and hypoxia-inducible factor-1 alpha, in the brain of *P. annectens* aestivating in air or in mud. There was a substantial decrease in the mRNA expression level of *glur1* in diencephalic areas of the brain in *P. annectens* after six months of aestivation in air. Although high levels of *heat shock protein 70* mRNA expression were observed in hypothalamic, mesencephalic and cerebellar areas of brains of fish after six months of aestivation in air or in mud, high levels of *hypoxia-inducible factor-1 alpha* and *heat shock protein 27* transcripts were detected only in the brain of fish aestivating in mud. Therefore, Giusi et al. (2012) suggested that high *glur2* mRNA levels in thalamic and mesencephalic regions might constitute ideal situations for the maintenance of long sleeping-like states of *P. annectens* during the induction and maintenance phase of aestivation.

On the other hand, Hiong et al. (2014) reported that the protein abundance of Na^+/K^+-ATPase α1 (Nkaα1) increased significantly in the brain of *P. annectens* after 12 days of aestivation (early maintenance phase). They (Hiong et al. 2014) surmised that there could be a decrease in the supply of blood to the brain during the maintenance phase of aestivation resulting from a decrease in blood pressure and a reduction in blood volume due to dehydration, both of which would affect osmoregulation in brain cells. Furthermore, the rates of urea synthesis and accumulation could be the greatest during the induction phase and the early maintenance phase when ammonia production rate had yet to be profoundly suppressed. In addition, the rate of water loss could be high during the induction phase when the mucus cocoon has not been formed completely. Taken together, it would imply that the blood osmolality would increase and affect osmotic balance in brain cells. Therefore, Hiong et al. (2014) concluded that it would be important for *P. annectens* to rectify the situation during the induction or early maintenance phases of aestivation by up-regulating the protein abundance of the ubiquitous Nkaα1, as Nka is known to be involved in volume regulation in all animal cells.

Metabolic Rate, Growth and Regeneration

Based mainly on results from *P. aethiopicus*, it has long been accepted that a profound decrease in metabolic rate occurs in African lungfishes during the maintenance phase of aestivation in a mud cocoon or an artificial substratum

(Smith 1930; Swan et al. 1968; Lahiri et al. 1970; DeLaney et al. 1974), although there was no reference to whether aestivation took place in hypoxia or normoxia. Aestivation in a mud cocoon or cloth bag could induce a reduction in respiration rate by up to 90%, depending on the duration of aestivation (Smith 1930; Swan and Hall 1966; Swan et al. 1968; Lahiri et al. 1970; DeLaney et al. 1974; Fishman et al. 1986; Hochachka and Guppy 1987). However, Perry et al. (2008) reported that *P. dolloi* exhibited constant rates of O_2 consumption before (0.95 ± 0.07 mmol kg^{-1} hr^{-1}), during (1.21 ± 0.32 mmol kg^{-1} hr^{-1}) and after (1.14 ± 0.14 mmol kg^{-1} hr^{-1}) extended periods (one-two months) of aestivation in a completely dried mucus cocoon in air (normoxia). Subsequently, Loong et al. (2008a,b) obtained results which suggested that metabolic depression in aestivating African lungfish was triggered by hypoxia and might not be an integral part of aestivation.

There are indications that the systemic response to aestivation in African lungfishes in mud cocoon is coordinated via an 'anti-metabolic' factor in addition to hormone- and paracrine factor-mediated extracellular signalling. Swan et al. (1968, 1969) reported the presence of an 'anti-metabolic' factor in the brain of aestivating *P. aethiopicus*, which, when injected intravenously, could lower O_2 consumption and body temperature of the rat. The 'anti-metabolic' factor appeared to be a peptide, because Reinhard (1981) reported that peptide extracts from the brains of aestivating African lungfish, when added to Chinese Hamster ovary cells or primary liver cells from the rat, caused a decrease in DNA synthesis up to 25% of control. The effect was nontoxic, reversible, and lasted about two hours. Brain extracts from aroused lungfish caused no effect. After gel filtration on Sephadex G-15, the activity was found in the molecular weight range of 400 to 1300 and could be destroyed by Pronase P (Reinhard 1981). With the production and release of the 'anti-metabolic' factor from the brain, it is logical to deduce that the growth must be retarded in African lungfishes during aestivation. Indeed, Conant (1976) reported that prolonged aestivation (> 17 months) in African lungfishes was associated with shrinkage of the body including fins and skeletal elements. Furthermore, the tail tip became noticeably blunter as the axis shortened and all limbs lost some of their length during aestivation. Blanc et al. (1956) reported that after 28 weeks of aestivation, one fish had lost 57 g out of 345 g originally, was 35 mm shorter than the original 400 mm snout-tail length, had pectoral limbs only 2/3 their initial length, and had pelvic limbs shortened by over 10%. Poll (1938) observed in Congo that aestivating individuals of *Protopterus* excavated near Lualaba had limbs reduced to as little as 1/3 of the length of those belonging to active fish taken from open water.

For African lungfishes aestivating in mud, metabolic depression during aestivation is not associated with a substantial decrease in extra- or intracellular pH. For *P. aethiopicus* aestivating in mud or a cloth bag, plasma pH only decreases by ~ 0.2 to 0.3 units (DeLaney et al. 1974; Land and Bernier 1995). Thus, pH-mediated down-regulation of enzymes for metabolic depression has not been demonstrated in African lungfishes. While the maximal activities

(V_{max}) of many metabolic enzymes are not affected by aestivation, reduced activity of cytochrome c oxidase may be an important way of down-regulating flux through all catabolic pathways in certain tissues (Frick et al. 2008a,b). Cytochrome c oxidase is the terminal enzyme in the electron transport chain and thus has the potential to reduce flux through pathways of aerobic amino acid, carbohydrate and lipid catabolism.

Many lower vertebrates can grow and/or regenerate despite adverse conditions. As early as 1909, Ellis showed that fasting tadpoles could regenerate at the same rate and to the same degree as well-fed counterparts. Moreover, Hui and Smith (1970) reported that larvae of *Ambystoma* (axolotl), which had not been fed for 10 weeks, regenerated hind limbs fully as rapidly as controls despite a 50% loss of body mass. However, African lungfishes would seem to be exceptions to this general rule. Conant (1973) examined limb and tail regeneration of *P. annectens* and *P. aethiopicus* under conditions of fasting or induced aestivation. For fish that were induced to aestivate after they had regenerated varying amounts of limb and tail tissues, aestivation sharply limited but did not halt further growth if the regenerate was in the latent phase. However, it did stop growth or even cause shrinkage of more mature regenerates. Short-term aestivation experiments revealed that the bulk of the growth took place during the induction phase preceding dry cocoon formation (Conant 1973). Thus, Conant (1973) concluded that deep aestivation had an inhibitory effect on tissue regeneration, which would imply that the fish must regain the ability of tissue regeneration upon arousal from aestivation. Whether the inhibitory effect on tissue regeneration is in any way related to the 'anti-metabolic factor' is uncertain at present.

Fasting and the Intestine

Icardo et al. (2012) described the structural modifications that occurred in the alimentary canal of *P. annectens* during aestivation. With fasting, all gut segments underwent structural modifications. The epithelium covering the intestinal vestibule appeared to have bursts of activation at four months of aestivation, but adopted a more quiescent appearance at six months. After four months of aestivation, the ridge area of the spiral intestine showed epithelial disintegration, cell desquamation, cell death, and loss of the freshwater phenotype. However, after six months of aestivation, the epithelium adopted a stratified appearance. Except for epithelial disintegration, the smooth portion of the spiral intestine followed a similar pattern of modifications than the ridge area. The entire epithelium of spiral intestine appeared to be renewed during the maintenance phase of aestivation. The presence of intraepithelial mast cells suggested that inflammation was part of the cellular response to aestivation. Since intestinal cells are known to undergo massive hypertrophy and apoptosis during long term fasting, and since feeding and drinking do not occur during the maintenance phase of aestivation which may last several years in African lungfishes, how the intestinal cells of African lungfishes avoid apoptotic cell death during aestivation is an enigma at present.

Torpor and Muscle Disuse

African lungfishes conserve energy during fasting by reducing locomotor activity. Conant (1973) reported that normally well-fed lungfishes (*P. annectens* and *P. aethiopicus*) showed limited nocturnal movement in the form of occasional episodes of swimming alternating with quiet periods. However, after two months of fasting, all such movements were abolished, and except for respiration, the fishes remained apparently motionless for days at a time. In comparison, African lungfishes would remain motionless inside cocoons in mud (DeLaney et al. 1974; Fishman et al. 1986; Sturla et al. 2002; Loong et al. 2008b) or in air (Chew et al. 2004; Ip et al. 2005a; Loong et al. 2008b) during the maintenance phase of aestivation. There is absolutely no locomotor activity, although the aestivating lungfish would continue to respond to sensory stimuli with bradycardia and altered breathing. In humans, disuse muscle atrophy occurs as a result of limb immobilization (stemming from bone fractures) or extended bed rest, or as a consequence of micro-gravity effects during prolonged space travel (Fitts et al. 2000, 2001). Aestivating African lungfishes represent a rare case of an animal that can ameliorate or suppress disuse muscle atrophy through a long period of inactivity. It has been demonstrated in mammals that disuse muscle atrophy occurs as a result of a decrease in muscle protein synthesis and an increase in muscle protein degradation (Booth and Seider 1979; Thomason et al. 1989). Although it is uncertain whether protein synthesis is suppressed during aestivation, no changes in expression of myosin isoforms in muscle of *P. annectens* have been found during aestivation (Chanoine et al. 1994). Enhanced protein breakdown may be expected but protease (cathepsin) activities in liver and muscle are the same among unfed, fed and aestivated *Protopterus* (Janssens 1964). Indeed, Smith (1930) reported that after 250 days of aestivation, African lungfish only lost approximately 9–18% of their original body mass. Furthermore, disuse muscle atrophy is not overtly noticeable in *P. dolloi* and *P. annectens* even after one year of aestivation in air in the laboratory, as they can struggle out of the mucus cocoon and swim to the water surface to breathe air within one-two hours upon arousal (Y.K.I. and S.F.C., unpubl. observations). In contrast to lungfish muscles, human muscles undergo rapid atrophy when not being used. At present, how aestivating African lungfish suppress disuse muscle atrophy is unclear. Chng et al. (2014) reported that the decrease in *argininosuccinate synthetase* mRNA expression level, accompanied with decreases in the concentrations of arginine and nitric oxide, in the skeletal muscle of aestivating *P. annectens* might ameliorate the potential of disuse muscle atrophy.

Respiration

In water, African lungfishes are bimodal breathers that use both gills and lungs for respiratory gas transfer (Burggren and Johansen 1986; Graham 1997); they surface every three–five minutes for a breath of fresh air. African lungfishes are obligatory air-breathers, because the vast majority of O_2 uptake of adult lungfish occurs over the lung (Lenfant and Johansen 1968; Lahiri et al. 1970; McMahon 1970; Johansen et al. 1976a), and they would succumb if denied access to air. On the other hand, CO_2 excretion normally occurs largely across the gills (Burggren and Johansen 1986) although the lung is the predominant route of CO_2 transfer in *P. dolloi* (Perry et al. 2005b). Ultsch (1996) reported that African lungfishes relied on their lungs for about 90% of their O_2 uptake and used the gills for 70–80% of their CO_2 excretion in normoxic water. For *P. amphibius*, the respiratory role of the gills is the greatest in smaller fish as compared with larger ones (Johansen et al. 1976a; Babiker 1979). In addition, the relative importance of aerial and aquatic respiration in *P. aethiopicus* changes with water O_2 levels. Up to 40% of the respiration is via the gills in normoxic water while in hypoxic water 90–100% of respiration is via the lungs with a constant metabolic rate being maintained as the proportions of aerial versus aquatic respiration change (Seifert and Chapman 2006). Field studies suggest that wild African lungfishes do not rely exclusively on aerial respiration (Mlewa et al. 2007). In the laboratory, *P. dolloi* uses the lung for the majority of both O_2 uptake (91.0 ± 2.9%) and CO_2 excretion (76.0 ± 6.6%) (Perry et al. 2005a,b). Perry et al. (2005a,b) also reported that aerial hypercapnia resulted in an increase in partial pressure of arterial blood CO_2 whereas aquatic hypercapnia had no effect, supporting the notion that the lung was the more important site of CO_2 excretion.

The mechanics of pulmonary ventilation in African lungfishes in water have been well characterized; inspiration is actively driven by a buccal force pump, while expiration involves passive elastic recoil (McMahon 1969). While the tongue blocks the movement of air through the open mouth, the buccal floor is lowered to create negative pressure within the buccal cavity. After the tongue is withdrawn, air entering the mouth is pumped into the lung by positive pressure when the buccal floor is raised and the glottis opens. In comparison, the mechanisms underlying pulmonary ventilation in aestivating lungfish are controversial. DuBois (1892) reported aspiration or 'suctional' breathing in aestivating *P. annectens*, and Lomholt et al. (1975) also concluded that aestivating *P. amphibius* relied on aspiration breathing. They (DuBois 1892; Lomholt et al. 1975) noted that the floor of the buccal cavity was motionless and that the mouth remained open during the entire respiratory cycle. Such observations, together with corroborating pressure measurements, were inconsistent with the operation of a positive pressure buccal force pump. However, more recent studies do not support the existence of aspiration breathing in aestivating African lungfishes. DeLaney and Fishman (1977) examined pulmonary breathing in *P. aethiopicus* in detail and concluded that filling of the lung during aestivation was driven by the same buccal force pump employed by aquatic lungfish. A difference between aquatic and aestivating animals is the involvement of rapid movements of the pectoral musculature to elicit positive pulmonary pressures to allow exhalation in aestivation; these movements are not required in aquatic animals, where exhalation is driven by hydrostatic forces (DeLaney and Fishman 1977). These exhalation/inhalation events are repeated in rapid succession in aestivating lungfish leading to periods of tachypnea lasting approximately 30 seconds followed by longer apnea (Lomholt et al. 1975; DeLaney and Fishman 1977; Lomholt 1993). For aestivating *P. dolloi*, Perry et al. (2008) observed a similar pattern of breathing where episodes consisting of 5–30 rapid breaths were interspersed with periods of apnea lasting approximately three minutes. As described for *P. aethiopicus* (DeLaney and Fishman 1977), each breathing episode was accompanied by shallow movements of the body wall in the vicinity of the posterior buccal musculature. These movements reflect the muscle contractions that generate positive intrapulmonary pressures to assist exhalation. The number of breaths per breathing period increased during the initial phase of aestivation, whereas the number of breathing episodes was the same as in the bimodal breathers. These data are consistent with the observation of a transient rise in breathing frequency during the first two weeks of aestivation in *P. aethiopicus* (DeLaney et al. 1974). In non-aestivating fish, the convective forces of ventilation cause a near complete tidal exchange of gases in the lung (Jesse et al. 1967; McMahon 1969), but the aestivating lungfish has a very small tidal volume and a large residual lung volume (Lomholt et al. 1975), making diffusion the principal means for renewal of gas inside the lung.

In fish that use the gas bladder for buoyancy, the cells of the gas gland are specialized to produce lactic acid to acidify the blood and release oxygen via the Root effect (Pelster 1995). For fish gas bladder, the ratio of the activity of lactate dehydrogenase to the activity of citrate synthase is very high (530 to 1,300) (Ewart and Driedzic 1990; Walsh and Milligan 1993), reflecting the role of lactate in 'salting out' O_2 from haemoglobin to fill the gas bladder. By contrast, Frick et al. (2008a,b) reported a ratio of ~ 50 for lactate dehydrogenase: citrate synthase from the lung of *P. dolloi*, which was comparable to those from the air-breathing organ of gar fish (*Lepisosteus platyrhincus*; Frick et al. 2007) and the mammalian lung (Murphy et al. 1980; Mirejovská et al. 1981). This ratio remained unchanged during aestivation when *P. dolloi* relied entirely on the lung for O_2 uptake (Frick et al. 2008a,b). Hence, Ballantyne and Frick (2010) proposed that when the gas bladder (as swim-bladder) was modified to act as an air-breathing organ, as in lungfish, the activity of lactate dehydrogenase was reduced. However, it is generally agreed that a respiratory lung, perhaps appearing first in early jawed vertebrates, was the ancestral organ (Romer 1966; Liem 1988; Graham 1997), and that the evolutionary transition from paired, ventral lungs to a dorsal single gas bladder occurred gradually (Graham 1997). Therefore, it is logical to conclude that the activity of lactate dehydrogenase increased substantially in the gas bladder during the lung to gas bladder transition.

Surfactants or specialized lipids are needed to facilitate the expansion and contraction of the lungs during respiration. It has been suggested (Daniels and Orgeig 2003) that the surfactant mixtures of primitive vertebrates contain higher cholesterol levels. The surfactants found in the lungs of lungfishes have been characterized (Orgeig and Daniels 1995; Power et al. 1999; Daniels and Orgeig 2003). The cholesterol/phospholipid ratio ($\mu g/\mu g$) of surfactants from the Australian lungfish *N. forsteri*, which cannot aestivate, is high (~ 0.30) and more closely resembles that of teleost fish gas bladders (~ 0.27) and lungs of air-breathing fishes (~ 0.20) (Daniels and Orgeig 2003). In comparison, the cholesterol/phospholipid ratio of surfactants from African lungfishes are low (~ 0.06) and more closely resemble those of amphibians, reptiles, birds and mammals (~ 0.05–0.075) (Daniels and Orgeig 2003). As with most non-mammalian vertebrates, it is possible that lungfish surfactant functions as anti-glue at low lung volumes or when the lungs are completely collapsed.

Circulation, Heart Rate, Blood Pressure and Haemoglobin

When an African lungfish aestivates during a dry spell, it has to rely entirely on its lungs for breathing. The collapse of the gill filaments could sharply reduce the blood flow through the gills, thereby directing most of the blood to the upper gill-less arches (and to the cranial and coronary arteries). In effect, the lungfish would become predominantly a heart-lung-brain preparation. Furthermore, blood supply to the kidney must decrease in order to impede urine production during aestivation. However, thrombosis would occur when blood circulation is restricted or slowed down. Thus, it is no coincidence that Hiong et al. (2015) reported that several genes related to blood coagulation, including *fibrinogen, apolipoprotein H* and *serine proteinase inhibitor clade C (antithrombin) member 1*, were down-regulated in the liver of fish after six months of aestivation, which could signify a decrease in clot formation.

Amelio et al. (2008) demonstrated that the cardiac *eNos* expression level increased in *P. dolloi* after six days of aestivation but decreased in those aestivated for 40 days. They (Amelio et al. 2008) concluded that nitric oxide contributed, probably in an autocrine-paracrine fashion, to cardiac readjustments during aestivation. During aestivation, heart rate and blood pressure decrease in African lungfishes (Lahiri et al. 1970; Fishman et al. 1986). DeLaney et al. (1974, 1976) reported that the mean arterial blood pressure in *P. aethiopicus* gradually decreased from control values of 20–28 mmHg to 14–18 mmHg during the first 30 days of aestivation in mud or in an artificial cloth-bag, whereas the heart rate decreased more gradually from 22–30 beats min^{-1} to 11–16 beats min^{-1} after 60 days of aestivation. Ventilation frequency increased 2- to 5-fold during the first 30 days of aestivation in the cloth-bag and then returned to the control range (2–10 per hour) within 45 days (DeLaney et al. 1974). The arterial P_{O_2} increased from the control level of 25–40 to 50–58 mmHg during the first 10 days of aestivation, and then returned to the control level (DeLaney et al. 1974). The arterial P_{CO_2} increased from control values of 25–30 mmHg to 45–70 mmHg, while arterial pH decreased concomitantly from 7.55–7.60 to 7.40–7.26 after the cocoon formation (DeLaney et al. 1974). On the other hand, plasma bicarbonate concentration increased from a low of 8.6 mmol l^{-1} in an active fish to 49.6 mmol l^{-1} in an aestivating fish (Johansen et al. 1976b). Therefore, it can be concluded that the sequential cardiopulmonary changes during the onset of aestivation are gradual.

Johansen et al. (1976b) studied blood respiratory properties in active and aestivating (28–30 months) *P. amphibius*, and reported that the haematocrit, O_2 capacity and blood haemoglobin concentration increased by about 50% during aestivation, probably due to the inevitable dehydration that accompanied aestivation. While the red cell haemoglobin concentration was unaltered, aestivation led to a 50% reduction in total blood concentration of nucleoside triphosphates, resulting from a prominent decrease (by ~ 80%) in the guanosine triphosphate/haemoglobin ratio with adenosine triphosphate concentration remaining largely unaltered (Johansen et al. 1976b). Guanosine triphosphate/ haemoglobin ratio changes were accompanied by marked changes in O_2 affinity (Johansen et al. 1976b). The P_{50} value in blood from active fish was 33 mmHg at pH 7.5 compared to 9 mmHg for the aestivating fish (Johansen et al. 1976b). The striking increase in O_2-haemoglobin affinity during aestivation could be an adaptation to a reduced alveolar O_2 availability associated with the hypoxic conditions of aestivating in a mud cocoon (Lomholt et al. 1975). Altered mechanics of pulmonary ventilation and reduced ambient O_2 availability caused by the hypoxic subterranean habitat could be the selection pressures for the development of the high affinity of haemoglobin to O_2. In addition, alteration in the haemoglobin O_2 affinity might also result from alteration in the haemoglobin system, although Weber et al. (1977) reported that the O_2-binding characteristics and the multiplicity of the stripped haemoglobin from active *P. amphibius* were the same as in specimens that had been aestivating for about 30 months (Johansen et al. 1976b). In fact, the findings of Weber et al. (1977) differ from those of DeLaney et al. (1976) who reported, besides increased haemoglobin concentration, marked changes in the relative concentrations of the haemoglobin components of *P. aethiopicus* during aestivation. DeLaney et al. (1976) reported the presence of four electrophoretically distinct types of haemoglobins in *P. aethiopicus*. These four types of haemoglobins (fractions I, II, III and IV) were present in both the aquatic and aestivating lungfish. They (DeLaney et al. 1976) reported a relative increase in the quantities of types II and IV within three months of aestivation, and these two types of haemoglobin persisted for the following 13 months. Recently, Loong et al. (2012b) demonstrated that the mRNA expression level of *haemoglobin beta* was up-regulated in the liver of *P. annectens* during the induction phase (six days) of aestivation, and it is possible that *haemoglobin beta* (Loong et al. 2012b) coded for one of the two or both types of haemoglobin (fractions II and IV, which are tetramers) that were reported by DeLaney et al. (1976). Hence, it is essential to re-examine the haemoglobin isoforms and their expression levels during the three phases of aestivation in African lungfishes. Of note, similar to elasmobranch haemoglobins (Bonaventura et al. 1974), the O_2 affinity of haemoglobin from lungfish has low sensitivity to urea, which appeared to be an adaptive response to high concentrations of urea accumulated during the maintenance phase of aestivation (Weber et al. 1977).

The chief haemocytopoietic organs in *P. aethiopicus* are the spleen, kidneys, and intestine (Jordon and Speidel 1931). Erythrocytes are formed in the splenic pulp, the usual mother cell being the lymphoid haemoblast with vesicular, sieve-like nucleus. During aestivation, degenerating granulocytes become more prominent throughout the tissues and in the circulation of *P. aethiopicus* (Jordon and Speidel 1931). Furthermore, erythrocytopoiesis practically ceases, with senile erythrocytes often present due to the appearance of intracellular nuclear degeneration.

Carbohydrate Metabolism

During the maintenance phase of aestivation, African lungfishes have absolutely no intake of food and water for an extended period. Therefore, they must rely on stored fuel for energy production. Depending on whether aestivation occurs in air (normoxia) or in mud or an artificial substratum (hypoxia), the substrates and the related metabolic pathways for energy metabolism naturally vary. In this regard, information in the literature is confusing. For instance, DeLaney et al. (1974) and Fishman et al. (1986) reported that the P_{O_2} of African lungfish aestivating in mud decreased from ~ 50 to 25 mmHg, suggesting that aestivating fish experienced hypoxia. If there was indeed a reduction in O_2 availability during aestivation, the aestivating African lungfish should have an increased reliance on anaerobic metabolism (Ballantyne and Frick 2010). However, there was not only a lack of accumulation of lactate in any tissue of *P. dolloi* during aestivation in cloth bags, but also a reduction in lactate concentration in the heart and muscle (Frick et al. 2008a). Therefore, Frick et al. (2008a) surmised that fish undergoing aestivation in cloth bags in their study were not in a hypoxic environment. However, they did not monitor the blood O_2 level in the aestivating fish, and hence the severity of hypoxia confronting those aestivating lungfish was uncertain. Because of this, it may not be valid to draw any correlation (Ballantyne and Frick 2010) between the blood O_2 results of DeLaney et al. (1974) and Fishman et al. (1986) and the biochemical results of Frick et al. (2008a).

Based on glycogen stores, African lungfishes appear to be well adapted for hypoxia. The amount of glycogen stored within a tissue indicates the tissue's anaerobic capacity (McDougal et al. 1968). In African lungfishes, the liver has the highest concentration of glycogen (Dunn et al. 1983; Frick et al. 2008a), and it supplies glucose to extrahepatic tissues during hypoxia. Fishes typically store small amounts of glycogen within the white muscle, but lungfishes have plenty of glycogen rosettes stored between myofibrils (Hochachka et al. 1978a; Hochachka 1980). These glycogen rosettes are present not only in red and white muscles, but also in the heart and liver. Lungfishes can use glycogen both aerobically and anaerobically (Hochachka and Hulbert 1978), but it is probably most advantageous to spare it for hypoxic episodes. Similar to brains of other anoxia tolerant species (McDougal et al. 1968; Kerem et al. 1973; Dunn et al. 1983), the brain glycogen contents in lungfishes are approximately two- to four-fold higher than those in most vertebrates. Furthermore, the lactate dehydrogenase/citrate synthase ratios are also high in tissues of *P. aethiopicus* and *P. dolloi* (Dunn et al. 1983; Frick et al. 2008a,b), which is a common characteristic of air-breathing fishes which inhabit hypoxic environments and require a large anaerobic potential (Dunn et al. 1983). In *P. dolloi* and *P. aethiopicus*, the highest activity of lactate dehydrogenase is found in the heart (Dunn et al. 1983; Frick et al. 2008a), and it is higher than levels reported for hearts of tuna, salmon and *Arapaima* (Hochachka et al. 1978b,c; Mommsen et al. 1980). Thus, based on both the high levels of lactate dehydrogenase and the large glycogen stores, Ballantyne and Frick (2010) concluded that hearts of African lungfishes display a high anaerobic potential. However, heart tissues are known to express the lactate dehydrogenase isoform (LDH-B) that turns lactate to pyruvate, which can then be shuttled into the tricarboxylic acid cycle for energy metabolism or into gluconeogenesis for glucose and glycogen formation. Therefore, it is rather logical to conclude that the hearts of African lungfishes have a high capacity for synthesizing glycogen from lactate or other substrates, instead of a high capacity for catabolizing glycogen for lactate production while aestivating in hypoxia.

Although a decrease in liver glycogen content (Janssens 1964; Babiker and El Hakeem 1979) with increased glycogen phosphorylase activity (Janssens 1965) has been observed in African lungfishes during the maintenance phase of aestivation, the muscle glycogen content either remains unchanged (Janssens 1964) or increases significantly in the aestivating fish as compared with the fed control (Babiker and El Hakeem 1979). On the other hand, glycogen concentrations in the liver and kidney of aestivating *P. dolloi* are higher than those of the fasting control, while the glycogen content in muscle and heart remain unchanged during aestivation (Frick et al. 2008a). Thus, African lungfishes appear to undergo continuous gluconeogenesis throughout aestivation (Janssens and Cohen 1968b). Indeed, *P. aethiopicus* (Dunn et al. 1983) and *P. dolloi* (Frick et al. 2008a) have a substantial capacity for glucose and glycogen synthesis, as reflected by their high activity (up to 100-fold greater than other fish species) of fructose 1,6-bisphosphatase, which is a key enzyme in the gluconeogenesis pathway, in the liver (Suarez and Mommsen 1987). Furthermore, Frick et al. (2008a) reported a lack of increases in activities of enzymes involved in glycolysis in *P. dolloi* during aestivation. Taken altogether, it can be concluded that glucose and glycogen are not the major fuels for energy metabolism in African lungfishes during the maintenance phase of aestivation, particularly in normoxia. Rather, the continuous build-up of urea in the body of aestivating African lungfishes (Janssens and Cohen 1968a,b; Chew et al. 2004; Frick et al. 2008a) suggests protein/amino acids as the major metabolic fuel during long-term aestivation. Furthermore, the relatively unchanged levels of glycogen (Janssens and Cohen 1968a,b; Frick et al. 2008a) indicate that gluconeogenesis could be supported by the carbon skeletons derived from amino acid catabolism.

Lipid Metabolism

African lungfishes have large lipid stores along the posterior third of their body; they are stored under the skin and not in the muscle. It has been postulated that these stores serve as important energy sources during periods of food deprivation (Hochachka and Hulbert 1978). Overall, the lipid content of muscle fibres is very low (Dunn et al. 1981), with a small amount of lipid present within the red muscle, but not in white muscle fibre. Only low levels of lipids are present in the plasma and tissues of *P. dolloi* (Frick et al. 2008b). Although the activity of 3-hydroxyacyl CoA dehydrogenase, an enzyme indicative of lipid catabolism, is ~ 10-fold higher in the liver of *P. dolloi* (Frick et al. 2008b) than in other primitive fishes (Speers-Roesch et al. 2006; Frick et al. 2007), the activities of other enzymes involved in lipid metabolism in *P. dolloi* are low. Specifically, carnitine palmitoyl CoA transferase is not detectable in tissues other than the liver and kidney (Frick et al. 2008b), where its activity is still very low (Speers-Roesch et al. 2006; Frick et al. 2007). Furthermore, the activity of malic enzyme, an enzyme involved in the generation of NADPH required for fatty acid synthesis (Henderson and Tocher 1987), is generally low compared to other fish species, suggesting the capacity for lipogenesis is limited in lungfishes (Frick et al. 2008b). Hence, although the total plasma lipids in lungfish aestivating for two years decreases to less than half of the levels found in control fish (Babiker and El Hakeem 1979), lipids cannot be a major substrate for energy metabolism in African lungfish during the maintenance phase of aestivation. Indeed, Smith (1931) reported that African lungfishes derived 50% of their energy requirements from protein even when lipid is available.

Ketone bodies (e.g., β-hydroxybutyrate and acetoacetate) can be used as oxidative fuels to replace carbohydrates during periods of starvation and as lipid precursors (Newsholme and Leech 1983). Ketogenesis occurs primarily in the liver, but judging by the activity of β-hydroxybutyrate dehydrogenase, the liver of *P. dolloi* has low capacity for ketone synthesis (Frick et al. 2008b). In extrahepatic tissues, β-hydroxybutyrate dehydrogenase would operate in the reverse direction, but the β-hydroxybutyrate dehydrogenase activities in extrahepatic tissues are low. Furthermore, succinyl coenzyme-A ketotransferase is low or undetectable in the white muscle, kidney and liver tissues (Frick et al. 2008b). Taken together, it can be concluded that ketone bodies do not act as a major substrate for energy metabolism in *P. dolloi* during the maintenance phase of aestivation.

Nitrogen Metabolism

Studies of *P. annectens* feeding on chironomid larvae indicate a nitrogen quotient of 0.27, which is typical of a very high reliance on protein/amino acids as aerobic fuels (Iftikar et al. 2008). Besides being the building blocks of proteins that are needed for survival, growth and development, amino acids have numerous functions. Under normal circumstances most animals take in amino acids in excess of what is needed to sustain growth and protein turnover. Unlike carbohydrates and lipids, which can be stored as glycogen and triglycerides, respectively, amino acids are not stored to any great extent and animals are not known to possess protein stores solely for the purpose of energy metabolism (Campbell 1991). Therefore, excess amino acids from diets are preferentially degraded, and their carbon skeletons are then channelled directly into the tricarboxylic acid cycle or converted to glucose through gluconeogenesis (Campbell 1991). The first step in amino acid catabolism involves the removal of the α-amino nitrogen as ammonia. For some amino acids, deamination involves specific deaminases, but many amino acids are deaminated through transdeamination (Campbell 1973, 1991). Transdeamination of amino acids usually occurs in the liver and requires an initial transamination of the amino acid with α-ketoglutarate in the cytosol to form glutamate, which then enters the mitochondria to be oxidatively deaminated by glutamate dehydrogenase. Glutamate dehydrogenase is therefore crucial to the regulation of amino acid catabolism, and hence ammonia production. It also plays an important role in integrating nitrogen and carbohydrate metabolism. Amino acid catabolism releases ammonia which, because of its toxicity, must be disposed of or detoxified. African lungfishes are normally ammonotelic, excreting more than 50% of the nitrogenous wastes as ammonia. However, they would turn ureotelic momentarily after a meal (Lim et al. 2004). In addition, they would detoxify endogenous ammonia to urea which accumulates to high levels during the maintenance phase of aestivation (Chew et al. 2004; Ip et al. 2005a; Loong et al. 2008a,b).

During aestivation, amino acids can be released from muscles and other tissues through increased protein catabolism. Indeed, Bakiker and El Hakeem (1979) obtained results which suggested that *P. annectens* probably utilized carbohydrates during the initial phase of aestivation only, and it relied almost entirely on protein during the maintenance phase of aestivation. Degradation of amino acids for energy metabolism releases ammonia. Ammonia is toxic (Cooper and Plum 1987); it acts on the central nervous system of vertebrates, including fish, causing hyperventilation, hyperexcitability, convulsions, coma, and finally death. Many air-breathing fishes have evolved mechanisms to defend against ammonia toxicity during aerial exposure (Ip et al. 2001, 2004a,b; Chew et al. 2006; Ip and Chew 2010; Chew and Ip 2014). Out of these different mechanisms, the two main ones adopted by African lungfishes to ameliorate ammonia toxicity during emersion or aestivation are increase in urea synthesis and suppression of ammonia production. Unlike most teleosts, African lungfishes are ureogenic and possess a full complement of ornithine-urea cycle enzymes (Janssens and Cohen 1966, 1968a; Mommsen and Walsh 1989; Chew et al. 2003, 2004; Loong et al. 2005), including Cps III (Chew et al. 2003; Loong et al. 2005, 2012a), in their liver. Because of this, previous studies focused mainly on the importance of increased urea synthesis to defend against ammonia toxicity in aestivating African lungfishes (Smith 1930, 1935; Janssens 1964; Janssens and Cohen 1968a,b), but the importance of decreased ammonia production was largely neglected until recently (Chew et al. 2003, 2004; Loong et al. 2005; Ip et al. 2005a). Chew et al. (2004) demonstrated that the rate of urea synthesis in *P. dolloi* increased 2.4- and 3.8-fold during six days and 40 days of aestivation in air, respectively. They (Chew et al. 2004) also discovered that the rate of ammonia production in *P. dolloi* decreased to 26 and 28% during the first six days and the following 34 days of aestivation, respectively. For *P. aethiopicus* that underwent 46 days of aestivation, a 20% decrease in the rate of ammonia production during the initial 12 days, and a profound decrease (96%) in ammonia production during the final 12 days of aestivation (day 34 to day 46) have been observed (Ip et al. 2005a). Hence, it becomes apparent that different African lungfish species exhibit different capacity of reduction in ammonia production.

Loong et al. (2008b) examined the importance of increased urea synthesis and decreased ammonia production in *P. annectens* aestivating in air versus in mud. Twelve days of aestivation in air led to significant increases in contents of urea, but not ammonia, in the tissues of *P. annectens*. The estimated rate of urea synthesis increased 2.7-fold despite the lack of changes in the activities of hepatic ornithine-urea cycle enzymes, but the change in rate of ammonia production was minor. After 46 days of aestivation in air, the ammonia concentration in the liver decreased significantly, but urea concentration increased in all tissues, indicating a combined strategy of increased urea synthesis (1.4-fold of the day 0 value) and decreased ammonia production (56% of the day 0 value) to defend against ammonia toxicity. By contrast, 12 days of aestivation in mud produced only minor increases in tissue urea contents, with ammonia contents remained unchanged (Loong et al. 2008b). Surprisingly, 46 days of aestivation in mud resulted in no changes in tissue urea contents, indicating a profound suppression of urea synthesis and ammonia production (2.6 and 1.2%, respectively, of the corresponding day 0 value) have occurred. Fish aestivated in mud could also be exposed to environmental hypoxia as reflected by the low blood Po_2 and muscle ATP content (Loong et al. 2008b), and hypoxia could have induced reductions in metabolic rate and ammonia production. Obviously, it is advantageous for fish aestivating in mud to have a lower dependency on increased urea synthesis to detoxify ammonia, because urea synthesis *de novo* is energy-intensive.

In another study, Loong et al. (2008a) discovered that the dependency of *P. annectens* on increased urea synthesis and decreased endogenous ammonia production to ameliorate ammonia toxicity was determined by whether aestivation occurred in normoxia or hypoxia. The rate of urea synthesis increased 2.4-fold, with only a 12% decrease in the rate of ammonia production in the normoxic fish. By contrast, the rate of ammonia production in the hypoxic fish decreased by 58%, with no increase in the rate of urea synthesis. Using *in vivo* ^{31}P NMR spectroscopy, Loong et al. (2008a) demonstrated that hypoxia led to significantly lower ATP concentrations on day 12 and significantly lower creatine phosphate concentrations on days 1, 6, 9 and 12 in the anterior region of the fish as compared with those of fish exposed to normoxia. Additionally, the hypoxic fish had a lower creatine phosphate concentration in the middle region than that in the normoxic fish on day 9. Hence, Loong et al. (2008a) concluded that lowering the dependency on increased urea synthesis to detoxify ammonia by reducing ammonia production could be a strategy to conserve cellular energy during aestivation in hypoxia.

Since glutamate dehydrogenase is in a crucial position to regulate ammonia production, Loong et al. (2008a) undertook a study to examine whether there would be changes in a specific activity and kinetic properties of glutamate dehydrogenase from the liver of *P. annectens* during the induction and the maintenance phases of aestivation, and whether these changes would be different between normoxic and hypoxic fish. They discovered that the activities of hepatic glutamate dehydrogenase, in the amination and deamination directions, remained relatively constant in fish exposed to normoxia during the induction phase (three or six days) of aestivation (Loong et al. 2008a). However, there was a significant increase in the glutamate dehydrogenase amination activity, with the deamination activity remained unchanged, in fish aestivating in normoxia on day 12. Hence, glutamate dehydrogenase would act less favourably in the deamination direction during the maintenance phase of aestivation to reduce the production of ammonia through transdeamination. Simultaneously, the hepatic glutamate dehydrogenase amination activity, but not the deamination activity, from fish aestivating in normoxia on day 12 became highly dependent on the availability of ADP. Results of Loong et al. (2008a) indicated that transdeamination of amino acids through the hepatic glutamate dehydrogenase became responsive mainly to the cellular energy status of the fish during the maintenance phase of aestivation (day 12) in normoxia. Since ammonia concentrations in various tissues of *P. annectens* exposed to normoxia (or hypoxia) remained relatively unchanged, Loong et al. (2008a) concluded that changes in the activity of hepatic glutamate dehydrogenase occurred primarily to reduce ammonia production, and not to detoxify ammonia during the maintenance phase of aestivation. In comparison, for fish exposed to hypoxia, significant increases in the hepatic glutamate dehydrogenase amination activity, the amination/deamination ratio, and the dependency of the amination activity on ADP activation, occurred much earlier (day 6) during the induction phase of aestivation (Loong et al. 2008a). Hence, decreased ammonia production through changes in the activity of hepatic glutamate dehydrogenase in *P. annectens* was more effectively induced and exacerbated by a combination of aestivation and hypoxia than aestivation alone (in normoxia). Apparently, glutamate dehydrogenase was critically regulated during the transition between the induction and the maintenance phases of aestivation in hypoxia, suppressing ammonia production to reduce the dependency on increased urea synthesis for the detoxification of ammonia.

In spite of suppressing ammonia production during the maintenance phase of aestivation, endogenous ammonia must be detoxified because of the complete impediment of ammonia excretion. African lungfishes detoxify ammonia to urea through the hepatic ornithine-urea cycle. By synthesizing and accumulating the moderately less toxic urea, aestivating African lungfishes can carry out protein catabolism for a longer period without being intoxicated by ammonia. Chew et al. (2004) reported that the rate of urea synthesis in *P. dolloi* increased 2.4-fold and 3.8-fold during the first six days and the subsequent 34 days of aestivation, respectively, and urea accumulated in various tissues and organs. Although activities of ornithine-urea cycle enzymes in fish aestivated for six days remained unchanged, the activities of several ornithine-urea cycle enzymes increased significantly in fish aestivated for 40 days. Previous works by Janssens and Cohen (1968a) also showed that urea accumulation occurred in *P. aethiopicus* aestivated for 78–129 days in an artificial mud cocoon, but they concluded that urea accumulation did not involve an increase in the rate of urea synthesis, even though the fish appeared to be in continuous gluconeogenesis throughout aestivation. Subsequently, Ip et al. (2005a) undertook a study to test the hypothesis that the urea synthesis rate in *P. aethiopicus* was up-regulated to detoxify ammonia during the initial period of aestivation (day 0 to day 12), and that a profound suppression of ammonia production occurred at a later period of aestivation (day 34 to day 46) which eliminated the need to sustain the increased rate of urea synthesis. Contrary to the report of Janssens and Cohen (1968a), Ip et al. (2005a) demonstrated a drastic increase in urea synthesis (three-fold) in *P. aethiopicus* during the initial 12 days of aestivation, although the magnitude of the increase in urea synthesis decreased over the next 34 days. Between day 34 and day 46 (12 days), the rate of urea synthesis decreased to 42% of the day 0 control value instead. There were significant increases in tissue urea contents and activities of some ornithine-urea cycle enzymes from the liver (Ip et al. 2005a). Since there was a meager 20% decrease in the rate of ammonia production in *P. aethiopicus* during the initial 12 days, as compared to a 96% decrease during the final 12 days of aestivation (day 34 to day 46), Ip et al. (2005a) concluded that *P. aethiopicus* depended mainly on increased urea synthesis to ameliorate ammonia toxicity during the initial period of aestivation, but it suppressed ammonia production profoundly during prolonged aestivation, eliminating the need to increase urea synthesis which is energy-intensive.

Urea, at high concentrations, can denature proteins and exerts disruptive effects on enzymes (Hochachka and Somero 1984), but there is no evidence of the presence of methylamines or other potential counteracting solutes in aestivating *P. dolloi* (Wilkie et al. 2007). It has been suggested that accumulated urea contributes to metabolic depression in dormant animals (Griffith 1991) by reversibly inhibiting key metabolic enzymes (Hand and Somero 1982; Yancey et al. 1982). Indeed, studies (Costanzo and Lee 2005; Muir et al. 2007) on hibernating wood frogs (*R. sylvatica*) suggest a link between urea accumulation and metabolic depression. Muir et al. (2008) measured aerobic metabolism of isolated organs from the wood frog in the presence or absence of elevated urea at various temperatures using frogs acclimatized to different seasons. When organs from winter frogs were tested at 10°C, metabolism was significantly decreased by ~ 15% in urea-treated liver and stomach and by ~ 50% in urea-treated skeletal muscle. Therefore, Muir et al. (2008) concluded that the presence of urea depressed the metabolism of living organs, and thereby reduced energy expenditure. However, no information is available on the possible role of urea in metabolic rate reduction in aestivating African lungfishes at present. It has been previously suggested that aestivation in air entails desiccation and increased tissue urea contents might serve the secondary function of facilitating water retention in tissues through vapour pressure depression (Withers 1998; Storey 2002). However, Machin (1975) established that urea concentration of 300 mmol l^{-1} had only minor contribution to the gradient for water movement between tissues and dry air. More importantly, Loong et al. (2008a) experimented on two groups of *P. annectens* that underwent aestivation in closed boxes with similar flow rates of air or 2% O$_2$ in N$_2$, and thereby experienced similar magnitudes of desiccation, but fish aestivating in hypoxia suppressed ammonia production and accumulated much less urea. Hence, it can be concluded that increased urea synthesis in *P. annectens* is an adaptation responding primarily to rates of protein degradation and amino acid catabolism instead of desiccation (Loong et al. 2008a).

Impediment of Urine Production and the Kidney

With a total absence of water intake through the digestive tract or cutaneous surface, it is absolutely essential for aestivating African lungfishes not to produce urine and to render the kidney non-functional during the maintenance phase of aestivation. Ojeda et al. (2008) studied the structural and lectin-binding modifications experienced by the renal corpuscle of *P. dolloi* during the maintenance phase of aestivation and observed that the renal corpuscle underwent a marked size reduction, affecting all the structural components. The parietal cells of Bowman's capsule lost their flattened appearance and adopted the organization of a stratified epithelium. The glomerular capillaries collapsed while the podocytes approached each other. The foot processes lost their regular arrangement, the filtration slits became unnoticeable, and the subpodocyte space disappeared. The glomerular basement membrane thickened enormously. On the whole, these modifications thickened the filtration barrier, thus reducing the filtration coefficient and allowing the aestivating lungfish to cope with dehydration. Hence, Ojeda et al. (2008) concluded that the renal corpuscle of *P. dolloi* was a highly dynamic structure capable of modifying its architecture in response to environmental changes. On the other hand, Amelio et al. (2008) demonstrated that both renal localization and expression of eNos increased in the kidney of *P. dolloi* during the maintenance phase of aestivation. They suggested that regulation of Na^+ and water reabsorption and glomerular filtration rate could involve eNos-produced nitric oxide in the kidney.

Oxidative Defence

Page et al. (2010) reported that most of the major intracellular antioxidant enzymes, including the mitochondrial superoxide dismutase, cytosolic superoxide dismutase, catalase, glutathione peroxidase and glutathione reductase, were up-regulated in the brain of *P. dolloi* after 60 days of aestivation. Aestivation also resulted in increases in activities of several of these enzymes in the heart. These changes were unrelated to fasting because similar phenomena were not observed in a group of fish that were deprived of food but maintained in water for the same period of time. Products of lipid peroxidation (4-hydroxynonenal adducts) and oxidative protein damage (carbonylation) were similar in control and aestivating lungfish, although protein nitrotyrosine levels were elevated in brain tissue of aestivators. Therefore, Page et al. (2010) concluded that aestivating *P. dolloi* experienced little oxidative damage in the brain and heart because of increased oxidative stress resistance in these organs resulting from increases in intracellular antioxidant capacity.

Ching et al. (2014) conducted experiments to elucidate the physiological significance of ascorbate synthesis in *P. annectens* in relation to the aestivation process. The concentrations of ascorbate, dehydroascorbate and total ascorbate + dehydroascorbate were quantified in the kidney and brain of *P. annectens* during the three phases of aestivation. Ascorbate levels decreased 99.9%, and dehydroascorbate increased 48.5% in the kidney of *P. annectens* after six months of aestivation in air. In addition, there were significant decreases in the mRNA expression level and the protein abundance of L-gulonolactone oxidase in the kidney (Ching et al. 2014). The sharp decrease in the mRNA expression level of *L-gulonolactone oxidase* could be related to the shut-down of renal function (DeLaney et al. 1974) and change in renal corpuscle structure during the prolonged maintenance phase of aestivation (Ojeda et al. 2008). As the kidney constitutes the major site of ascorbate synthesis in lungfishes, the decline of L-gulonolactone oxidase activity and hence ascorbate production, would expose the lungfish to potential oxidative stress during aestivation. However, there was only a minor decrease in ascorbate levels in the brain of *P. annectens* after six months of aestivation. Although the brain ascorbate concentration fluctuated during the aestivation process, it was still maintained at a relatively high level. The expression of *L-gulonolactone oxidase* and the presence of the L-gulonolactone oxidase enzyme activity in the brain of *P. annectens* (Ching et al. 2014) could well explain the unusual phenomenon of maintaining relatively high ascorbate concentrations in the brain during the maintenance phase of aestivation, in spite of a drastic decrease in the L-gulonolactone oxidase activity in, and a reduction in the ascorbate output from, the kidney.

Arousal from Aestivation

External Signals and Internal Cues for Arousal

In the authors' laboratories, African lungfishes aestivating in air have been routinely aroused from aestivation by re-immersion. While it is certain that water availability constitutes the major external signal to arouse African lungfishes from aestivation (Ip and Chew 2010; Hiong et al. 2014, 2015), the sensing mechanisms, the internal cues and the signal transduction mechanisms involved have not been examined. It is uncertain whether the skin and gills are involved in sensing the return of water, as they are covered by the dried cocoon which is presumably water-impermeable. In particular, the gills of *P. annectens* are still covered with remnants of cocoon material several days after arousal from aestivation (J.M. Icardo, pers. comm.), which would render them ineffective in osmosensing and/or respiration. Of note, African lungfishes depend entirely on pulmonary respiration during the maintenance phase of aestivation in air or in mud. However, arousal prescribes the fish being submerged in water, and the arousal process would take at least several hours. During this period, the mouth would be filled with water, and the fish would undergo transient asphyxia due to the blockage of air intake into its lungs, the lack of ventilation movements, and the non-functional gills. The same would apply to the arousal of fish aestivating in mud as the air channel in the mud leading to the mouth would be filled up with water. Whether the sensing mechanism lies in the mouth and whether transient asphyxia constitutes part of the arousal mechanism are uncertain at present.

For *P. annectens* that were aroused from aestivation, Giusi et al. (2011) detected increases in the mRNA expression level of *gabaar α1* subunit in the telencephalon and various cerebellum neuronal fields, which are areas commonly associated with GABAergic-dependent plasticity and motor properties in rodents (Fritschy and Panzanelli 2006; Li et al. 2009). Thus, Giusi et al. (2011) proposed that Gabaar α1 might have a major role in the correct arrangement of the other Gabaergic subunits, in the recovery of physiological functions during arousal, and in the early activation of protective pathways against ischemic-like stress episodes in the brain of *P. annectens* as in mammals (Zepeda et al. 2004).

Icardo et al. (2008) reported that the heart of *P. dolloi* had high capacity for functional recovery during the maintenance and arousal phases of aestivation. They proposed that the drastic reduction in the amounts of urea accumulated in the body tissues upon arousal (Wood et al. 2005) might produce an osmotic imbalance that eventually resulted in rupture of the membranes and the massive accumulation of vacuolized cytoplasm components in the septal myocytes of the heart (Icardo et al. 2008). Subsequently, these areas attract the macrophages which are involved in debris clearance, and such a process may facilitate tissue regeneration. Hence, it is logical to deduce that the increased excretion of urea during rehydration and the resulting changes in osmolality of tissue fluids could be an important internal cue for tissue regeneration upon arousal from aestivation, the confirmation of which awaits future studies.

Rehydration

During arousal in water, water absorption must precede urea excretion because urea is crucial to this osmotic phenomenon. Wilkie et al. (2007) studied *P. dolloi* during prolonged (five months) exposure to air, but because the fish used in their experiment did not really undergo aestivation, their results actually offered insights into what would happen during the arousal phase when water becomes once again available to aestivating African lungfishes. The results of Wilkie et al. (2007) indirectly support the proposition of Riddle (1983), who proposed that urea accumulated during the maintenance phase of aestivation could facilitate water uptake from the environment upon rehydration during arousal. Based on results of Wilkie et al. (2007), the 13-fold increase in muscle urea content was the likely explanation for the 56% increase in muscle water content observed after five months of air exposure. However, the phenomena reported by Wilkie et al. (2007) that muscle acted as a 'water reservoir' during air exposure and that the body mass decreased by 20% during subsequent re-immersion in water did not reflect the real situation of arousal from aestivation. Firstly, the level of urea accumulated in the body during the normally short period of induction phase of aestivation would not lead to such a large increase in muscle water content. Secondly, after a long period of aestivation, there should be a decrease, and not an increase, in the muscle water content. Thirdly, during arousal, it is essential for the fish to gain water from instead of losing it to the environment. Although it is unlikely that fish can accumulate such high levels of urea during a six–eight day induction period, a similar magnitude of increase in urea content can be expected to have occurred in fish that underwent an equivalent period (i.e., five months) of aestivation. Therefore, it can be deduced from results reported by Wilkie et al. (2007) that water absorption occur through the cutaneous surface of African lungfishes when water becomes available during arousal.

There could be a sharp decrease in plasma osmolality due to decreases in plasma Na^+ and urea concentrations in relation to rehydration and increased urea excretion during the first 48 hours of arousal. Hence, the aestivating lungfish would be confronted with a profound osmotic stress during the early phase of arousal from aestivation. Since Nka is known to interact with aquaporin which is involved in the permeation of water through biomembranes, the down-regulation in the mRNA expression levels of not only *nkaα1*, but also *nkaα2* and *nkaα3*, in the brain of *P. annectens* during the arousal phase (Hiong et al. 2014) indicate a possible reduction in the interaction between Nka and Aqp4 and Na_x channels which would in principle prevent the occurrence of brain edema during arousal.

Increased Blood Circulation and Oxidative Stress

When African lungfishes enter into aestivation, blood flow is shunted away from splanchnic organs and returned during arousal. The heart beat rate decreases during the maintenance phase of aestivation but recovers upon arousal. This increases the potential for transient ischemia-reperfusion events. Intracellular alterations following ischemic and/or reperfusion have the potential to induce physiological and oxidative stress and perhaps even apoptosis in sensitive organs. How African lungfishes defend against oxidative stress upon arousal is uncertain at present. Upon arousal from aestivation, there is a general recovery of the haemocytopoietic tissues (Jordon and Speidel 1931). Erythrocyte amitoses may become very numerous. Amitosis has also been observed in the case of thrombocytes, granulocytes monocytes and lymphoid haemoblasts (Jordon and Speidel 1931).

Excretion of Accumulated Urea

Upon arousal in water, African lungfishes can efficiently excrete the excess urea accumulated in the body during the maintenance phase of aestivation (Smith 1930; Janssens 1964). Working on *P. dolloi* exposed to terrestrial conditions for six days, Chew et al. (2003) demonstrated that its urea excretion rate increased 22-fold during re-immersion as compared to the control specimen. This is the greatest increase in urea excretion reported for fishes during emersion-immersion transition, and suggests that *P. dolloi* possesses transporters which facilitate the excretion of urea after arousal from aestivation. Subsequently, Wood et al. (2005) reported that after 21–30 days of aestivation in air or exposed to air without aestivation, the urea excretion rate increased dramatically, reaching 2000–6000 μmol-N hr^{-1} kg^{-1} at 10–24 hours, in *P. dolloi* during re-immersion. The skin appeared to be an important site of urea-N excretion, because 72% of the urea-N efflux occurred through the posterior 85% of the body with minimal involvement of the kidney. Indeed, Hung et al. (2009) obtained the full length cDNA sequence of a putative urea transporter (UT-A) from *P. annectens*. The putative urea transporter was expressed in gills, kidney, liver, skeletal muscle, and skin. Upon re-immersion in water after 33 days of air exposure, *P. annectens* exhibited a massive rise in urea-N excretion which peaked at 12–30 hours and persisted until 70 hours. Quantitative real-time-PCR revealed significant elevation in the mRNA expression level of the putative urea transporter in the skin between hour 14 and 48 of re-immersion. Thus, the increase in urea excretion was attributable to the transcriptional activation of the putative urea transporter in the skin of African lungfishes during the arousal phase of aestivation. During arousal, increased urea excretion in *P. dolloi* occurred in pulses (Wood et al. 2005). Ip and Chew (2010) proposed that it could be an adaptation to ensure complete rehydration, which was dependent on tissue urea content, and to minimize instantaneous osmotic shock to cells, upon arousal. Through the injection of NH_4Cl + urea, Ip et al. (2005c) concluded that excretion of accumulated urea in *P. dolloi* was regulated by the level of internal ammonia. Hence, it is possible that an increase in ammonia production occurred through increased amino acid catabolism upon arousal, and the increased production of endogenous ammonia acted as a signal to enhance urea excretion.

Urine Production and Kidney Regeneration

Ojeda et al. (2008) reported that all the modifications which rendered the kidney non-functional during the maintenance phase of aestivation were partially reversed during the first few days of arousal with the return of water. How the kidney nephrons undergo regeneration has not been explored.

Metabolic Rate and Energy Metabolism

No information is available on the metabolic rate of African lungfish after arousal from long periods of aestivation. However, it is logical to deduce that arousal would lead to an increase in the metabolic rate for structural and functional reconstruction of cells and tissues. Aroused fish would only start to feed after 7–10 days, and the fuel for ATP production to support locomotor activity and tissue reconstruction/regeneration is unclear at present. It is probable that there could be a prominent increase in the mobilization of muscle protein during the arousal phase of aestivation, but such a possibility has not been explored.

Feeding, Tissue Regeneration and Protein Synthesis

African lungfishes would initiate feeding only after 7–10 days of arousal from aestivation, indicating that some restructuring of the intestinal epithelium is a prerequisite to feeding. Icardo et al. (2012) reported that after arousal from six months of aestivation, cell phenotypes in the digestive tract of *P. annectens* were restored in about six days, but full structural recovery was not attained during the experimental period (15 days post-aestivation). Hence, the initial recovery of the intestinal epithelium was independent of food intake. Since structural changes would require increased syntheses of certain proteins, which occurred before re-feeding, it would imply the mobilization of amino acids of endogenous origin. However, at present, there is a dearth of knowledge on protein degradation and synthesis during the arousal phase of aestivation. Furthermore, no information is available presently on the postprandial nitrogen metabolism and excretion in African lungfishes a short period after arousal from aestivation.

Perspective

In this chapter, we have presented up-to-date information on aestivation in African lungfishes. Special efforts were made to address some controversial and enigmatic issues pertaining to the induction, maintenance and arousal phases of aestivation. Adaptive responses differ during the three phases, and they are essential to the understanding of the overall aestivation process. However, many of these adaptive responses have not been studied in detail. Specifically, limited information is available on processes related to the induction or arousal phases of aestivation. Therefore, future efforts should be made to identify adaptive responses particular to each of the three phases of aestivation in African lungfishes.

NOTE: Two different types of abbreviations are adopted in this chapter because the standard abbreviations of genes/proteins of fishes (http://zfin.org/cgi-bin/webdriver?MIval=aa-ZDB_home.apg) are different from those of human/non-human primates ("http://www.genenames.org/" http://www.genenames.org). Specifically, for fishes, gene symbols are italicized, all in lower case, and protein designations are the same as the gene symbol, but not italicized with the first letter in upper case.

References

Amelio, D., F. Garofalo, E. Brunelli, A.M. Loong, W.P. Wong, Y.K. Ip, B. Tota and M.C. Cerra. 2008. Differential NOS expression in the freshwater and aestivating *Protopterus dolloi* (lungfish): heart versus kidney readjustments. Nitric Oxide 18: 1–10.

Amemiya, C.T., J. Alföldi, A.P. Lee, S. Fan, H. Philippe, I. MacCallum, I. Braasch, T. Manousaki, I. Schneider, N. Rohner, C. Organ, D. Chalopin, J.J. Smith, M. Robinson, R.A. Dorrington, M. Gerdol, B. Aken, M.A. Biscotti, M. Barucca, D. Baurain, A.M. Berlin, G.L. Blatch, F. Buonocore, T. Burmester, M.S. Campbell, A. Canapa, J.P. Cannon, A. Christoffels, G. De Moro, A.L. Edkins, L. Fan, A.M. Fausto, N. Feiner, M. Forconi, J. Gamieldien, S. Gnerre, A. Gnirke, J.V. Goldstone, W. Haerty, M.E. Hahn, U. Hesse, S. Hoffmann, J. Johnson, S.I. Karchner, S. Kuraku, M. Lara, J.Z. Levin, G.W. Litman, E. Mauceli, T. Miyake, M.G. Mueller, D.R. Nelson, A. Nitsche, E. Olmo, T. Ota, A. Pallavicini, S. Panji, B. Picone, C.P. Ponting, S.J. Prohaska, D. Przybylski, N.R. Saha, V. Ravi, F.J. Ribeiro, T. Sauka-Spengler,

G. Scapigliati, S.M.J. Searle, T. Sharpe, O. Simakov, P.F. Stadler, J.J. Stegeman, K. Sumiyama, D. Tabbaa, H. Tafer, J. Turner-Maier, P. van Heusden, S. White, L. Williams, M. Yandell, H. Brinkmann, J.-N. Volff, C.J. Tabin, N. Shubin, M. Schartl, D.B. Jaffe, J.H. Postlethwait, B. Venkatesh, F. Di Palma, E.S. Lander, A. Meyer and K. Lindblad-Toh. 2013. The African coelacanth genome provides insights into tetrapod evolution. Nature 496: 311–316.

Ayson, F.G., T. Kaneko, M. Tagawa, S. Hasegawa, E.G. Grau, R.S. Nishioka, D.S. King, H.A. Bern and T. Hirano. 1993. Effects of acclimation to hypertonic environment on plasma and pituitary levels of two prolactins and growth hormone in two species of tilapia, *Oreochromis mossambicus* and *Oreochromis niloticus*. Gen. Comp. Endocrinol. 89: 138–148.

Babiker, M.M. 1979. Respiratory behaviour, oxygen consumption and relative dependence on aerial respiration in the African lungfish (*Protopterus annectens*, Owen) and an airbreathing teleost (*Clarias lazera*, C.). Hydrobiologia 65: 177–187.

Babiker, M.M. and O.H. El Hakeem. 1979. Changes in blood characteristics and constituents associated with aestivation in the African lungfish, *Protopterus annectens* Owen. Zool. Anz 202: 9–16.

Ballantyne, J.S. and N.T. Frick. 2010. Lungfish metabolism. pp. 305–340. *In*: J.M. Jorgensen and J. Joss (eds.). The Biology of Lungfishes. Science Publishers, New Hampshire.

Bayomy, M.F.F., A.G. Shalan, S.D. Bradshaw, P.C. Withers, T. Stewart and G. Thompson. 2002. Water content, body weight and acid mucopolysaccharides, hyaluronidase and beta-glucuronidase in response to aestivation in Australian desert frogs. Comp. Biochem. Physiol. A 131: 881–892.

Belkin, D.A. 1965. Reduction of metabolic rate in response to starvation in the turtle, *Sternotherus minor*. Copeia 1965: 367–368.

Blanc, M., F. D'Aubenton and Y. Plessis. 1956. Étude de l'enkystement de *Protopterus annectens*. Bull. Inst. Fondam. 18A: 843–854.

Bole-Feysot, C., V. Goffin, M. Edery, N. Binart and P.A. Kelly. 1998. Prolactin (PRL) and its receptor: actions, signal transduction pathways and phenotypes observed in PRL receptor knockout mice. Endocr. Rev. 19: 225–268.

Bonaventura, J., C. Bonaventura and B. Sullivan. 1974. Urea tolerance as a molecular adaptation of elasmobranch hemoglobins. Science 186: 57–59.

Booth, F.W. and M.J. Seider. 1979. Early change in skeletal muscle protein synthesis after limb immobilization of rats. J. Appl. Physiol. 47: 974–977.

Brien, P., M. Poll and J. Boiullon. 1959. Ethologie de la reproduction de *Protopterus dolloi*. Ann. Mus. R. Congo. Belge. 71: 3–21.

Brinkmann, H., B. Venkatesh, S. Brenner and A. Meyer. 2004. Nuclear protein-coding genes support lungfish and not the coelacanth as the closest living relatives of land vertebrates. Proc. Natl. Acad. Sci. USA 101: 4900–4905.

Burggren, W.W. and K. Johansen. 1986. Circulation and respiration in lungfishes (Dipnoi). J. Morphol. 190: 217–236.

Campbell, J.W. 1973. Nitrogen excretion. pp. 279–316. *In*: C.L. Prosser (ed.). Comparative Animal Physiology (3rd ed.). Saunders College Publishing, Philadelphia.

Campbell, J.W. 1991. Excretory nitrogen metabolism. pp. 277–324. *In*: C.L. Prosser (ed.). Comparative Animal Physiology (4th ed.) Environmental and Metabolic Animal Physiology. Wiley-Liss Inc., New York.

Carlisky, N.J. and A. Barrio. 1972. Nitrogen metabolism of the South American lungfish *Lepidosiren paradoxa*. Comp. Biochem. Physiol. B 41: 857–873.

Chanoine, C., A. El-Attari, M. Guyotlenfant, L. Ouedraogo and C.-L. Gallien. 1994. Myosin isoforms and their subunits in the lungfish *Protopterus annectens:* changes during development and the annual cycle. J. Exp. Zool. 269A: 413–421.

Chew, S.F. and Y.K. Ip. 2014. Excretory nitrogen metabolism and defense against ammonia toxicity in air-breathing fishes. J. Fish Biol. 84: 603–638.

Chew, S.F., T.F. Ong, L. Ho, W.L. Tam, A.M. Loong, K.C. Hiong, W.P. Wong and Y.K. Ip. 2003. Urea synthesis in the African lungfish *Protopterus dolloi*: hepatic carbamoyl phosphate synthetase III and glutamine synthetase are upregulated by 6 days of aerial exposure. J. Exp. Biol. 206: 3615–3624.

Chew, S.F., N.K.Y. Chan, W.L. Tam, A.M. Loong, K.C. Hiong and Y.K. Ip. 2004. Nitrogen metabolism in the African lungfish (*Protopterus dolloi*) aestivating in a mucus cocoon on land. J. Exp. Biol. 207: 777–786.

Chew, S.F., L. Ho, T.F. Ong, W.P. Wong and Y.K. Ip. 2005. The African lungfish, *Protopterus dolloi*, detoxifies ammonia to urea during environmental ammonia exposure. Physiol. Biochem. Zool. 78: 31–39.

Chew, S.F., J.M. Wilson, Y.K. Ip and D.J. Randall. 2006. Nitrogenous excretion and defense against ammonia toxicity. pp. 307–395. *In*: A. Val, V. Almedia-Val and D.J. Randall (eds.). The Physiology of Tropical Fishes. Fish Physiology, 21. Academic Press, New York.

Childs, S.G. 2003. Muscle wasting. Ortho. Nurs. 22: 251–257.

Ching, B., S.F. Chew, W.P. Wong and Y.K. Ip. 2014. L-gulono-γ-lactone oxidase and ascorbate synthesis in the brain of the African lungfish *Protopterus annectens*. FASEB J. 28: 3506–3517.

Chng, Y.R., J.L.Y. Chew, B. Ching, X.L. Chen, W.P. Wong, S.F. Chew and Y.K. Ip. 2014. Molecular characterization of *argininosuccinate synthase* and *argininosuccinate lyase* from the liver of the African lungfish *Protopterus annectens*, and their mRNA expression levels in the liver, kidney, brain and skeletal muscle during aestivation. J. Comp. Physiol. B. 184: 835–853.

Conant, E.B. 1973. Regeneration in the African lungfish, *Protopterus*. III. Regeneration during fasting and estivation. Biol. Bull. 144: 248–261.

Conant, E.B. 1976. Urea accumulation and other effects of estivation in the African lungfish, *Protopterus*. Virginia J. Sci. 27: 42.

Cooper, J.L. and F. Plum. 1987. Biochemistry and physiology of brain ammonia. Physiol. Rev. 67: 440–519.

Costanzo, J.P. and R.E. Lee. 2005. Cryoprotection by urea in a terrestrially hibernating frog. J. Exp. Biol. 208: 4079–4089.

Daniels, C.B. and S. Orgeig. 2003. Pulmonary surfactant: the key to the evolution of air breathing. News Physiol. Sci. 18: 151–157.

DeLaney, R.G. and A.P. Fishman. 1977. Analysis of lung ventilation in the aestivating lungfish *Protopterus aethiopicus*. Am. J. Physiol. Regul. Integr. Comp. Physiol. 233: R181–R187.

DeLaney, R.G., S. Lahiri and A.P. Fishman. 1974. Aestivation of the African lungfish *Protopterus aethiopicus*: Cardiovascular and respiratory functions. J. Exp. Biol. 61: 111–128.

DeLaney, R.G., C. Shub and A.P. Fishman. 1976. Hematologic observations on the aquatic and aestivating African lungfish *Protopterus aethiopicus*. Copeia 1976: 423–434.

DeLaney, R.G., S. Lahiri, R. Hamilton and A.P. Fishman. 1977. Acid base balance and plasma composition in the aestivating lungfish (*Protopterus*). Am. J. Physiol. Regul. Integr. Comp. Physiol. 232: R10–R17.

Doldán, M.J., B. Prego, B.I. Holmqvist and E. de Miguel. 1999. Distribution of GABA-immunolabeling in the early zebrafish (*Danio rerio*) brain. Eur. J. Morphol. 37: 126–129.

DuBoiS, R. 1892. Contribution a l'etude du mecanisme respiratoire des Dipnolques. Ann. Soc. Lhm. Lyon 39: 65–72.

Dunn, J.F., W. Davison, G.M.O. Maloiy, P.W. Hochachka and M. Guppy. 1981. An ultrastructural and histochemical study of the axial musculature in the African lungfish. Cell Tissue Res. 220: 599–609.

Dunn, J.F., P.W. Hochachka, W. Davidson and M. Guppy. 1983. Metabolic adjustments to diving and recovery in the African lungfish. Am. J. Physiol. Regul. Integr. Comp. Physiol. 245: R651–R657.

Ebert, B., K.A. Wafford and S. Deacon. 2006. Treating insomnia: current and investigational pharmacological approaches. Pharmacol. Ther. 112: 612–629.

Ellefsen, S., K.O. Stensløkken, C.E. Fagernes, T.A. Kristensen and G.E. Nilsson. 2009. Expression of genes involved in GABAergic neurotransmission in anoxic crucian carp brain (*Carassius carassius*). Physiol. Genomics 36: 61–68.

Ellis, M.M. 1909. The relation of the amount of tail regenerated to the amount removed in tadpoles of *Rana clansitans*. J. Exp. Zool. 7: 421–455.

Evans, T.G. and G.N. Somero. 2008. A microarray-based transcriptomic time-course of hyper- and hypo-osmotic stress signaling events in the euryhaline fish *Gillichthys mirabilis*: osmosensors to effectors. J. Exp. Biol. 211: 3636–3649.

Evans, D.H., P.M. Piermarini and K.P. Choe. 2005. The multifunctional fish gill: dominant site of gas exchange, osmoregulation, acid-base regulation, and excretion of nitrogenous waste. Physiol. Rev. 85: 97–177.

Ewart, H.S. and W.R. Driedzic. 1990. Enzyme activity levels underestimate lactate production rates in cod (*Gadus morhua*) gas gland. Can. J. Zool. 68: 193–197.

Facciolo, R.M., M. Crudo, G. Giusi and M. Canonaco. 2010. GABAergic influences on ORX receptor-dependent abnormal motor behaviors and neurodegenerative events in fish. Toxicol. Appl. Pharmacol. 243: 77–86.

Fiol, D.F. and D. Kültz. 2007. Osmotic stress sensing and signaling in fishes. FEBS J. 274: 5790–5798.

Fishman, A.P., A.I. Pack, R.G. DeLaney and R.J. Galante. 1986. Estivation in *Protopterus*. J. Morphol. Suppl. 1: 237–248.

Fishman, A.P., R.J. Galante, A. Winokur and A.I. Pack. 1992. Estivation in the African lungfish. Proc. Am. Philos. Soc. 136: 61–72.

Fitts, R.H., D.R. Riley and J.J. Widrick. 2000. Physiology of a microgravity environment invited review: microgravity and skeletal muscle. J. Appl. Physiol. 89: 823–839.

Fitts, R.H., D.R. Riley and J.J. Widrick. 2001. Functional and structural adaptations of skeletal muscle to microgravity. J. Exp. Biol. 204: 3201–3208.

Forster, R.P. and L. Goldstein. 1966. Urea synthesis in the lungfish: relative importance of purine and ornithine cycle pathways. Science 153: 1650–1652.

Fraser, E.J., P.T. Bosma, V.L. Trudeau and K. Docherty. 2002. The effect of water temperature on the GABAergic and reproductive systems in female and male goldfish (*Carassius auratus*). Gen. Comp. Endocrinol. 125: 163–175.

Frick, N.T., J.S. Bystriansky and J.S. Ballantyne. 2007. The metabolic organization of a primitive air-breathing fish, the Florida gar (*Lepisosteus platyrhincus*). J. Exp. Zool. 307A: 7–17.

Frick, N.T., J.S. Bystriansky, Y.K. Ip, S.F. Chew and J.S. Ballantyne. 2008a. Carbohydrate and amino acid metabolism in fasting and aestivating African lungfish (*Protopterus dolloi*). Comp. Biochem. Physiol. A 151: 85–92.

Frick, N.T., J.S. Bystriansky, Y.K. Ip, S.F. Chew and J.S. Ballantyne. 2008b. Lipid, ketone body and oxidative metabolism in the African lungfish, *Protopterus dolloi*, following 60 days of fasting and aestivation. Comp. Biochem. Physiol. A 151: 93–101.

Fritschy, J.M. and P. Panzanelli. 2006. Molecular and synaptic organization of $GABA_A$ receptors in the cerebellum: effects of targeted subunit gene deletions. Cerebellum 5: 275–285.

Giusi, G., M. Crudo, A. Di Vito, R.M. Facciolo, F. Garofalo, S.F. Chew, Y.K. Ip and M. Canonaco. 2011. Lungfish aestivating activities are locked in distinct encephalic gamma-aminobutyric acid type A receptor alpha subunits. J. Neurosci. Res. 89: 418–428.

Giusi, G., M. Zizza, R.M. Facciolo, S.F. Chew, Y.K. Ip and M. Canonaco. 2012. Aestivation and hypoxia-related events share common silent neuron trafficking processes. BMC Neurosci. 13: 39.

Godet, R. 1959. Le role de la posthypophyse dans le passage de la phase aquatique a la phase terrestre chez le Protoptere du Senegal. C. R. Soc. Biol. Paris 153: 691–693.

Godet, R. 1962. Evolution de la pars intermedia hypophysaire chez le Protoptere au conrs de l'epreuve de deshydration. C. R. Soc. Biol. Paris 156: 148–150.

Godet, R., R. Michel and M. Dupe. 1964. Metabolisme d'l 131 chez le Protopere (*Protopterus annectens* Owen) en phase seche (cocoon experimental). C. R. Soc. Biol. Paris 158: 1236–1241.

Graham, J.B. 1997. Air-breathing Fishes: Evolution, Diversity and Adaptation. Academic Press, San Diego.

Greenwood, P.H. 1986. The natural history of the African lungfishes. J. Morphol. Suppl. 1: 163–179.

Gregory, P.T. 1982. Reptilian hibernation. pp. 53–154. *In*: C. Gans and F.H. Pough (eds.). Biology of the Reptilia, Vol. 13. Physiology and Physiological Ecology. Academic Press, London.

Griffith, R.W. 1991. Guppies, toadfish, lungfish, coelacanths and frogs—A scenario for the evolution of urea retention in fishes. Envir. Biol. Fish. 32: 1–4.

Guppy, M. and P. Withers. 1999. Metabolic depression in animals: physiological perspectives and biochemical generalizations. Biol. Rev. 74: 1–40.

Hailey, A. and J.P. Loveridge. 1997. Metabolic depression during dormancy in the African tortoise *Kinixys spekii*. Can. J. Zool. 75: 1328–1335.

Hand, S.C. and G.N. Somero. 1982. Urea and methylamine effects on rabbit muscle phosphofructokinase. J. Biol. Chem. 257: 734–741.

Henderson, R.J. and D.R. Tocher. 1987. The lipid composition and biochemistry of freshwater fish. Prog. Lipid Res. 26: 281–347.

Hiong, K.C., Y.K. Ip, W.P. Wong and S.F. Chew. 2013. Differential gene expression in the brain of the African lungfish, *Protopterus annectens*, after six days or six months of aestivation in air. PLoS One. PLoS ONE 8(8): e71205. doi:10.1371/journal.pone.0071205.

Hiong, K.C., Y.K. Ip, W.P. Wong and S.F. Chew. 2014. Molecular characterization of three Na^+/K^+-ATPase α-subunit isoforms from, and their mRNA expression and protein abundance in, the brain of the African lungfish, *Protopterus annectens*, during three phases of aestivation in air. J. Comp. Physiol. B 184: 571–587.

Hiong, K.C., Y.K. Ip, W.P. Wong and S.F. Chew. 2015. Differential gene expression in the liver of the African lungfish, *Protopterus annectens*, after 6 months of aestivation in air or 1 day of arousal from 6 months of aestivation. PLoS One 10(3): e0121224. doi: 10.1371/journal.pone.0121224.

Hochachka, P.W. 1980. Living Without Oxygen. Harvard University Press, Cambridge.

Hochachka, P.W. and W.C. Hulbert. 1978. Glycogen 'seas', glycogen bodies, and glycogen granules in heart and skeletal muscle of two air-breathing, burrowing fishes. Can. J. Zool. 56: 774–786.

Hochachka, P.W. and G.N. Somero. 1984. Biochemical Adaptation. Princeton University Press, Princeton.

Hochachka, P.W. and M. Guppy. 1987. Estivators. pp. 101–112. *In*: P.W. Hochachka and M. Guppy (eds.). Metabolic Arrest and the Control of Biological Time. Harvard University Press, Cambridge.

Hochachka, P.W., M. Guppy, H.E. Guderley, K.B. Storey and W.C. Hulbert. 1978a. Metabolic biochemistry of water vs. air breathing fishes: muscle enzymes and ultra-structure. Can. J. Zool. 56: 736–750.

Hochachka, P.W., M. Guppy, H.E. Guderley, K.B. Storey and W.C. Hulbert. 1978b. Metabolic biochemistry of water vs. air breathing osteoglossids: heart enzymes and ultrastructure. Can. J. Zool. 56: 759–768.

Hochachka, P.W., W.C. Hulbert and M. Guppy. 1978c. The tuna power plant and furnace. pp. 153–174. *In*: G.D. Sharp and A.E. Dizon (eds.). The Physiological Ecology of Tunas. Academic Press, New York.

Holl, R.W., M.L. Hartman, J.D. Veldhuis, W.M. Taylor and M.O. Thorner. 1991. Thirty-second sampling of plasma growth hormone in man: correlation with sleep stages. J. Clin. Endocrinol. Metab. 72: 854–861.

Hui, F. and A. Smith. 1970. Regeneration of the amputated amphibian limb: retardation by hemicholinium-3. Science 170: 1313–1314.

Hung, C.Y.C., F. Galvez, Y.K. Ip and C.M. Wood. 2009. A facilitated diffusion urea transporter in the skin of the African lungfish, *Protopterus annectens*. J. Exp. Biol. 212: 1202–1211.

Hwang, P.P. and T.H. Lee. 2007. New insights into fish ion regulation and mitochondrion-rich cells. Comp. Biochem. Physiol. A 148: 479–497.

Hwang, P.P., T.H. Lee and L.Y. Lin. 2011. Ion regulation in fish gills: recent progress in the cellular and molecular mechanisms. Am. J. Physiol. Regul. Integr. Comp. Physiol. 301: R28–R47.

Icardo, J.M., D. Amelio, F. Garofalo, E. Colvee, M.C. Cerra, W.P. Wong, B. Tota and Y.K. Ip. 2008. The structural characteristics of the heart ventricle of the African lungfish *Protopterus dolloi*: freshwater and aestivation. J. Anat. 213: 106–119.

Icardo, J.M., A.M. Loong, E. Colvee, W.P. Wong and Y.K. Ip. 2012. The alimentary canal of the African lungfish *Protopterus annectens* during aestivation and after arousal. Anat. Rec. 295: 60–72.

Iftikar, F.I., M. Patel, Y.K. Ip and C.M. Wood. 2008. The influence of feeding on aerial and aquatic oxygen consumption, nitrogenous waste excretion, and metabolic fuel usage in the African lungfish, *Protopterus annectens*. Can. J. Zool. 86: 790–800.

Ip, Y.K. and S.F. Chew. 2010. Nitrogen metabolism and excretion during aestivation. pp. 63–93. *In*: C.A. Navas and J.E. Carvalho (eds.). Aestivation: Molecular and Physiological Aspects, Progress in Molecular and Subcellular Biology, Vol. 49. Springer-Verlag, Berlin.

Ip, Y.K., S.F. Chew and D.J. Randall. 2001. Ammonia toxicity, tolerance and excretion. pp. 109–148. *In*: P.A. Wright and P.M. Anderson (eds.). Fish Physiology Vol. 19, Nitrogen Excretion. Academic Press, New York.

Ip, Y.K., S.F. Chew and D.J. Randall. 2004a. Five tropical air-breathing fishes, six different strategies to defend against ammonia toxicity on land. Physiol. Biochem. Zool. 77: 768–782.

Ip, Y.K., S.F. Chew, J.M. Wilson and D.J. Randall. 2004b. Defences against ammonia toxicity in tropical air-breathing fishes exposed to high concentrations of environmental ammonia: a review. J. Comp. Physiol. B 174: 565–575.

Ip, Y.K., P.J. Yeo, A.M. Loong, K.C. Hiong, W.P. Wong and S.F. Chew. 2005a. The interplay of increased urea synthesis and reduced ammonia production in the African lungfish *Protopterus aethiopicus* during 46 days of aestivation in a mucus cocoon on land. J. Exp. Zool. 303A: 1054–1065.

Ip, Y.K., B.K. Peh, W.L. Tam, S.L.M. Lee and S.F. Chew. 2005b. Changes in salinity and ionic compositions act as environmental signals to induce a reduction in ammonia production in the African lungfish Protopterus dolloi. J Exp. Zool. 307A: 456–463.

Ip, Y.K., B.K. Peh, W.L. Tam, W.P. Wong and S.F. Chew. 2005c. Effects of intra-peritoneal injection with NH_4Cl, urea or NH_4Cl+urea on nitrogen excretion and metabolism in the African lungfish *Protopterus dolloi*. J. Exp. Zool. 303A: 272–282.

Ip, Y.K., A.M. Loong, Y.R. Chng, K.C. Hiong and S.F. Chew. 2012. Hepatic carbamoyl phosphate synthetase (CPS) I and urea contents in the hylid tree frog, *Litoria caerulea*: transition from CPS III to CPS I. J. Comp. Physiol. B 182: 1081–1094.

Janssens, P.A. 1964. The metabolism of the aestivating African lungfish. Comp. Biochem. Physiol. 11: 105–117.

Janssens, P.A. 1965. Phosphorylase and glucose-6-phosphatase in the African lungfish. Comp. Biochem. Physiol. 16: 317–319.

Janssens, P.A. and P.P. Cohen. 1966. Ornithine-urea cycle enzymes in the African lungfish *Protopterus aethiopicus*. Science 152: 358–359.

Janssens, P.A. and P.P. Cohen. 1968a. Biosynthesis of urea in the estivating African lungfish and in *Xenopus laevis* under conditions of water-shortage. Comp. Biochem. Physiol. 24: 887–898.

Janssens, P.A. and P.P. Cohen. 1968b. Nitrogen metabolism in the African lungfish. Comp. Biochem. Physiol. 24: 879–886.

Jesse, M.J., C. Shub and A.P. Fishman. 1967. Lung and gill ventilation of the African lungfish. Respir. Physiol. 3: 267–287.

Johansen, K., J.P. Lomholt and G.M.O. Maloiy. 1976a. Importance of air and water breathing in relation to size of the African lungfish *Protopterus amphibius* Peters. J. Exp. Biol. 65: 395–399.

Johansen, K., G. Lykkeboe, R.E. Weber and G.M.O. Maloiy. 1976b. Respiratory properties of blood in awake and aestivating lungfish, *Protopterus amphibius*. Respir. Physiol. 27: 335–345.

Johnels, A.G. and G.S.O. Svensson. 1954. On the biology of *Protopterus annectens* (Owen). Arkiv. Zool. 7: 131–158.

Jordon, H.E. and C.C. Speidel. 1931. Blood formation in the African lungfish, under normal conditions and under conditions of prolonged estivation and recovery. J. Morphol. 51: 319–371.

Kerem, D., D. Hammond and R. Elsner. 1973. Tissue glycogen levels in the Weddell seal *Leptonychotes weddelli*: a possible adaptation to asphyxia hypoxia. Comp. Biochem. Physiol. A 45: 731–736.

Kreider, M.S., A. Winokur, A.I. Pack and A.P. Fishman. 1990. Reduction of thyrotropin-releasing hormone concentrations in central nervous system of African lungfish during estivation. Gen. Comp. Endocrinol. 77: 435–441.

Kültz, D., D. Fiol, N. Valkova, S. Gomez-Jimenez, S.Y. Chan and J. Lee. 2007. Functional genomics and proteomics of the cellular osmotic stress response in 'non-model' organisms. J. Exp. Biol. 210: 1593–1601.

Laberge, T. and P.J. Walsh. 2011. Phylogenetic aspects of carbamoyl phosphate synthetase in lungfish: a transitional enzyme in transitional fishes. Comp. Biochem. Physiol. D 6: 187–194.

Lahiri, S., J.P. Szidon and A.P. Fishman. 1970. Potential respiratory and circulatory adjustments to hypoxia in the African lungfish. Fed. Proc. 29: 1141–1148.

Land, S.C. and N.J. Bernier. 1995. Estivation: mechanisms and control of metabolic suppression. pp. 381–412. *In*: P.W. Hochachka and T.P. Mommsen (eds.). Biochemistry and Molecular Biology of Fishes, Vol. 5. Elsevier, Amsterdam.

Leloup, J. 1958. Influence de la température sur le fonctionnement thyroïdien de l'anguille normale. Comptes Rendus des Séances de l'Académie des Sciences 247: 2454–2456.

Lenfant, C. and K. Johansen. 1968. Respiration in the African lungfish, *Protopterus aethiopicus*. I. Respiratory properties of blood and normal patterns of breathing and gas exchange. J. Exp. Biol. 49: 437–452.

Li, P., U. Rudolph and M.M. Huntsman. 2009. Long-term sensory deprivation selectively rearranges functional inhibitory circuits in mouse barrel cortex. Proc. Natl. Acad. Sci. USA 106: 12156–12161.

Liem, K.F. 1988. Form and function of lungs: the evolution of air-breathing mechanisms. Am. Zool. 28: 739–759.

Lillehoj, E.P., K. Kato, W. Lu and K.C. Kim. 2013. Cellular and molecular biology of airway mucins. Int. Rev. Cell Mol. Biol. 303: 139–202.

Lim, C.K., W.P. Wong, S.M.L. Lee, S.F. Chew and Y.K. Ip. 2004. The ammonotelic African lungfish, *Protopterus dolloi*, increases the rate of urea synthesis and becomes ureotelic after feeding. J. Comp. Physiol. B 174: 555–564.

Lomholt, J.P. 1993. Breathing in the aestivating African lungfish, *Protopterus amphibius*. pp. 17–34. *In*: B.R. Singh (ed.). Advances in Fish Research. Narendra, New Delhi.

Lomholt, J.P., K. Johansen and G.M.O. Maloiy. 1975. Is the aestivating lungfish the first vertebrate with suctional breathing? Nature 257: 787–788.

Loong, A.M., K.C. Hiong, S.M.L. Lee, W.P. Wong, S.F. Chew and Y.K. Ip. 2005. Ornithine-urea cycle and urea synthesis in African lungfishes, *Protopterus aethiopicus* and *Protopterus annectens*, exposed to terrestrial conditions for 6 days. J. Exp. Zool. 303A: 354–365.

Loong, A.M., J.Y. Tan, K.C. Hiong, W.P. Wong, S.F. Chew and Y.K. Ip. 2007. Defense against environmental ammonia toxicity in the African lungfish, *Protopterus aethiopicus*: Bimodal breathing, skin ammonia permeability and urea synthesis. Aquat. Toxicol. 85: 76–86.

Loong, A.M., S.F. Ang, W.P. Wong, H.O. Pörtner, C. Bock, R. Wittig, C.R. Bridges, S.F. Chew and Y.K. Ip. 2008a. Effects of hypoxia on the energy status and nitrogen metabolism of African lungfish during aestivation in a mucus cocoon. J. Comp. Physiol. B 178: 853–865.

Loong, A.M., C.Y. Pang, K.C. Hiong, W.P. Wong, S.F. Chew and Y.K. Ip. 2008b. Increased urea synthesis and/or suppressed ammonia production in the African lungfish, *Protopterus annectens*: aestivation in air versus aestivation in mud. J. Comp. Physiol. B 178: 351–363.

Loong, A.M., Y.R. Chng, S.F. Chew, W.P. Wong and Y.K. Ip. 2012a. Molecular characterization and mRNA expression of carbamoyl phosphate synthetase III in the liver of the African lungfish, *Protopterus annectens*, during aestivation or exposure to ammonia. J. Comp. Physiol. B 182: 367–379.

Loong, A.M., K.C. Hiong, W.P. Wong, S.F. Chew and Y.K. Ip. 2012b. Differential gene expression in the liver of the African lungfish, *Protopterus annectens*, after 6 days of estivation in air. J. Comp. Physiol. B 182: 231–245.

Machin, J. 1975. Water relationships. pp. 105–163. *In*: V. Fretter and J. Peake (eds.). The Pulmonates, Vol. 1. Academic Press, New York.

Macpherson, A.J., M.B. Geuking and K.D. McCoy. 2005. Immune responses that adapt the intestinal mucosa to commensal intestinal bacteria. Immunology 115: 153–162.

Manzon, L.A. 2002. The role of prolactin in fish osmoregulation: a review. Gen. Comp. Endocrinol. 125: 291–310.

Martínez-Antón, A., C. Debolós, M. Garrido, J. Roca-Ferrer, C. Barranco, I. Alobid, A. Xaubet, C. Picado and J. Mullol. 2006. Mucin genes have different expression patterns in healthy and diseased upper airway mucosa. Clin. Exp. Allergy 36: 448–457.

Martyniuk, C.J., A.B. Crawford, N.S. Hogan and V.L. Trudeau. 2005. GABAergic modulation of the expression of genes involved in GABA synaptic transmission and stress in the hypothalamus and telencephalon of the female goldfish (*Carassius auratus*). J. Neuroendocrinol. 17: 269–275.

Masini, M.A., M. Sturla, M. Pestarino, F. Facchinetti, A. Gallinelli and B.M. Uva. 1999. Proopiomelanocortin (POMC) mRNA and POMC-derived peptides immunolocalization in the skin of *Protopterus annectens*, an African lungfish. Peptides 20: 87–91.

McDougal, D.B., J. Holowach, M.C. Howe, E.M. Jones and C.A. Thomas. 1968. The effects of anoxia upon energy sources and selected metabolic intermediates in the brains of fish, frog and turtle. J. Neurochem. 15: 577–588.

McMahon, B.R. 1969. A functional analysis of the aquatic and aerial respiratory movements of an African lungfish, *Protopterus aethiopicus*, with reference to the evolution of the lung-ventilation mechanism in vertebrates. J. Exp. Biol. 51: 407–430.

McMahon, B.R. 1970. The relative efficiency of gaseous exchange across the lungs and gills of an African lungfish *Protopterus aethiopicus*. J. Exp. Biol. 52: 1–15.

Mirejovská, E., A. Bass, J. Hurych and J. Teisinger. 1981. Enzyme changes during experimental silicotic fibrosis. II. Intermediary metabolism enzymes of the lungs. Environ. Res. 25: 434–440.

Mlewa, C.M., J.M. Green and R. Dumbrack. 2007. Are wild lungfish obligate air breathers? Some evidence from radioteletry. Afr. Zool. 42: 131–134.

Mommsen, T.P. and P.J. Walsh. 1989. Evolution of urea synthesis in vertebrates: the piscine connection. Science 243: 72–75.

Mommsen, T.P. and P.J. Walsh. 1991. Urea synthesis in fishes: evolutionary and biochemical perspectives. pp. 137–163. *In*: P.W. Hochachka and T.P. Mommsen (eds.). Biochemistry and Molecular Biology of Fishes, Vol. I, Phylogenetic and Biochemical Perspectives. Elsevier, New York.

Mommsen, T.P., C.J. French and P.W. Hochachka. 1980. Sites and patterns of protein and amino acid utilization during the spawning migration of salmon. Can. J. Zool. 66: 1059–1068.

Mora-Ferrer, C. and C. Neumeyer. 2009. Neuropharmacology of vision in goldfish: a review. Vision Res. 49: 960–969.

Muir, T.J., J.P. Costanzo and R.E. Lee, Jr. 2007. Osmotic and metabolic responses to dehydration and urea-loading in a dormant, terrestrially-hibernating frog. J. Comp. Physiol. B 177: 917–926.

Muir, T.J., J.P. Costanzo and R.E. Lee, Jr. 2008. Metabolic depression induced by urea in organs of the wood frog, *Rana sylvatica*: Effects of season and temperature. J. Exp. Zool. 309A: 111–116.

Murphy, B., W.M. Zapol and P.W. Hochachka. 1980. Metabolic activities of heart, lung and brain during diving and recovery in the Weddell seal. J. Appl. Physiol. 48: 596–605.

Newsholme, E.A. and A.R. Leech. 1983. Biochemistry for the Medical Sciences. John Wiley & Sons, New York.

Ojeda, J.L., W.P. Wong, Y.K. Ip and J.M. Icardo. 2008. Renal corpuscle of the African lungfish *Protopterus dolloi*: Structural and histochemical modification during aestivation. Anat. Rec. 291: 1156–1172.

Olsen, R.W. and W. Sieghart. 2009. GABA A receptors: subtypes provide diversity of function and pharmacology. Neuropharmacology 56: 141–148.

Orgeig, S. and C.B. Daniels. 1995. The evolutionary significance of pulmonary surfactant in lungfish (Dipnoi). Am. J. Respir. Cell Mol. Biol. 13: 161–166.

Otero, O. 2011. Current knowledge and new assumptions on the evolutionary history of the African lungfish, *Protopterus*, based on a review of its fossil record. Fish Fisheries 12: 235–255.

Page, M.M., K.D. Salway, Y.K. Ip, S.F. Chew, S.A. Warren, J.S. Ballantyne and J.A. Stuart. 2010. Upregulation of intracellular antioxidant enzymes in brain and heart during estivation in the African lungfish *Protopterus dolloi*. J. Comp. Physiol. B 180: 361–369.

Pelster, B. 1995. Metabolism of the swimbladder tissue. pp. 101–118. *In*: P.W. Hochachka and T.P. Mommsen (eds.). Biochemistry and Molecular Biology of Fishes, Vol. 4. Elsevier Science, Amsterdam.

Perry, S.F., K.M. Gilmour, B. Vulesevic, B. McNeil, S.F. Chew and Y.K. Ip. 2005a. Circulating catecholamines and cardiorespiratory responses in hypoxic lungfish (*Protopterus dolloi*): a comparison of aquatic and aerial hypoxia. Physiol. Biochem. Zool. 78: 325–334.

Perry, S.F., K.M. Gilmour, E.R. Swenson, B. Vulesevic, S.F. Chew and Y.K. Ip. 2005b. An investigation of the role of carbonic anhydrase in aquatic and aerial gas transfer in the African lungfish *Protopterus dolloi*. J. Exp. Biol. 208: 3805–3815.

Perry, S.F., R. Euverman, T. Wang, A.M. Loong, S.F. Chew, Y.K. Ip and K.M. Gilmour. 2008. Control of breathing in African lungfish (*Protopterus dolloi*): A comparison of aquatic and cocooned (terrestrialized) animals. Resp. Physiol. Neurobiol. 160: 8–17.

Peterson, C.C. and P.A. Stone. 2000. Physiological capacity for estivation of the Sonoran mud turtle, *Kinosternon sonoriense*. Copeia 2000: 684–700.

Poll, M. 1938. Poissons du Katanga (basin du Congo) récoltés par le professor Paul Brien. Rec. Zool. Bot. Afr. 30: 389–423.

Poll, M. and H. Damas. 1939. Poissons. Exploration du Parc National Albert. Mission H. Damas (1935–36). Institut des Parcs Nationaux du Congo belge, Bruxelles, Fasc. 6: 1–73.

Power, J.H., I.R. Doyle, K. Davidson and T.E. Nicholas. 1999. Ultrastructural and protein analysis of surfactant in the Australian lungfish *Neoceratodus forsteri*: evidence for conservation of composition for 300 million years. J. Exp. Biol. 202: 2543–2550.

Rand-Weaver, M. and H. Kawauchi. 1993. Growth hormone, prolactin and somatolactin: a structural overview. pp. 39–56. *In*: P.W. Hochachka and T.P. Mommsen (eds.). The Biochemistry and Molecular Biology of Fishes, Vol. 2. Elsevier, Amsterdam.

Reinhard, F.G. 1981. Suppression of DNA synthesis by peptide(s) from the brain of estivating lungfish *Protopterus annectens* Owen. Cryobiology 18: 103.

Riddle, W.A. 1983. Physiological ecology of land snails and slugs. pp. 431–461. *In*: W.D. Russell-Hunter (ed.). The Mollusca, Vol. 6. Academic Press, New York.

Romer, A.S. 1966. Vertebrate Paleontology. University of Chicago Press, Chicago.

Sakamoto, T., K. Iwata and M. Ando. 2002. Growth hormone and prolactin expression during environmental adaptation of gobies. Fish. Sci. 68: 757–760.

Sandovici, M., R.H. Henning, R.A. Hut, A.M. Strijkstra, A.H. Epema, H. van Goor and L.E. Deelman. 2004. Differential regulation of glomerular and interstitial endothelial nitric oxide synthase expression in the kidney of hibernating ground squirrel. Nitric Oxide 11: 194–200.

Seidel, M.E. 1978. Terrestrial dormancy in the turtle *Kinosternon flavescens*: respiratory metabolism and dehydration. Comp. Biochem. Physiol. A 61: 1–4.

Seifert, A.W. and L.J. Chapman. 2006. Respiratory allocation and standard rate of metabolism in the African lungfish, *Protopterus aethiopicus*. Comp. Biochem. Physiol. A 143: 142–148.

Shephard, K.L. 1994. Functions for fish mucus. Rev. Fish Biol. Fish. 4: 401–429.

Sievert, L.M., G.A. Sievert and P.V. Cupp, Jr. 1988. Metabolic rates of feeding and fasting juvenile midland painted turtles, *Chrysemys picta marginata*. Comp. Biochem. Physiol. A 90: 157–159.

Smith, G.M. and C.W. Coates. 1937. Memoirs: On the Histology of the Skin of the Lungfish *Protopterus annectens* after experimentally induced Aestivation. Q. J. Microsc. Sci. 2: 487–491.

Smith, H.W. 1930. Metabolism of the lungfish *Protopterus aethiopicus*. J. Biol. Chem. 88: 97–130.

Smith, H.W. 1931. Observations on the African lungfish, *Protopterus aethiopicus*, and on evolution from water to land environments. Ecology 12: 164–181.

Smith, H.W. 1935. The metabolism of the lungfish II. Effect of feeding meat on metabolic rate. J. Cell. Comp. Physiol. 6: 335–349.

Specker, J.L., D.S. King, R.S. Nishioka, K. Shirahata, K. Yamaguchi and H.A. Bern. 1985. Isolation and partial characterization of a pair of prolactins released *in vitro* by the pituitary of a cichlid fish, *Oreochromis mossambicus*. Proc. Natl. Acad. Sci. USA 82: 7490–7494.

Speers-Roesch, B., J.W. Robinson and J.S. Ballantyne. 2006. Metabolic organization of the spotted ratfish, *Hydrolagus colliei* (Holocephali: Chimaeriformes): insight into the evolution of energy metabolism in the chondrichthyan fishes. J. Exp. Zool. 305A: 631–644.

Staples, J.F., M. Kajimura, C.M. Wood, M. Patel, Y.K. Ip and G.B. McClelland. 2008. Enzymatic and mitochondrial responses to five months of aerial exposure in the slender lungfish (*Protopterus dolloi*). J. Fish Biol. 73: 608–622.

Storey, K.B. 2002. Life in the slow lane: molecular mechanisms of estivation. Comp. Biochem. Physiol. A 133: 733–754.

Storey, K.B. and J.M. Storey. 1990. Metabolic rate depression and biochemical adaptation in anaerobiosis, hibernation, and estivation. Q. Rev. Biol. 65: 145–174.

Sturla, M., M.A. Masini, P. Prato, C. Grattarola and B. Uva. 2001. Mitochondria-rich cells in gills and skin of an African lungfish, *Protopterus annectens*. Cell Tissue Res. 303: 351–358.

Sturla, M., P. Prato, C. Grattarola, M.A. Masini and B.M. Uva. 2002. Effects of induced aestivation in *Protopterus annectens*: a histomorphological study. J. Exp. Zool. 292: 26–31.

Suarez, R.K. and T.P. Mommsen. 1987. Gluconeogenesis in teleost fish. Can. J. Zool. 65: 1869–1882.

Swan, H. and F.G. Hall. 1966. Oxygen-hemoglobin dissociation in *Protopterus aethiopicus*. Am. J. Physiol. 210: 487–489.

Swan, H., D. Jenkins and K. Knox. 1968. Anti-metabolic extract from the brain of *Protopterus aethiopicus*. Nature 217: 671.

Swan, H., D. Jenkins and K. Knox. 1969. Metabolic torpor in *Protopterus aethiopicus*: an antimetabolic agent from the brain. Am. Nat. 103: 247–258.

Takezaki, N., F. Figueroa, Z. Zaleska-Rutczynska, N. Takahata and J. Klein. 2004. The phylogenetic relationship of tetrapod, coelacanth, and lungfish revealed by the sequences of forty-four nuclear genes. Mol. Biol. Evol. 8: 1512–1524.

Thomason, D.B., R.B. Biggs and F.W. Booth. 1989. Protein metabolism and beta-myosin heavy-chain mRNA in unweighed soleus muscle. Am. J. Physiol. 257: R300–R305.

Thornton, D.J. and J.K. Sheehan. 2004. From mucins to mucus: toward a more coherent understanding of this essential barrier. Proc. Am. Thorac. Soc. 1: 54–61.

Touhata, K., H. Toyohara, T. Mitani, M. Kinoshita, M. Satou and M. Sakaguchi. 1995. Distribution of L-gulono-1,4-lactone oxidase among fishes. Fisheries Sci. 61: 729–730.

Ultsch, G.R. 1989. Ecology and physiology of hibernation and overwintering among freshwater fishes, turtles, and snakes. Biol. Rev. 64: 435–516.

Ultsch, G.R. 1996. Gas exchange, hypercarbia and acid-base balance, paleoecology, and the evolutionary transition from water-breathing to air-breathing among vertebrates. Palaeogeogr. Palaeoclimatol. Palaeoecol. 123: 1–27.

Walsh, P.J. and C.L. Milligan. 1993. Roles of buffering capacity and pentose phosphate pathway activity in the gas gland of the gulf toadfish *Opsanus beta*. J. Exp. Biol. 176: 311–316.

Wasawo, D.P.S. 1959. A dry season burrow of *Protopterus aethiopicus*. Heckel. Rev. Zool. Bot. Afr. 60: 65–70.

Weber, R.E., K. Johansen, G. Lykkeboe and G.M.O. Maloiy. 1977. Oxygen-binding properties of hemoglobins from aestivating and active African lungfish. J. Exp. Zool. 199: 85–96.

Wilkie, M.P., T.P. Morgan, F. Galvez, R. Smith, M. Kajimura, Y.K. Ip and C.M. Wood. 2007. The African lungfish (*Protopterus dolloi*): Ionoregulation and osmoregulation in a fish out of water. Physiol. Biochem. Zool. 80: 99–112.

Winsky-Sommerer, R. 2009. Role of GABAA receptors in the physiology and pharmacology of sleep. Eur. J. Neurosci. 29: 1779–1794.

Withers, P.C. 1998. Urea: Diverse functions of a 'waste' product. Clin. Exp. Pharmacol. Physiol. 25: 722–727.

Wood, C.M., P.J. Walsh, S.F. Chew and Y.K. Ip. 2005. Greatly elevated urea excretion after air exposure appears to be carrier mediated in the slender lungfish (*Protopterus dolloi*). Physiol. Biochem. Zool. 78: 893–907.

Yancey, P.H., M.E. Clark, S.C. Hand, R.D. Bowlus and G.N. Somero. 1982. Living with water stress: evolution of osmolyte systems. Science 217: 1214–1222.

Zepeda, A., F. Sengpiel, M.A. Guagnelli, L. Vaca and C. Arias. 2004. Functional reorganization of visual cortex maps after ischemic lesions is accompanied by changes in expression of cytoskeletal proteins and NMDA and $GABA_A$ receptor subunits. J. Neurosci. 24: 1812–1821.

6

Anatomy of the Heart and Circulation in Lungfishes

José M. Icardo,[1,]* *Bruno Tota*[2] *and Yuen K. Ip*[3]

Introduction

Lungfish are air-breathing fishes that possess true lungs. This remarkable feature has often obscured the fact that lungfish are bimodal breathers that obtain oxygen both through gills and through lungs. On the other hand, the aerial dependence to meet the oxygen demands is quite diverse. The African (genus *Protopterus*) and the South American (*Lepidosiren paradoxa*) lungfish are considered to be obligate air-breathers (see Mlewa et al. 2011), whereas the Australian lungfish (*Neoceratodus forsteri*) only surface to breath air in situations of fatigue or when the oxygen content of the water descends to below critical levels (Kind 2011). It is unclear whether these behavioral differences may be related to lung anatomy. The African and the South American species have two lungs fused cranially whereas a single lung appears to be present in the Australian species. In any case, aquatic lungfish depend on water respiration since they use the gills for carbon dioxide exchange (Graham 1997).

Another important characteristic of lungfish is their ability to aestivate during periods of seasonal drying. Aestivation is a state of dormancy or torpor that presents, among other characteristics, muscle inactivity, reduced metabolic rates, cardiac and respiratory depression, prolonged fasting and urine suppression (see Jorgensen and Joss 2011). Changes in organ physiology

[1] Department of Anatomy and Cell Biology, University of Cantabria, 39011-Santander, Spain.
[2] Department of Cell Biology, University of Calabria, 87030-Cosenza, Italy.
[3] Department of Biological Sciences, National University of Singapore, Singapore 117543, Republic of Singapore.
* Corresponding author: icardojm@unican.es

are accompanied by structural changes that depend on the extent of the functional modifications. For instance, the decrease in heart function occurs without any significant modification of the gross heart structure (Icardo et al. 2008), whereas the kidney (Ojeda et al. 2008) and the gut (Icardo et al. 2012a) undergo significant structural modifications. Structural changes in the lungs and gills (Sturla et al. 2002), and in lymphoid organs (Icardo et al. 2012b, 2014), have also been reported during aestivation. Of note, these studies have been performed in different species of *Protopterus*. Although aestivation of the South American lungfish presents many similarities to that of the African species (Almeida-Val et al. 2011), no structural studies appear to have been done in *L. paradoxa*. On the other hand, *N. forsteri* does not aestivate.

Aquatic lungfish employ a diverse combination of gill and air-breathing strategies to satisfy their oxygen demands (changes in hemoglobin oxygen affinity are not considered here). On the other hand, aestivating lungfish rely exclusively on air breathing. For the lungs and the heart to cope with any of those situations, a complex circulation has been established. It involves the presence of pulmonary arteries reaching the lungs, and of a single pulmonary vein extended between the lungs and the caudal portion of the heart (Fig. 1). In addition, the heart has developed partial chamber partitionings and several other specialized features.

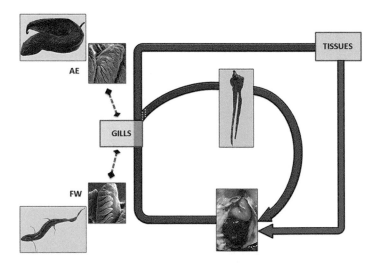

Fig. 1. Schematic representation of the lungfish circulation. The paired lung, the heart and the gills are depicted. In freshwater (FW) conditions most of the blood passes through the gills and into the dorsal aorta for tissue supply. The darkening of the tract between the heart and the gills reflects the existence of some right-left blood mixing in the heart. In aestivation (AE) the lamellae are collapsed, the filament circulation bypassed, and the venous blood is directed to the lungs. The white dotted lines represent the vasomotor segment of the pulmonary arteries and indicate the capability for blood shunting. The upper and lower left panels show images of *P. annectens* in aestivating and in aquatic conditions, respectively.

The Caudal Pole of the Heart

Lung ventilation implies the production of oxygen-rich blood. This oxygenated blood reaches the caudal end of the heart through a single pulmonary vein. At the same time, oxygen-poor blood from the systemic circulation also reaches the caudal end of the heart (Fig. 1). Should these blood streams mix in any significant proportion, the benefits of the existence of the lungs would have been obviated. To avoid this, the lungfish heart has developed complex specializations that ensure separation of the two streams as they pass through the different heart chambers (Bugge 1961; Burggren 1988; Icardo et al. 2005a,b).

The single pulmonary vein reaches the heart, enters the sinus venosus and attaches to its dorsal wall. Then, it runs forward toward the atrioventricular (AV) area (Fig. 2). The AV region of the lungfish heart is strikingly different from the same region in other fish (see Icardo and Colvee 2010). It does not have well-defined boundaries and lacks AV valves. Instead, it contains a large piece of tissue, the AV plug (Fig. 3). The bulky AV plug contains a core of hyaline cartilage and sits on top of the dorsal side of the ventricular septum (Fig. 4). The dorsal wall of the plug is attached to the sinus venosus (Fig. 3), but its ventral, lateral and apical surfaces are free (Fig. 4). When the pulmonary vein runs within the sinus venosus towards the AV plug, the vein meets a complex fold of membranous tissue named the pulmonalis fold (Fig. 3). This fold extends between the sinus, the atrial wall and the AV plug. The wall of the pulmonary vein then fuses with the pulmonalis fold, and the vein disappears as an anatomic entity. The oxygenated blood is next directed toward the left in a muscular-membranous duct, the pulmonary channel, until it reaches the left side of the atrium (Fig. 3) (for details, see Icardo et al. 2005a). The

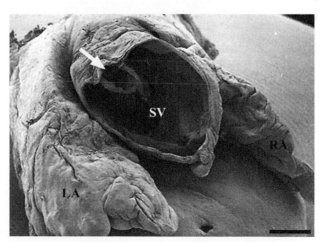

Fig. 2. *P. dolloi.* Fresh water. Dorsal view of the caudal pole of the heart. The pulmonary vein (arrow) is attached to the dorsal wall of the sinus venosus (SV). LA, left atrium. RA, right atrium. Scale bar: 500 μm (from Icardo et al. 2005a, J. Morphol. 263: 30–38).

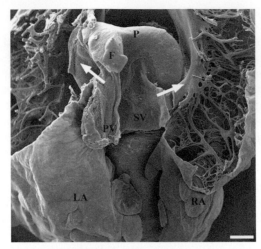

Fig. 3. *P. dolloi.* Fresh water. Dissection of the dorsal side of the heart. The sinus venosus (SV) has been opened, and the pulmonary vein (PV) has been dissected up to the pulmonalis fold (F). This exposes the pathways conveying the oxygenated blood into the left side of atrium (LA) and the oxygen-poor blood into the right side (RA) of atrium. P, atrioventricular plug. Scale bar: 500 μm (from Icardo et al. 2005a, J. Morphol. 263: 30–38).

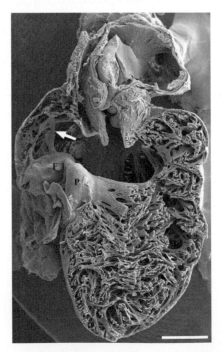

Fig. 4. *P. dolloi.* Fresh water. Sagittal section of the heart. Left side of the specimen, seen from the right. The plane of section passes through the right side of the ventricular septum. Note the trabecular structure of the ventricle. Arrow indicates the atrial septum. P, atrioventricular plug. F, pulmonalis fold. Scale bar: 1 mm (from Icardo et al. 2005a, J. Morphol. 263: 30–38).

systemic blood also enters the large sinus venosus. However, the formation of the pulmonary channel has closed the left side of the sinus venosus in such a way that the oxygen-poor blood can only be directed toward the right side of the atrium (Fig. 3). This situation is interesting from a phylogenetic point of view. In birds and mammals, the sinus venosus (or its derivative) conveys oxygen-poor blood into the right atrium, while the pulmonary vein conveys oxygenated blood into the left atrium.

Once separation of the blood streams is achieved in the sinus venosus area, other heart specializations help to maintain this separation throughout the heart. First, the single atrium shows an incipient muscular septum located in a dorsal and cranial position (Fig. 4). This partial septum is formed by the accretion of pectinate muscles (muscular trabeculae) and shows a crescent shape. Its presence contributes to prevent mixing of blood streams. On the other hand, the large AV plug obstructs the AV communication (Fig. 5). It is not difficult to envision the plug functioning as a valve and contributing to convey separated blood streams into the ventricular chamber.

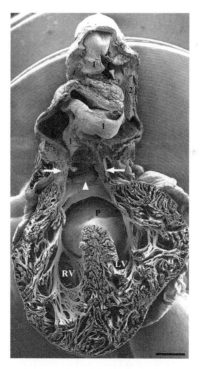

Fig. 5. *P. dolloi.* Fresh water. Frontal section of the heart (dorsal segment, anterior view). The ventricle is partially divided into left (LV) and right (RV) chambers. The atrioventricular plug (P) sits on top of the ventricular septum. The entire ventricle, including the septum, is trabeculated. The outflow tract (OFT) starts above the ventricular septum. The OFT lumen is occupied by the spiral (1) and bulbar (2) folds. The proximal OFT part (arrows) is the conus. It has a muscular structure and supports the vestigial conus valves (arrowhead). The rest of the OFT, the bulbus, has an arterial-like structure. Scale bar: 500 μm (from Icardo et al. 2005b, J. Morphol. 265: 43–51).

The Ventricle

The lungfish ventricle has a saccular appearance and a round apex (Fig. 4). Its ventral wall is attached to the pericardium by a thick tendon, the *gubernaculum cordis*, whose functional significance is still a matter of discussion (see Icardo et al. 2005b). Internally, the most distinct feature is the presence of an antero-posterior septum (Figs. 4, 5). The ventricular septum occupies more than half of the ventricular length and shows a smooth, free border (Figs. 4, 5). The septum incompletely divides the ventricle into right and left cavities (Figs. 4, 5). The ventricular free wall is continuous with the septum. The two muscular compartments are, at least in *Protopterus* and *Lepidosiren*, entirely trabeculated. Due to the existence of the septum, systolic contraction of the ventricle propels separated blood streams cranially, into the outflow tract (OFT) segment.

The Outflow Tract

A distinct characteristic of the lungfish OFT is that it follows a very tortuous course, extending from above the ventricular septum up to the origin of the branchial arteries that supply the gills. First, the OFT runs cranially; then, it bends ventrally and to the left; finally, it bends again and follows a nearly straight cranial course (Fig. 1). The presence of two bending points divides the OFT into proximal, middle and distal portions. From the structural point of view, the proximal portion is homologous to the conus arteriosus of other fishes, while the middle and distal portions correspond to the bulbus arteriosus.

The proximal part, the conus, is endowed with a layer of compact, well-vascularized myocardium (Fig. 6). This portion, as in any other fish, supports the conus valves. In *Protopterus*, these valves are situated in the ventral and dorsal walls of the OFT, and are organized into two to three rows (Icardo et al. 2005b). This situation appears to be more complex in *N. forsteri*, where several valve rows have been described (see Burggren 1988). It should be underscored that conus valves in *Protopterus* are vestigial, being reduced in most cases to mere transverse protrusions with no signs of valve excavation (Fig. 5). Thus, it is unlikely that they play a significant role in heart dynamics. The rest of the OFT wall, the bulbus, shows an arterial-like structure formed by elastic and collagenous fibers oriented along the main OFT axis (Icardo et al. 2005b). In addition, the bulbus is covered externally by a thin layer of myocardium which is continuous with the conus myocardium (Fig. 6).

A striking characteristic of the lumen of the lungfish OFT is the presence of two masses of loose connective tissue collectively termed bulbar folds (Figs. 5, 6). The more prominent of these two masses is the spiral fold. It runs along most of the length of the OFT, spiraling from the ventral wall of the proximal OFT via the dorsal wall of the middle OFT to the right side of the distal OFT (Bugge 1961; Icardo et al. 2005b). The other tissue mass is the bulbar fold. It is much shorter and occupies part of the middle and distal OFT. The two folds show a similar structure. They are covered by endocardium and

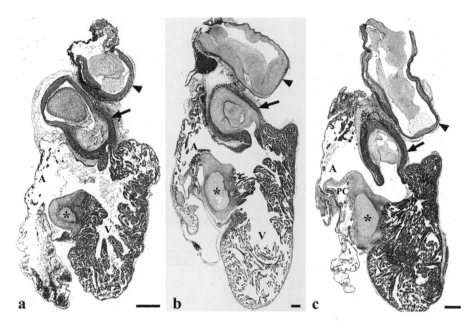

Fig. 6. Composite showing parasagittal sections of different lungfishes in freshwater and in aestivation. A, atrium. V, ventricle. PC, pulmonary channel. Arrows indicate the conus myocardium. Arrowheads indicate myocardial layer of bulbus. Asterisks, cartilage core of the atrioventricular plug. The structure is very similar in all cases. Note a better alignment of the spiral and bulbar folds during aestivation (c) than in fresh water (a). a: *P. dolloi*. Fresh water. Hematoxylin-eosin. b: *L. paradoxa*. Fresh water. Martin's trichrome. c: *P. dolloi*. Six months of aestivation. Martin's trichrome. Scale bars: a–c: 1 mm (a: from Icardo et al. 2005b, J. Morphol. 265: 43–51).

are composed of a clear matrix that contains small amounts of collagen and elastin, and sparsely distributed fibroblast-like cells (Fig. 6). The two folds show a narrow base of insertion into the OFT wall, have a bulky appearance, and fuse at their distal ends. They establish an incomplete, spiraling partitioning of the OFT lumen (Fig. 6). It has been hypothesized that the folds contribute to separate the blood streams in this segment of the heart, forming ventral and dorsal channels (Szidon et al. 1969). A ventral aorta is lacking in lungfish. Instead, three pairs of branchial arches arise directly from the OFT and end into the gill system.

Blood Streams Through the Heart

The inner side of the lungfish heart is provided with a succession of partial divisions that should maintain the separation between the oxygenated and deoxygenated blood streams. The fact that the heart septa are incomplete indicates the possibility of blood mixing. Indeed, mixing occurs throughout the heart and in the area of communication of the efferent gill vessels. Despite this, oxygenated blood flows preferentially into the dorsal aorta whereas oxygen-poor blood moves preferentially into the pulmonary arteries (Szidon et al. 1969; Burggren and Johansen 1986). Early studies indicate that when the pulmonary vein is injected with an opaque medium, the contrast fills the left side of the ventricle, the ventral channel of the OFT, and the ventral branchial arches. Conversely, when the contrast is injected into the vena cava, it fills the right side of the ventricle, the dorsal channel of the OFT, and the posterior branchial arches (Szidon et al. 1969). Of note, the anterior arches do not contain gill filaments. They are the origin of the circulation to the brain and to the heart. They also form the systemic (dorsal) aorta. The absence of filaments in the anterior arches minimizes possible oxygen losses to stagnant, oxygen-poor water. The posterior arches contain well-developed gill filaments and are the origin of the pulmonary arteries. The aorta and the pulmonary arteries remain communicated by ducts or vessels (Szidon et al. 1969). More specifically, a persistent *ductus arteriosus*, located between the origin of the aorta and that of the pulmonary arteries, shunts the blood towards the systemic or towards the pulmonary circulation (Laurent 1996). It should also be underscored that the oxygen content of the water may vary widely. Low oxygen levels can be compensated for by an increase in the number of lung respiratory cycles, meaning that more blood has to be directed towards the pulmonary arteries. The wall of these arteries contains specialized vasomotor segments consisting of thickenings of the media. Numerous branches of the vagus nerve, and numerous cells with granular content, populate the vasomotor segments. Thus, blood shunting between systemic and pulmonary systems (Fig. 1) depends on the existence of receptors and is adjusted by nervous and endocrine factors (Laurent 1996; Burggren et al. 1997).

Aestivation

The process of aestivation modifies the circulatory parameters greatly. Concomitant with the reduction of the metabolic rate, the blood pressure and the heart rate decrease by 50% (DeLaney et al. 1974). Cardiac output decreases similarly. Yet, the heart does not undergo significant anatomical or histological modifications (Icardo et al. 2008). This should not come as a surprise since the heart, albeit at a low pace, is still working. The single modification observed relates to the relative position of the bulbar folds. During aestivation, the folds undergo a small reduction in size, probably as a result of general tissue dehydration. This results in a better alignment of the folds and in a better definition of the blood flow pathways (Fig. 6c). Although this modification is apparently insignificant, it would minimize blood mixing through the OFT and would improve the quality of the blood reaching the aorta and the systemic circulation.

Despite the absence of gross morphological modifications, several changes occur at the cellular and subcellular levels. In aquatic conditions, the cardiomyocytes of the lungfish ventricle show large cytoplasmic areas occupied by a material of low electron density (Fig. 7). This material appears pale-gray in semi-thin sections (Fig. 7a), and may surround myofibrils and mitochondriae (Fig. 7b) (Chopin and Bennett 1995; Icardo et al. 2008). In contrast, myocardial cells during aestivation show partial or total loss of that material (Fig. 7c), and many of the cells appear to have empty cytoplasms (Fig. 7d). A large number of residual bodies, linked to cellular digestion processes, are also characteristic of aestivation (Icardo et al. 2008). It has been hypothesized that the pale material contains glycogen (or glycosaminoglycan) deposits that could be used as a source of water and food supply during long periods of aestivation (Icardo et al. 2008). The characteristics of the residual bodies are consistent with that hypothesis. The use of the periodic acid of Schiff (PAS) reaction clearly demonstrates that myocardial cells store large amounts of glycogen (Fig. 8a), and that a large part of that glycogen is depleted after six months of aestivation (Fig. 8b). The physiological significance of this observation is described below.

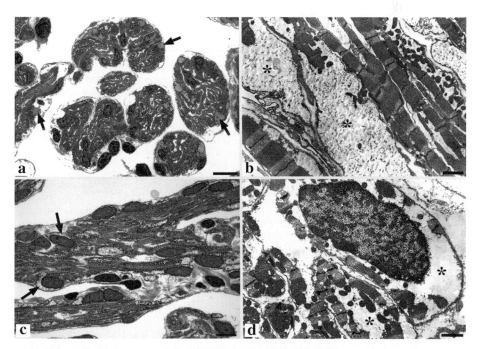

Fig. 7. Composite illustrating changes in myocardial ultrastructure during aestivation. *P. dolloi.* (a) Fresh water. Myocardial cells show large cytoplasmic areas occupied by a material that appears pale gray (arrows) in black and white pictures. Most nuclei show a heterochromatin pattern. (b) Fresh water. Large areas of the cytoplasm (asterisk) are occupied by low electron density material. (c) Six months of aestivation. Many cells show clear cytoplasms (arrows). Most nuclei show a euchromatin pattern. (d) Six months of aestivation. A large myocardial cell shows a few myofibrils and an 'empty' cytoplasm (asterisks). Scale bars: a, 20 μm; b, 1 μm; c, 20 μm; d, 1 μm (a, c, d: from Icardo et al. 2008, J. Anat. 213: 106–119).

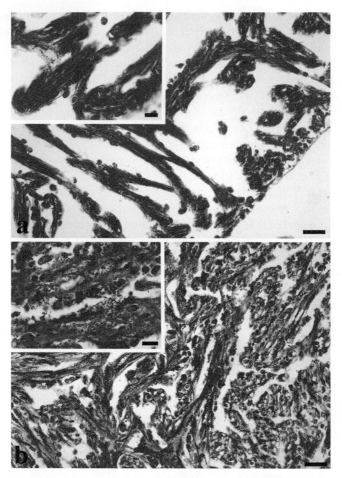

Fig. 8. Ventricular myocardium of *P. annectens.* (a) Fresh water. PAS staining. Myocardial ventricular cells are intensely positive. Inset: Detail of myocardial trabeculae. (b) Six months of aestivation. PAS positivity becomes less intense, granular and patchy, indicating glycogen depletion. Inset: detail of myocardial trabeculae. Scale bars: a, 50 μm; inset of a, 20 μm; b, 50 μm; inset of b, 20 μm.

Another intriguing feature of the lungfish heart is the occurrence of cell death in the ventricular septum. In *P. dolloi*, necrotic cell death is a specific feature of the cardiomyocytes located in the ventricular septum (Icardo et al. 2008). In fact, the ventricular septum and the free ventricular wall appear to constitute different compartments from the structural and functional points of view. Curiously, the number of dead cells increases greatly after six months of aestivation. A satisfactory explanation for the presence of myocardial necrosis is still awaited. In *P. annectens*, myocardial cell death in the ventricular septum is accompanied by the presence of apoptotic cell fragments (unpubl. observations in aquatic animals). In this species, however, TUNEL assays have not detected differences in the number of apoptotic nuclei between the ventricles of freshwater and aestivating animals (Amelio et al. 2013). Control of apoptotic myocardial remodeling in *P. annectens* is considered below.

Aestivation is also accompanied by a shift in the distribution of chromatin in the nuclei of the myocardial cells. Most of the cardiomyocytes in freshwater animals show a heterochromatin pattern, typical of a low-transcription, low-energy state (Fig. 7a). In contrast, and contrary to what might have been expected, most myocardial cells show a euchromatin pattern during aestivation (Fig. 7c), probably reflecting an increase in metabolic activity (Icardo et al. 2008). Although the real significance of this feature is still unclear, aestivation cannot be looked upon as a simple reduction of functional and metabolic activities, including the restriction of gene activities (Icardo et al. 2008, 2012a). Observations in several organs indicate that aestivation involves the interplay between the downregulation and the upregulation of diverse activities, from the cell and tissue levels (Icardo et al. 2008, 2012a,b; Ojeda et al. 2008) to the gene level (Hiong et al. 2013). The interplay regulates the adaptive responses of the different organs to aestivation.

In addition to heart structural and functional modifications, the aestivation process involves the isolation of the gills from the environment and the establishment of exclusive lung ventilation. To this end, the gills become surrounded by a protective layer of mucus. Furthermore, the secondary lamellae appear collapsed (Fig. 1). Thus, the lungs become the only source for oxygen supply. This involves blood flow redistribution at the branchial arches level (see above). In addition, shunt vessels connecting the afferent and efferent gill arteries bypass the lamellae circulation (Laurent 1996). With the gills closed, there is a preferential distribution of the systemic blood to the lungs while oxygenated blood is mainly directed toward the systemic circulation. Independent regulation of the amount of blood that each system receives is fundamental in the adaptation of lungfish to the diverse environmental conditions (Burggren et al. 1997). It should be underscored that aestivation also induces lung modifications. During aestivation the lungs are better vascularized, appear expanded, and their inner structure, formed by ridges and air spaces, becomes visible to the naked eye (Sturla et al. 2002, and pers. observations). Physiological hypertrophy of the lung is another adaptive response of the organism and reflects the exclusive air-breathing mode.

Energy Metabolism

It is currently assumed that, during aestivation, the heart undergoes metabolic modifications related to a decrease of the general metabolism. However, the way in which the heart may cope with a shortage of basic supplies, as well as with a decrease in oxygen availability, is unclear. In fact, a situation of metabolic shortage is, in itself, unclear. Uncertainties stem from the difficulty to interpret data obtained from different species, from the use of different methods and periods of aestivation, and from the lack of information about the blood transport of energy substrates during the aestivation period.

During aestivation, *P. aethiopicus* experiences a significant decrease in heart rate (from ~ 25 to 15 beats min^{-1}) and in arterial pO_2 (from ~ 50 to 25 mmHg) (Delaney et al. 1974; Fishman et al. 1986). Under these conditions, delivery of oxygen to the peripheral tissues is reduced, potentially resulting in functional hypoxia (Belkin 1968). In addition, the blood oxygen tension reaches much lower levels as the blood approaches the venous side of the vascular system. If the oxygen tension decreases to very low levels, the heart and other organs may shift from aerobic to anaerobic metabolism to generate the energy needed for survival.

Can these parameters be applied to the heart (or, for that matter, to the entire body) of aestivating lungfish? First, it is unclear whether the heart experiences a situation of hypoxia during aestivation. It is true that cell death increases in the ventricular septum of *P. annectens* during aestivation (see above), and that this feature can be explained by an increased sensitivity to reduced oxygen levels. However, the heart is the first organ to receive blood from the lungs (Fig. 1), taking the oxygen directly from the newly oxygenated blood. In addition, reduction of heart activity would decrease both the metabolic rate and the cellular demands for oxygen. Therefore, heart hypoxia during aestivation is unlikely. In this regard, the effect of the mode of aestivation on oxygen tension has not been investigated. It is possible that the availability of oxygen is different if aestivation takes place in air (Chew et al. 2004), in cloth bags (Delaney et al. 1974), or in deep holes excavated in the mud (Janssens 1964), where a relatively long, thin duct separates the animal nest from the surface. Also, the resilience of the species may vary. For instance, *P. aethiopicus* appears to show a significant tolerance to hypoxia (Dunn et al. 1983).

On the other hand, the heart is a very aerobic organ that obtains most of its required energy from glucose released by the liver, and from the cytoplasmic reserves of glycogen. Glycogen is a very efficient storage form of glucose, can be quickly mobilized, and is used as a metabolic fuel by organisms during both bursts of activity and quiescent periods. Lungfishes of the genus *Protopterus* and *Lepidosiren* are known to have large glycogen stores, which may act as an important energy source during aestivation. However, the concentration of glycogen in the heart of *P. dolloi* remains constant after a period of aestivation of two months (Frick et al. 2008a). Stable levels of glycogen do not fit well with the structural depletion of the glycogen reserves that are observed after six months of aestivation (Fig. 8). It should be underscored that several organs of the aestivating lungfish require a period of adaptation to the new physiological situation. This period has been set at two months in the case of the heart (Delaney et al. 1974; Icardo et al. 2008), and at four months in the case of the gut (Icardo et al. 2012a). After this period of adaptation, morphology and function may change significantly. Thus, it is possible that myocardial glycogen is mobilized after periods of aestivation longer than two months, progressively exhausting the energy reserves. In any case, the possible contribution of glucose stored in the liver to cardiomyocyte metabolism does not appear to have been considered.

It was recently reported that the heart of the aestivating *P. dolloi* shows a down-regulation of citrate synthase, which is a key enzyme in the tricarboxylic acid cycle (Krebs cycle) (Frick et al. 2008a). Would this indicate hypoxic stress and a shift towards anaerobic glycolysis? In other fish species, reductions in arterial pO_2 to 20 and 26 mmHg activate anaerobic glycolysis leading to the accumulation of lactate (Burggren and Cameron 1980; Boutilier et al. 1988). However, the whole body of the African lungfish does not display an increased reliance on anaerobic metabolism during aestivation. In addition, the heart concentration of lactate decreases during this period (Frick et al. 2008a).

The type of energy substrates feeding into the glycolytic pathways is also unclear. Aestivation leads to a significant decrease in hexokinase activity in the heart, although pyruvate kinase activity remains unchanged. The significant negative correlation between the activity of hexokinase and glucose levels in the heart of the aestivating *P. dolloi* suggests that hexokinase activity might be regulated by glucose concentrations (Frick et al. 2008a). Furthermore, citrate synthase, which is one of the flux-generating enzymes of the Krebs cycle, is downregulated in the aestivating heart (Frick et al. 2008a). A reduction in the carbon flux through the Krebs cycle would result in the accumulation of citrate which, in turn, would inhibit phosphofructose kinase activity (Randle et al. 1964). This would result in the accumulation of fructose-6-phosphate and, ultimately, of glucose-6-phosphate within the cell. High glucose-6-phosphate concentrations would inhibit hexokinase activity, resulting in glucose build up which, in turn, reduces glucose uptake by the cell (Randle et al. 1964). It was proposed that the purpose of this regulatory mechanism is to conserve glucose in the heart and muscle tissues, utilizing ketone bodies as an alternative fuel source during aestivation (Frick et al. 2008a). However, a later study indicated that ketone bodies do not contribute substantially to the energy requirements of *P. dolloi* during aestivation (Frick et al. 2008b).

Beta-oxidation of fatty acid molecules could be an alternative source of energy during aestivation. Food deprivation, however, causes a depletion of oxaloacetate, which is an intermediate of the Krebs cycle. The depletion of oxaloacetate slows the Krebs cycle and impedes the entry into the cycle of acetyl-CoA produced from lipid β-oxidation. Under these circumstances, acetyl-CoA is converted into acetoacetate via succinyl Co-A keto-transferase. Thus, aestivation and fasting would induce the accumulation of acetoacetate, indicating a decrease in lipid catabolism. Strikingly, the amount of acetoacetate decreases in the heart of *P. dolloi* during aestivation (Frick et al. 2008b). Does this mean an increase in lipid mobilization in the heart cells? Not necessarily. The decrease in the amount of acetoacetate can be explained if it is being used as fuel during aestivation, or if its production is reduced (Frick et al. 2008b). In fact, the levels of succinyl Co-A keto-transferase, a key enzyme in the synthesis of acetoacetate, were significantly lower in the heart of aestivating fish. The reduction in acetoacetate production appears to be linked to a downregulation of β-oxidation of lipids during aestivation. This fits well with the decrease in activity of 3-hydroxyacyl CoA dehydrogenase, an enzyme involved in fatty acid β-oxidation (Frick et al. 2008b). In all, the decrease in the activity of these two key enzymes suggests a downregulation of lipid metabolism in the aestivating heart. This decrease does not need to be an active feature. Rather, it could simply be due to reduced cardiac activity. The use of proteins and amino acids in the heart and other organs for energy metabolism is discussed in another chapter of this volume.

Cellular Stress during Aestivation

Metabolic changes, modifications of cellular and tissue activities, and modifications in the availability and/or use of oxygen, are conditions that induce cellular stress (Carey et al. 2003). All of these modifications appear to occur during aestivation. Cells respond to stress by the activation of signaling pathways that defend the cells against insults. If the damage cannot be repaired, activation of the death signaling pathways leads to cell destruction (Fulda et al. 2010). The heat shock response (active against temperature elevations, oxidative stress and toxic insults), the unfolded protein response (associated with the accumulation of unfolded proteins in the endoplasmic reticulum) and the DNA damage response are among the well-known protective responses. While little is known about the activity of these different pathways in the lungfish heart, more information is available on the response to oxidative stress.

Cell survival requires the presence of molecular oxygen. However, the reactive products of oxygen are amongst the most deleterious cell agents. The accumulation of these oxygen by-products, such as hydrogen peroxide and free radical intermediates, is counterbalanced by the presence of antioxidative agents including catalase, glutathione peroxidase and superoxide dismutases. For instance, overexpression of the mitochondrial superoxide dismutase (MnSOD) and/or of the cytosolic superoxide dismutase (CuZnSOD) is known to increase oxidative stress resistance (Murakami et al. 1997; Shan et al. 2007; Jang et al. 2009). Overexpression of MnSOD, CuZnSOD, glutathione peroxidase and catalase also acts to protect against the ischemia associated with the release of cytochrome c from mitochondria. This limits the incidence of apoptotic cell death that may occur in species that are not particularly tolerant to oxidative or other stressors (Zemlyak et al. 2009). It was initially suggested that the upregulation of intracellular antioxidant enzymes would protect the cells against stress-related injury during aestivation (Hermes-Lima and Zenteno-Savín 2002; Carey et al. 2003). More recently, the presence of the superoxide dismutases MnSOD and CuZnSOD has been reported in the heart of *P. dolloi* (Page et al. 2010). Curiously, the protein levels of these enzymes are not modified during aestivation. By contrast, aestivation leads to a significant increase in catalase and glutathione peroxidase activity. Since

Page et al. (2010) found little evidence of tissue oxidative damage in the heart of *P. dolloi* that had aestivated for 60 days, they concluded that a significant upregulation of intracellular antioxidant capacity increased oxidative stress resistance in this organ.

On the other hand, aestivation and arousal involve multilevel rearrangements and compromises between often opposite tissue/organ-specific requirements. However, nothing is known regarding the mechanisms that orchestrate these multilevel rearrangements. A major modulating role may be played by the Nitric Oxide (NO)/Nitric Oxide Synthase (NOS) system. NO is a universal, multi-faceted regulator of the aerobic biome (redox and energy balance). It is also crucial in cardio-circulatory homeostasis in mammals (Shen et al. 2001) and fish (Imbrogno et al. 2011). Perhaps, the best known function of NOS is to regulate blood flow through the production of endothelium-dependent vasodilation. NO is produced by different NOS isoforms that have ubiquitous tissue locations (including the heart and the skeletal muscle) and specialized subcellular compartments.

In the mammalian heart (Shen et al. 2001) and skeletal (Shen et al. 2000) muscle, endogenous physiological NO concentrations modulate mitochondrial respiration by reducing myocardial O_2 consumption (MVO_2) without changing the rate of ATP synthesis. At the same time, endogenous NO also affects substrate selection and utilization since its inhibition results in increased glucose and lactate uptake and decreased fatty acid utilization (Shen et al. 2001). The coupling between MVO_2 and ATP synthesis, as well as the coupling between MVO_2 and cardiac performance, results in a finely tuned regulation of metabolic needs/demands. Therefore, it is not surprising that endogenous NO appears implicated in stress-induced compensations such as hibernation and aestivation (Kaminski and Andrade 2001; Amelio et al. 2008).

It has already been indicated that, during aestivation, vascular resistance, blood pressure, heart rate and cardiac output decrease significantly (DeLaney et al. 1974). This has to be accompanied by modifications in the vascular supply of many organs. For instance, quiescent muscular masses or non-functional kidneys may undergo drastic reductions in blood supply. Modulation of cardiovascular parameters and blood shunting are prime functional targets of NO. Conceivably, modifications of NO activity may be of extreme beneficial influence during both the adaptive and the maintenance phases of aestivation. In freshwater *Protopterus*, the epicardium, the endocardial endothelium and the cardiomyocytes of both atria and ventricle are positive for the endothelial isoform (eNOS) of NOS (Amelio et al. 2008). eNOS localizes to the cell plasmalemma and, to a lesser degree, to the cytoplasm. In addition, low levels of the neural isoform (nNOS) of NOS appear at the level of the endocardium whereas the inducible isoform (iNOS) was undetectable. Of note, the cardiac distribution of NOS is shared by modern fish indicating the conservative nature of NO biosynthesis (Tota et al. 2005). Significantly, the levels of eNOS increase greatly during the beginning of aestivation in *P. dolloi*, returning to freshwater levels after 40 days of aestivation (Amelio et al. 2008). It has been suggested that eNOS acts in a paracrine fashion to protect the myocardium against insults and helps to adapt heart metabolism and performance to reduced levels of flow (Amelio et al. 2008). However, the reasons why the cardiac levels of eNOS are similar in freshwater conditions and long-term aestivation remain unclear.

Very recently, the localization and activity of the eNOS isoform and its partners, Akt and Hsp-90, have been studied in the cardiac and skeletal muscles of *P. annectens* (Amelio et al. 2013). Western blotting and immunofluorescence microscopy have shown tissue-specific modulation of the expression of these mediators. During aestivation, phospho-eNOS/eNOS and phospho-Akt/Akt ratios increased in the heart (whose pump function is continuous) but decreased in the skeletal muscle (which ceases to function during aestivation but can immediately contract after arousal). By contrast, Hsp-90 increased in the two muscle types during aestivation. On the other hand, an increased rate of apoptotic cell death (TUNEL assay) was evident in the skeletal muscle after aestivation, while the myocardial apoptotic rate remained unchanged as compared with freshwater controls. In agreement with the maintenance of heart activity during aestivation, the heart expression of the apoptosis repressor ARC also remained unchanged through the aestivating and arousal periods as compared with the freshwater controls. In contrast, ARC expression was strongly reduced in the skeletal muscle of the aestivating *P. annectens* (Amelio et al. 2013). Taken together, the changes in the eNOS/NO system and cell turnover strongly suggest that NO is implicated in the morpho-functional readjustments that occur in the cardiac and skeletal muscles of lungfish during the switch from freshwater to aestivation, and between the maintenance and arousal phases of aestivation.

Acknowledgements

This work was supported by the "Ministerio de Ciencia y Tecnología", Spain (Grant Number CGL2008-04559/BOS) and by the Ministry of Education of the Republic of Singapore (Grant Number R-154-000-429-112).

References

Almeida-Val, V.M.F., S.R. Nozawa, N.P. Lopes, P.H.R. Aride, L.S. Mesquita-Saad, M.N.P. Silva, R.T. Honda, M.S. Ferreira-Nozawa and A.L. Val. 2011. Biology of the South American lungfish, *Lepidosiren paradoxa*. pp. 129–147. *In*: J.M. Jorgensen and J. Joss (eds.). The Biology of Lungfishes. Science Publishers, Enfield.

Amelio, D., F. Garofalo, E. Brunelli, A.M. Loong, W.P. Wong, Y.K. Ip, B. Tota and M.C. Cerra. 2008. Differential NOS expression in freshwater and aestivating *Protopterus dolloi* (lungfish): heart vs. kidney readjustments. Nitric Oxide 18: 1–10.

Amelio, D., F. Garofalo, W.P. Wong, S.F. Chew, Y.K. Ip, M.C. Cerra and B. Tota. 2013. Nitric oxide synthase-dependent "On/Off" switch and apoptosis in freshwater and aestivating lungfish, *Protopterus annectens*: Skeletal muscle versus cardiac muscle. Nitric Oxide 32: 1–12.

Belkin, D.A. 1968. Anaerobic brain function: effects of stagnant and anoxic anoxia on persistence of breathing in reptiles. Science 162: 1017–1018.

Boutilier, R.G., G. Dobson, U. Hoeger and D.J. Randal. 1988. Acute exposure to graded levels of hypoxia in rainbow trout (*Salmo gairdneri*): metabolic and respiratory adaptations. Respir. Physiol. 71: 69–82.

Bugge, J. 1961. The heart of the African lungfish, *Protopterus*. Vidensk Meddr. Dansk natuhr. Foren. 123: 193–210.

Burggren, W.W. 1988. Cardiac design in lower vertebrates: what can phylogeny reveal about ontogeny. Experientia 144: 919–930.

Burggren, W.W. and N.J. Cameron. 1980. Anaerobic metabolism, gas exchange, and acid basebalance during hypoxic exposure in the channel catfish, *Ictalurus punctatus*. J. Exp. Zool. 213: 405–416.

Burggren, W.W. and K. Johansen. 1986. Circulation and respiration in lungfishes (Dipnoi). J. Morphol. Suppl. 1: 217–236.

Burggren, W.W., A. Farrell and H. Lillywhite. 1997. Vertebrate cardiovascular systems. pp. 215–308. *In*: W.H. Dantzler (ed.). Handbook of Physiology, Sect. 13, Comparative Physiology, Vol. 1. Oxford University Press, New York.

Carey, H.V., M.T. Andrews and S.L. Martin. 2003. Mammalian hibernation: cellular and molecular responses to depressed metabolism and low temperature. Physiol. Rev. 83: 1153–1181.

Chew, S.F., N.K.Y. Chan, A.M. Loong, K.C. Hiong, W.L. Tam and Y.K. Ip. 2004. Nitrogen metabolism in the African lungfish (*Protopterus dolloi*) aestivating in a mucus cocoon on land. J. Exp. Biol. 207: 777–786.

Chopin, L.K. and M.B. Bennett. 1995. Cellular ultrastructure and catecholamine histofluorescence of the heart of the Australian lungfish, *Neoceratodus forsteri*. J. Morphol. 223: 191–201.

Delaney, R.G., S. Lahiri and A.P. Fishman. 1974. Aestivation of the African lungfish *Protopterus aethiopicus*: cardiovascular and respiratory functions. J. Exp. Biol. 61: 111–128.

Dunn, J.F., P.W. Hochachka, W. Davison and M. Guppy. 1983. Metabolic adjustments to diving and recovery in the African lungfish. Am. J. Physiol. 245: R651–R657.

Fishman, A.P., A.I. Pack, R.G. Delaney and R.J. Galante. 1986. Aestivation in *Protopterus*. J. Morph. Suppl. 1: 237–248.

Frick, N.T., J.S. Bystriansky, Y.K. Ip, S.F. Chew and J.S. Ballantyne. 2008a. Carbohydrate and amino acid metabolism in fasting and aestivating African lungfish (*Protopterus dolloi*). Comp. Biochem. Physiol. A 151: 85–92.

Frick, N.T., J.S. Bystriansky, Y.K. Ip, S.F. Chew and J.S. Ballantyne. 2008b. Lipid, ketone body and oxidative metabolism in the African lungfish, *Protopterus dolloi* following 60 days of fasting and aestivation. Comp. Biochem. Physiol. A 151: 93–101.

Fulda, S., A.M. Gorman, O. Hori and A. Samali. 2010. Cellular stress responses: cell survival and cell death. Int. J. Cell Biol. 2010:214074 (doi:10.1155/2010/214074).

Graham, J.B. 1997. Air-breathing Fishes. Evolution, Diversity and Adaptation. Academic Press, San Diego.

Hermes-Lima, M. and T. Zenteno-Savín. 2002. Animal response to drastic changes in oxygen availability and physiological oxidative stress. Comp. Biochem. Physiol. C 133: 537–556.

Hiong, K.C., Y.K. Ip, W.P. Wong and S.F. Chew. 2013. Differential gene expression in the brain of the African lungfish, *Protopterus annectens*, after six days or six months of aestivation in air. PLoS One 8: e71205.

Icardo, J.M., J.L. Ojeda, E. Colvee, B. Tota, W.P. Wong and Y.K. Ip. 2005a. The heart inflow tract of the African lungfish *Protopterus dolloi*. J. Morphol. 263: 30–38.

Icardo, J.M., E. Brunelli, I. Perrotta, E. Colvée, W.P. Wong and Y.K. IP. 2005b. Ventricle and outflow tract of the African lungfish *Protopterus dolloi*. J. Morphol. 265: 43–51.

Icardo, J.M., D. Amelio, F. Garofalo, E. Colvee, M.C. Cerra, W.P. Wong, B. Tota and Y.K. Ip. 2008. The structural characteristics of the heart ventricle of the African lungfish *Protopterus dolloi*: freshwater and aestivation. J. Anat. 213: 106–119.

Icardo, J.M., W.P. Wong, E. Colvee, A.M. Loong and Y.K. Ip. 2012a. The alimentary canal of the African lungfish *Protopterus annectens* during aestivation and after arousal. Anat. Rec. 295: 60–72.

Icardo, J.M., W.P. Wong, E. Colvee, A.M. Loong and Y.K. Ip. 2012b. The spleen of the African lungfish *Protopterus annectens*: freshwater and aestivation. Cell Tissue Res. 350: 143–156.

Icardo, J.M., W.P. Wong, E. Colvee, A.M. Loong, A.G. Zapata and Y.K. Ip. 2014. Lympho-granulocytic tissue associated with the wall of the spiral valve in the African lungfish *Protopterus annectens*. Cell Tissue Res. 355: 397–407.

Imbrogno, S., B. Tota and A. Gattuso. 2011. The evolutionary functions of cardiac NOS/NO in vertebrates tracked by fish and amphibian paradigms. Nitric Oxide. 25: 1–10.

Jang, Y.C., V.I. Perez, W. Song, M.S. Lustgarten, A.B. Salmon, J. Mele, W. Qi, Y. Liu, H. Liang, A. Chaudhuri, Y. Ikeno, C.J. Epstein, H. Van Remmen and A. Richardson. 2009. Overexpression of Mn superoxide dismutase does not increase life span in mice. J. Gerontol. A Biol. Sci. Med. Sci. 64: 1114–1125.

Janssens, P.A. 1964. The metabolism of the aestivating African lungfish. Comp. Biochem. Physiol. 11: 105–117.

Jorgensen, J.M. and J. Joss. 2011. The Biology of Lungfishes. Science Publishers, Enfield.

Kaminski, H.J. and F.H. Andrade. 2001. Nitric oxide: biologic effects on muscle and role in muscle diseases. Neuromuscul. Disord. 11: 517–524.

Kind, P.K. 2011. The natural history of the Australian lungfish *Neoceratodus forsteri* (Kreft 1870). pp. 61–95. *In*: J.M. Jorgensen and J. Joss (eds.). The Biology of Lungfishes. Science Publishers, Enfield.

Laurent, P. 1996. Vascular organization of lungfish, a landmark in ontogeny and phylogeny of air-breathers. pp. 47–58. *In*: J.S.D. Munshi and H.M. Dutta (eds.). Fish Morphology. Horizon of New Research. A.A. Balkema, Rotterdam.

Mlewa, C.M., J.M. Gree and R.L. Dunbrack. 2011. The general natural history of the African lungfishes. pp. 97–127. *In*: J.M. Jorgensen and J. Joss (eds.). The Biology of Lungfishes. Science Publishers, Enfield.

Murakami, K., T. Kondo, C.J. Epstein and P.H. Chan. 1997. Overexpression of CuZn-superoxide dismutase reduces hippocampal injury after global ischemia in transgenic mice. Stroke 28: 1797–1804.

Ojeda, J.L., W.P. Wong, Y.K. Ip and J.M. Icardo. 2008. Renal corpuscle of the African lungfish *Protopterus dolloi*: structural and histochemical modifications during aestivation. Anat. Rec. 291: 1156–1172.

Page, M.M., K.D. Salway, Y.K. Ip, S.F. Chew, S.A. Warren, J.S. Ballantyne and J.A. Stuart. 2010. Upregulation of intracellular antioxidant enzymes in brain and heart during estivation in the African lungfish *Protopterus dolloi*. J. Comp. Physiol. B 180: 361–369.

Randle, P.J., E.A. Newsholme and P.B. Garland. 1964. Regulation of glucose uptake by muscle: effects of fatty acids, ketone bodies and pyruvate, and alloxan-diabetes and starvation on the uptake and metabolic fate of glucose in rat heart and diaphragm muscles. Biochem. J. 93: 652–665.

Shan, X., L. Chi, C. Luo, S. Qian, D. Gozal and R. Liu. 2007. Manganese superoxide dismutase protects mouse cortical neurons from chronic intermittent hypoxia-mediated oxidative damage. Neurobiol. Dis. 28: 206–215.

Shen, W., X. Xu, M. Ochoa, G. Zhao, R.D. Bernstein, P. Forfia and T.H. Hintze. 2000. Endogenous nitric oxide in the control of skeletal muscle oxygen extraction during exercise. Acta Physiol. Scand. 168: 675–86.

Shen, W., R. Tian, K.W. Saupe, M. Spindler and J.S. Ingwall. 2001. Endogenous nitric oxide enhances coupling between O_2 consumption and ATP synthesis in guinea pig hearts. Am. J. Physiol. Heart Circ. Physiol. 281: H838–846.

Sturla, M., P. Paola, G. Carlo, M.M. Angela and U.B. Maria. 2002. Effects of induced aestivation in *Protopterus annectens*: a histomorphological study. J. Exp. Zool. 292: 26–31.

Szidon, J.P., S. Lahiri, M. Lev and A.P. Fishman. 1969. Heart and circulation of the African lungfish. Circ. Res. 25: 23–38.

Tota, B., D. Amelio, D. Pellegrino, Y.K. Ip and M.C. Cerra. 2005. NO modulation of myocardial performance in fish hearts. Comp. Biochem. Phsyol. A Mol. Integr. Physiol. 142: 164–177.

Zemlyak, I., S.M. Brooke, M.H. Singh and R.M. Sapolsky. 2009. Effects of overexpression of antioxidants on the release of cytochrome *c* and apoptosis-inducing factor in the model of ischemia. Neurosci. Lett. 453: 182–185.

7

The Cardiac Outflow Tract of Primitive Fishes

Adrian C. Grimes

Introduction

At the cranial tip of the heart of all vertebrates is a structure known as the cardiac outflow tract (OFT), through which blood is pumped into the arterial system. In phylogenetically basal animals, the OFT is a tubular structure that connects the heart to a single systemic circulation, whereas in more apical animals, a more complex OFT is either partially or fully septated and acts as a conduit through which blood also passes to a secondary, pulmonary circulation. All fishes possess a heart that consists of a single atrium and a single ventricle, and most fish species lack this secondary pulmonary circulation so there is no requirement to prevent the intermixing of oxygenated and deoxygenated blood. Their heart pumps blood through a single, unseptated OFT to the gills where gas exchange occurs. Even in the three genera of lungfishes (Dipnoi), which show a gradation of cardiac specialization that correlates with each species' dependence upon atmospheric oxygen (Johansen and Lenfant 1968; Farmer 1999), a single ventricle pumps blood from the heart, and it is primarily adaptations of the venous (inflow) and arterial (outflow) poles that allow a partial separation of lung-oxygenated and system-deoxygenated blood. Thus, one should take care not to consider the morphology and evolution of the OFT in isolation, as specialization in this region may have been driven by coincident or subsequent adaptations in other cardiovascular structures.

Department of Medicine, Wisconsin Institutes of Medical Research Rm 8453, 1111 Highland Ave, Madison, WI 5370, USA.
E-mail: agrimes@pediatrics.wisc.edu

In all fishes, the OFT is interposed between the ventricle and the ventral aorta—the vessel lying just outside the pericardial cavity and which takes blood exiting the heart to the gills and then on to the systemic circulation. In the lungfishes, an additional vessel, the pulmonary artery, also lies just outside the pericardial cavity and transports blood to either a single lung in the case of the Australian lungfish (*Neoceratodus forsteri*), or two lungs in the African lungfishes (four species in the genus *Protopterus*) and in the South American lungfish (*Lepidosiren paradoxa*).

While this chapter is concerned with the morphology of the OFT in fully developed (juvenile and/or adult) species, it should be noted that some of the component structures of the OFT may be developmentally transient and can undergo complex morphological changes during embryogenesis. As will be discussed later, the myocardial component of the OFT is almost entirely myocardial in early development as a result of the cardiogenic programs and processes that help construct the heart. After an initial heart tube is formed at the embryonic midline, the OFT is added secondarily by an accretion of cells from a population of cardiac progenitors occupying a proliferating growth center known as the secondary heart field (Waldo et al. 2005) and located in the caudal coelomic wall (Soufan et al. 2006; van den Berg et al. 2009). These cells appear to be incorporated into the OFT such that the majority of cells initially differentiate as myocardium but, after a short developmental period, a few cells at the lumenal surface differentiate as smooth muscle. As development proceeds, and the OFT is extended, the ratio of lumenal smooth muscle to surrounding myocardium adjusts such that eventually the distal OFT possesses an entirely smooth muscle phenotype. This phenomenon was originally demonstrated experimentally in chick and mouse—vertebrate models possessing a four-chambered heart—but has now been observed in several fish species, such as the gilthead sea bream (Icardo et al. 2003), the sturgeon (Guerrero et al. 2004) and the zebrafish (Grimes et al. 2006; Hami et al. 2011; Guner-Ataman et al. 2013). It is these cardiogenic processes that ultimately determine the relative contribution the proximal myocardial and distal smooth muscle components will make to the OFT in adult species.

Generally, it has been proposed that the hearts of fishes can be divided approximately into four main types of increasing complexity (Tota et al. 1983; Tota 1989; Davie and Farrell 1991) based primarily on ventricular structure. However, the heart has been studied comprehensively in only a very small number of the approximately 33,000 formally described species of fishes (Eschmeyer and Fong 2014—see also Fishbase.org), the vast majority of which belong to the modern teleost group within the actinopterygians (ray-finned fishes). Consequently, the complexity and diversity of the OFT has been largely overlooked. Indeed, in many text books, just two basic OFT forms are proposed for the fishes: a generalized chondrichthyan form, in which a prominent myocardial conus arteriosus persists into adulthood, and a generalized teleost form, in which the conus arteriosus is said to have been 'replaced' by the smooth muscle or elastic structure known as the bulbus arteriosus (e.g., Helfman et al. 1989; Kapoor and Khanna 2004; Schmidt-Nielsen 2006; Hill et al. 2008). Some work has given consideration to the OFT of 'primitive' fishes (e.g., Guerrero et al. 2004; Icardo et al. 2004; Farrell 2007; Zaccone et al. 2011; Durán et al. 2014), but descriptions are limited to very few of these evolutionarily important species. As a result, at least 28 species of Acipenseriformes (sturgeons and paddlefishes), 12 species of Polypteriformes (bichirs and reedfish), seven species of Lepisosteiformes (gars) and the bowfin *Amia calva*, are largely misrepresented or ignored. In addition to the basal actinopterygians, many orders of true teleosts are also considered to be phylogenetically more ancient, and the structure of the cardiac OFT in representative species of these orders has also been largely unaddressed. Considered among the most basal of the teleosts are the superorders Osteoglossomorpha and Elopomorpha (e.g., Patterson and Rosen 1977; Lauder and Liem 1983; Nelson 2006), and a prominent conus arteriosus has been described in the OFT of several of these fishes, such as the African bonytongue *Heterotis niloticus*, the silver arowana *Osteoglossum bicirrhosum* and in some *Notopterus* species (Smith 1918; Grimes et al. 2010). Also, while not much is known about the OFT of approximately 1700 species of Elopomorphs, an extended conus arteriosus with two tiers of outflow valves is found in the bonefish *Albula vulpes*, the Japanese gissu *Pterothrissus gissu*, the Pacific tarpon *Megalops cyprinoides* and the Atlantic tarpon *M. atlanticus* (Senior 1907a,b; Smith 1918; Parsons 1929; Goodrich 1930).

Much has been learned about cardiac form and function from the widely-studied zebrafish (*Danio rerio*), a modern true teleost in the Cyprinidae family (order Cypriniformes); but, like all fishes, the zebrafish possesses a single atrium and a single ventricle, and the septation processes that divide the OFT of more apical vertebrates into separate aortic and pulmonary outflows, do not occur in such species. Further, the zebrafish is highly derived and possesses a prominent intrapericardial smooth muscle structure, the bulbus arteriosus, which had, until recently, been considered an evolutionary apomorphy (Durán et al. 2008). Nonetheless, the zebrafish does possess a myocardial component to the OFT—a conus arteriosus—that is quite obvious during embryogenesis (Grimes et al. 2006), but which becomes far less prominent in adult life after the bulbus arteriosus has been fully formed. Extant actinopterygians and, in particular, the living true teleosts belong to a phylogenetically advanced lineage that diverged from the tetrapod lineage at least 450 million years ago. Durán et al. (2008) suggest that the chondrichthyan heart, rather than that of the actinopterygians, should be regarded as the faithful reflection of the primitive gnathostomate heart, but it would clearly be advantageous to our understanding of both phylogeny and ontogeny of the OFT if the primitive actinopterygian species were studied in finer detail. Indeed, Icardo et al. (2004) proposed that the sturgeon may be an excellent model for studying particular processes of cardiogenesis such as heart looping. Moreover, several authors have recognized the likely homology between the conus arteriosus of primitive fishes and the region of the mammalian embryonic outflow that ultimately gives rise to the right ventricle (e.g., Holmes 1975; Romer and Parsons 1985; Moorman et al. 2007). It therefore seems only prudent to follow such observations with more detailed research of relevant species. The following summary descriptions present the most recent information on the morphology of the OFT of ancient fish species.

The Jawless Fishes—Agnathans

The group Craniata (animals with skulls) can be divided into two subphyla: the basal hagfishes (or Hyperotreti—meaning 'perforated palate') and the Vertebrata (animals with a backbone composed of cartilaginous or bony vertebrae). Current consensus, based primarily on studies of anatomy and morphology, has it that the lampreys (Hyperoartia—meaning 'complete palate') are found at the base of the vertebrate subphylum as a sister group to the jawed vertebrates, the gnathostomes, making hagfishes and lampreys paraphyletic (e.g., Janvier 1996). However, this consensus view is by no means unanimous and new molecular data appear to support an alternative hypothesis that these two groups are monophyletic and are more closely related to each other than to the jawed vertebrates (e.g., Delarbre et al. 2002; Kuraku et al. 2009; Heimberg et al. 2010). Regardless of this ongoing debate, together, the 77 nominal species of hagfishes and 46 species of lamprey (Eschmeyer and Fong 2014) are the jawless fishes, the agnathans, and in terms of cardiovascular morphology they are highly specialized, possessing a distinguishing feature of a 'partially open' circulatory system. This system lacks the capillary beds of most other vertebrates but, rather, has several open sinuses providing the function of venous drainage. As may be expected from the morphological and anatomical view of their phylogenetic position, it appears that the hagfishes possess a more primitive circulatory system and have a greater number of sinuses than the lampreys, along with extremely low blood pressure. The heart of all agnathans comprises a single atrium and a single ventricle, but in addition to the branchial heart hagfishes possess five accessory pumps situated within the systemic circulation that aid in venous return. Both the branchial heart and one of the accessory pumps, the portal heart, are invested in myocardium in the hagfishes, although they beat asynchronously (Jensen 1965). The hagfish heart was classically considered aneural (e.g., Jensen 1965), but several studies have demonstrated that the myocardium of both the portal and branchial hearts contains a plexus of course and fine argyrophilic nerve fibers (i.e., stained by silver) and that myelinated nerve fibers and ganglion cells can be found in the vicinity of the heart (Hirsch et al. 1964; Yamauchi 1980). However, physiological studies have found the heart to be remarkably insensitive to acetylcholine and the hagfish heart is thus considered to be at least 'functionally aneural'. In contrast, there are large numbers of catecholamine-containing chromaffin cells (i.e., stained by chromium salts) located in the endocardium of the hagfish heart, and the release of these catecholamine stores (primarily epinephrine and norepinephrine) is cardioexcitatory. Chromaffin cells have been described in the endocardial lining of the OFT in both hagfish and lampreys (Dahl et al. 1971; Wright 1984).

Although several bodies of work have stated that there is no morphologically obvious conus or bulbus arteriosus in the hagfish (Randall and Davie 1980; Farrell 2007) an intra-pericardial fibrous OFT can be seen clearly under low-magnification microscopic observation and this has been variously described as either the conus arteriosus or bulbus arteriosus (Wright 1984). Confusingly, a slight thickening of the proximal ventral aorta has also been described as the bulbus (Jones and Braun 2011), but as this feature is outside the pericardial cavity, it should not be considered part of the OFT. The fibrous OFT structure that *is* intrapericardial does not appear to be homologous to the conus arteriosus of phylogenetically advanced species, as it contains no myocardium, but has three distinct tissue layers described using blood vessel nomenclature (Wright 1984): the inner-most endothelium; a tunica media composed of scattered smooth muscle cells arranged either circumferentially or longitudinally and characterized ultrastructurally by the presence of numerous, tightly packed microfibrils within a fibrous matrix; and an outer adventitia of dense connective tissue. Because of the absence of myocardium and the presence of smooth muscle cells, the author considers that the structure should more properly be called the bulbus arteriosus. Based on evidence from the Pacific hagfish *Eptatretus stoutii*, the bulbus possesses a single bicuspid valve (Chapman et al. 1963; Farrell 2007; Grimes and Kirby 2009). Of note is the fact that the valve sinuses (the supporting bases of the leaflets) lie within the pericardial cavity and are supported by myocardium (Grimes and Kirby 2009). Although Wright (1984) observed that the valve arises from the myocardium, this is generally not well documented, but may indicate both myocardial and smooth muscle components of the outflow tract in these species.

In contrast, the lampreys have a single, more typically piscine heart that is innervated by a branch of the vagus nerve that enters the heart either along the jugular vein and into the sinus venosus at the inflow or passes along, and terminates at, the OFT. Interestingly, this vagal stimulation is excitatory, whereas it is inhibitory in all other vertebrates. Again, while it has been stated that there is no morphologically obvious conus or bulbus arteriosus in the lamprey (Randall and Davie 1980; Farrell 2007) a short but obvious bulb-like structure that has been called the bulbus cordis (e.g., Nakao et al. 1981) or the conus arteriosus (e.g., Wright 1984) arises directly from the anterior portion of the ventricle and supports a semilunar outflow valve (Fig. 1). This OFT structure is innervated by "large-cored-vesicle-containing nerve fibers and endings" (Nakao et al. 1981). Similar to the hagfish, the outflow valve is supported by myocardium that may be more compact than the remainder of the ventricle (Grimes and Kirby 2009), which suggests the possibility of both myocardial and smooth muscle OFT components. However, because

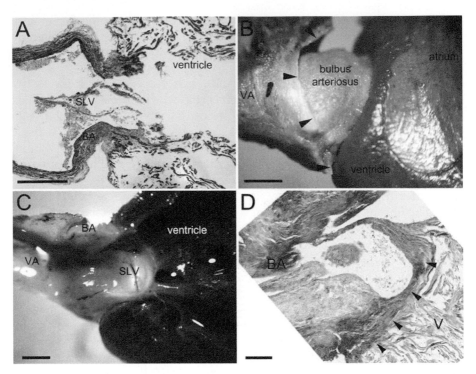

Fig. 1. The intrapericardial OFT of agnathans. (A) longitudinal section through the heart of a hagfish, trichrome stained, showing the bulbus arteriosus (BA) and that the semilunar outflow valves (SLV) arise from the ventricular myocardium (adapted from Grimes and Kirby 2009). (B) The OFT of the heart of an 18″ lamprey specimen, showing an obvious bulbus arteriosus within the pericardial cavity. Arrowheads delineate the pericardial sac. The ventral aorta (VA) lies outside the pericardium. (C) The same lamprey heart cut in half sagittally, showing the semilunar valve (SLV) sinuses surrounded by myocardium. (D) Longitudinal section of a lamprey heart, trichrome stained, confirming the base of the valve arising from ventricular myocardium. BA = bulbus arteriosus; V = ventricle. Scale bars (A–C) = 1 mm, D = 250 μm.

the OFT as described in adult specimens is not myocardial, it is likely not homologous to the conus arteriosus of more apical vertebrates and would be better named the bulbus arteriosus. Its walls are much thicker than in the hagfish but have similar layers (Wright 1984), with the tunica media containing distinct concentric layers of fibrous material and smooth muscle cells that are oriented both longitudinally and circumferentially. Similar to the hagfish, the tunica media contains numerous, tightly packed microfibrils. Of note, because neither the hagfish nor the lamprey appear to express elastin, it has been proposed that the likely function of microfibrils in the OFT tunica media is similar to that performed by elastin in the teleost bulbus arteriosus (Wright 1984). The ventral walls of the distal lamprey OFT also contain large distensions that protrude into the lumen and partially occlude it. These distensions are composed of irregularly shaped cells, fibrous material and collagen fibers, and, in concert with the elastic myofibrils, may mitigate fluctuations in blood pressure and ensure a constant flow of blood to the gills. Similar structures are seen in more advanced teleosts, such as the dolphinfish *Coryphaena hippurus* (the author's own observations). Because the hagfish has considerably lower blood pressure, such adaptations may be unnecessary and could explain why the walls of its OFT are thinner.

None of the agnathans has a coronary blood supply (Farrell et al. 2012) and the heart in these species are thus considered to be avascular. Coronary vessels are an adaptation that does not appear until the chondrichthyans.

It is worth emphasizing that the 'primitive' nature of the hearts of hagfish and lampreys is, in fact, the result of around 500 million years of evolution and adaptation. These are highly specialized animals and it is likely that many of their features—including perhaps the lack of a jaw—may be the result of degenerative evolutionary processes, meaning that the most recent common ancestor of the agnathans and jawed vertebrates may have had phenotypic characters that were far more complex than those possessed by modern hagfish and lampreys (see Heimberg et al. 2010). The fact that the heart of the hagfish is 'functionally aneural' in spite of the presence of nerve fibers and ganglion cells may be exemplary of those gradual degenerative processes—these species simply no longer have a requirement for neural stimulation of the heart within the niche they have so successfully exploited. Moreover, when considering the OFT of the agnathans, the lack of a myocardial component in the adult form may also be a degenerated state. While the concept of ontogeny recapitulating phylogeny is no longer considered literally and universally valid, perhaps future studies on the embryologic development of these fascinating species may provide clues about earlier phenotypic characters and reveal a myocardial component—a conus arteriosus—that regresses towards the ventricle during development, as has been observed in more apical species such as the sturgeon (Guerrero et al. 2004) the shark (Rodríguez et al. 2013) and zebrafish (Grimes et al. 2006).

Chondrichthyes—Sharks, Skates, Rays and Chimaeras

The Chondrichthyes (cartilaginous fishes) can be divided into two subclasses—the Elasmobranchii (sharks, skates and rays) and the Holocephali (chimaeras). There are 1328 nominal species of sharks, skates and rays, and 49 species of chimaeras (Eschmeyer and Fong 2014) and the heart of these cartilaginous fishes is considerably more complex than that of the agnathans. This is the most basal group in which, distal to the ventricle, a distinct myocardial conus arteriosus arises, and which persists into adulthood (Fig. 2). This pulsatile, myocardial, cylindrical structure supports several rows of pocket valves—the number being dependent upon species—and extends cranially to a smooth muscle, distal OFT component—the bulbus arteriosus—that lies

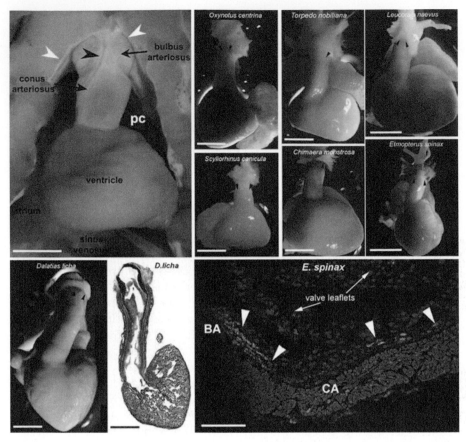

Fig. 2. The OFT of chondrichthyans. In the top right panel (adapted from Durán et al. 2008), in the heart of *Galeus atlanticus*, white arrowheads highlight the boundary of the pericardial cavity (pc). The black arrowhead indicates a coronary artery. Two components of the OFT—the proximal, myocardial conus arteriosus and the distal, smooth muscle bulbus arteriosus—can be seen. Scale bar = 5 mm. Note that the relative contribution each of these structures makes to the OFT is highly variable dependent upon species (see main text for details). Seven other species are shown to illustrate the diversity of OFT morphology, but that the two OFT components appear to be ubiquitous among chondrichthyans. Black arrowheads in these panels highlight the boundary between the conus arteriosus and the bulbus arteriosus. Scale bars = 5 mm. In the bottom left panels (wholemount and trichrome-stained sagittal section) is the heart of *Dalatius licha*, which has a conus arteriosus that extends for approximately 85% of the OFT and contains six tiers of valves. Scale bars = 5 mm. The bottom right panel shows detail of a sagittal section of the OFT of *Eptmopterus spinax*, labeled with antibodies against tropomyosin (myocardium – red) and transgelin (smooth muscle—green). DAPI stains the nuclei blue. This image shows the overlapping region of the OFT between the myocardial and smooth muscle components. Scale bar = 250 μm. BA = bulbus arteriosus; CA = conus arteriosus; a = atrium (Images of sections adapted from Grimes et al. 2010).

proximal to the pericardial wall (Sans-Coma 2007; Durán et al. 2008). During embryogenesis, the conus has four longitudinal, endocardial ridges, that are arranged dorsally, ventrally and laterally, and which can occlude the lumen during contraction. It is from these ridges that the valves arise in transverse tiers (Parsons 1929; Goodrich 1930; Farrell and Jones 1992). In the Chimaeridae, only two tiers are present but as many as six tiers may be present in some sharks (e.g., *Dalatias licha*—see Fig. 2).

The conus arteriosus is described using cardiac muscle terminology and has three layers: the outermost epicardium, the medial layer of myocardium and, at the luminal side, the endocardium. Thus, situated between endocardial and epicardial layers, and having the appearance of the ventricular compact layer, the intermediate layer of the walls of the elasmobranch conus contain circumferentially arranged cardiac myocytes, collagen and capillaries (Zummo and Farina 1989). The function of the conus is still not well defined, but its contractions apparently extend ventricular systole marginally (Johansen et al. 1966) and may allow the structure to act as an additional pump, ensuring a constant flow of blood into the ventral aorta (Kardong 2006). However, the relatively small lumenal volume of the conus relative to the ventricle means that this proposed 'auxiliary pump' is unlikely to be able to enhance flow considerably. Alternatively, because only the distal-most valve is able to occlude the lumen in the absence of contraction, the function of the contractile myocardium may primarily be in the active closure of the remainder of the conus, which would thus prevent backflow (Satchell and Jones 1967). The presence of nodal tissue at the junction of the ventricle, which sets up a delay between ventricular ejection and valve closure, provides additional evidence for this role (Satchell 1991). Another alternative hypothesis is that the contractions of the conus and subsequent valve closure are able to regulate blood flow into distinct pathways through the gills. This phenomenon is exemplified by the big skate *Raja binoculata* and rough skate *Dipturus nasutus*, in which the most anterior rows of gills are ventilated predominantly by the spiracle, and the posterior gills ventilated by the mouth. During early systole, the row of conus valves most distal to the ventricle may occlude the opening to the arteries that perfuse the posterior gills (Satchell 1976).

Although the entire OFT is myocardial in early embryogenesis (Rodriguez et al. 2013), at the distal limit of the adult chondrichthyan OFT, but still within the pericardial cavity, is a distinct, non-myocardial structure interposed between the conus and the ventral aorta (Sans-Coma 2007; Durán et al. 2008) that may be morphogenetically equivalent to the actinopterygian bulbus arteriosus (Durán et al. 2008). Although described in classical literature (e.g., Parsons 1929), this structure appears to have been overlooked in many descriptions of the chondrichthyan OFT over almost a century, but recent studies in 10 species from seven families have confirmed its presence (Durán et al. 2008; Grimes and Kirby 2009; Grimes et al. 2010). Although its walls have arterial characteristics, they are crossed by coronary arteries and are histologically quite different from the walls of the ventral aorta. The length of this portion of the chondrichthyan OFT is variable, dependent upon species, and is even highly variable among different specimens of the same species (noted by Parsons in 1929). It comprises as much as half of the entire OFT in the black-mouthed dogfish *Galeus melanostomus* and, in the dogfish *Scyliorhinus canicula*, up to one-third of the length of the intrapericardial outflow tract is composed of this 'unstriped' muscle (Parsons 1929). This interspecies size variability may correlate with the life-style and activity of the animal, as the bulbus arteriosus appears to be less developed in slow swimmers (Sans-Coma 2007).

In addition to the gross morphological distinction between the proximal myocardial and distal smooth muscle components of the chondrichthyan OFT, it has been proposed that a third component can be distinguished ultrastructurally or histologically. This third component is the region that forms an overlapping seam at the junction between myocardial and smooth muscle components. Smooth muscle cells that are continuous with the smooth muscle of the bulbus arteriosus and labeled with various immunohistochemical markers can be seen extending down within the sub-endocardial wall at the luminal side of the myocardial conus arteriosus, often to at least the most distal outflow valve and, in some cases, well beyond it (Grimes and Kirby 2009; Grimes et al. 2010; Rodríguez et al. 2013). Thus, the myocardium of this third OFT region forms a collar around a ring of smooth muscle around, or just efferent to, the valves. As can be deduced from descriptions below, this appears to be a common feature among the ancient fishes and may even be present in all vertebrates.

There is considerable interspecies variability with regard to coronary arteries, but chondrichthyans generally have a cephalad coronary supply that approaches the heart along the OFT in two main configurations described as dorsoventral and lateral (De Andres et al. 1990, 1992). The dorsoventral arrangement, where two coronary trunks pass, respectively, along the dorsal and ventral aspects of the OFT, is shared by lamnoid and advanced carcharhinid sharks; whereas the lateral arrangement, where the two trunks pass along the lateral aspect of the OFT, is found in hexanchoid and squaloid sharks. An intermediate arrangement has been observed in scyliorhinid species such as the small-spotted catshark *Scyliorhinus canicula* and the blackmouth catshark *Galeus melastomus*. A dorsoventral arrangement for the sand tiger shark *Carcharias taurus*, but a lateral arrangement in a batoid, the little skate *Raja erinacea* have also been detailed (Parker and Davis 1899). The coronary vessels of all chondrichthyans invade the compact myocardium of both the conus arteriosus and the ventricle and, in many cases, even the spongy myocardium of the ventricle is vascularized (Farrell et al. 2012; De Andres et al. 1992).

The chondrichthyans have both sympathetic and parasympathetic nervous systems. However, the sympathetic system does not reach the heart. Although lacking sympathetic innervation, the chondrichthyan heart receives inhibitory cholinergic stimulation from two distinct branches of the vagus nerve. In some batoids (skates and rays), such as the little skate *Raja erinacea* and the starry ray *Raja radiata*, nerve fibers showing neuropeptide Y-like and bombesin-like immunoreactivity have been identified in the conus arteriosus and the coronary vessels, but it is unclear what function they fulfill. Circulating catecholamines released by chromaffin cells in the blood vessel endothelium and cardiac endocardium provide cardioexcitatory cholinergic stimulation. Chromaffin cells in the lumenal lining of the conus of some sharks may play a role in this phenomenon.

Osteichthyes—Ray-Finned and Lobe-Finned Fishes

The Osteichthyes is a sister clade of the Chondrichthyes that includes the Actinopterygii (ray-finned fishes) and the Sarcopterygii (the lobe-finned fishes and the land-based tetrapods). The Actinopterygii can be further sub-divided into the Cladistia (bichirs), the Chondrostei (sturgeons and paddlefish), and the Halostei (the gars, the bowfin and the modern teleosts). Within the Osteichthyes, there was an at least partial genome duplication that occurred after the divergence of the sarcopterygian and actinopterygian lineages. The resulting genetic complexity has been proposed as an explanation for the wide diversity and success of fishes (Meyer and Schartl 1999) and it is interesting that the sarcopterygians and basal actinopterygians have maintained an elongated conus arteriosus, whereas this structure is far less prominent in most of the teleosts. The following descriptions are of the basal actinopterygians up to and including the bowfin, and exclude the modern teleosts.

Although there is considerable variation among the basal ray-finned fishes in terms of OFT structure, it follows a similar general pattern to that of the chondrichthyans, with two gross morphological components and a third ultrastructural region where the two major components overlap. This arrangement persists into adulthood. The relative contribution of the myocardial and smooth muscle components varies dependent upon family, where the conus arteriosus becomes reduced and the bulbus arteriosus is added distally, appearing to arise from a thickening of the arterial trunk within the pericardial cavity. It has been suggested that the reduction of the myocardial conus provides "snapshots in the development of form and function that stretches from the most ancient to most advanced fishes" (Satchell 1991; Jones and Braun 2011). However, while there is a greater myocardial contribution to the OFT in the bichirs than in the bowfin, for example, there is no obvious scale to suggest that the myocardial component has been reduced over evolutionary time. The relationships between the gars, the bowfin and the teleosts are somewhat controversial (see Janvier 2011), but it is generally accepted using morphological, paleontological and molecular criteria that the gars are more closely related to the bowfin than to the bichirs. Nonetheless, the extent of the conus arteriosus (and the number of outflow valves) in the gars is more similar to that found in the bichirs, from whom they diverged much earlier. Moreover, the presence of a relatively prominent conus arteriosus in many modern teleosts gives no clue to their phylogenetic position or their relationships to other species in which the conus appears more vestigial (e.g., compare the snakehead *Channa argus* [Anabantiformes] to the zebrafish *Danio rerio* [Cypriniformes]).

Actinopterygii

Cladistia—The Bichirs and Reedfish

The 12 extant species of bichir and reedfishes are generally considered to form a sister group to all other actinopterygians (Noack et al. 1996; Kikugawa et al. 2004; Nelson 2006; Takeuchi et al. 2009). However, the fossil record for these species is poor and molecular data has produced conflicting results, so their position is by no means firmly established (see Janvier 2011, for details). All 12 species are found in the freshwater river systems of tropical Africa, primarily in shallow or swampy floodplains and estuaries. They are obligate air-breathers and possess highly vascularized, paired lungs. However, unlike the Dipnoi and the land-based vertebrates, their lungs are smooth sacs and do not have alveolae. Paired pulmonary arteries situated outside the pericardial cavity direct deoxygenated blood from the heart to the lungs via the fourth gill arch.

Early anatomical studies on the polypterids described the OFT in terms of the number, shape and size of the OFT valves (Müller 1846; Boas 1880; Gegenbaur 1891, 1901). In a wider study of the conus arteriosus of several species of fish, Parsons (1929) stated that the myocardial component of the OFT of *Polypterus* "extends forwards over the conus to the pericardiac boundary". However, more recent work using histology and immunohistochemistry has provided much greater detail and identified a smooth muscle component to the OFT in these species (Zaccone et al. 2009a,b; Grimes et al. 2010; Durán et al. 2014). It is now clear that, similar to that found in the chondrichthyans, the cladistian OFT consists of two main components, a proximal myocardial conus arteriosus and a distal smooth muscle bulbus arteriosus. The conus arteriosus is a tubular structure that extends around 85–90% of the total length of the OFT (Fig. 3) and has three histological layers: an inner endocardial layer that possesses a relatively thick subendocardium containing several wavy sheets of elastin, interlaced by finer elastic fibers; a middle myocardial layer consisting of compact cardiac muscle arranged in well-organized, circumferential layers, between which there are fibers of type III collagen; and an outer epicardial layer, lined by epicardial cells, and with a thick subepicardium rich in type I collagen and crossed by coronary arteries whose branches penetrate the myocardium. At the most ventral aspect of the conus wall, several myocardial bundles are arranged longitudinally or helically.

Gegenbaur (1901) described within the polypterid conus arteriosus "nine circles of valves arranged in six longitudinal rows each one of which is made up of three large and three small units". This arrangement has recently been

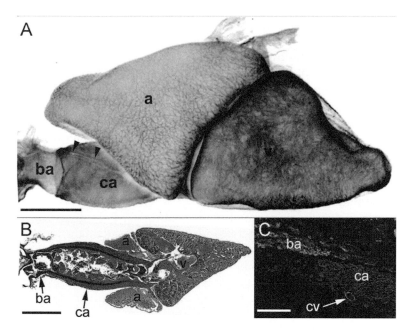

Fig. 3. The Cladistian OFT. (A) Lateral view of the heart of *Polypterus senegalus*. The conus arteriosus (ca) extends for around 85% of the OFT. ba = bulbus arteriosus; a = atrium; v = ventricle. Arrowheads highlight a coronary artery. Scale bar = 1 mm. (B) Longitudinal section through the heart of *P. senegalus*. Elastic stain identifies abundant elastic proteins extending deep into the conus arteriosus and supporting the nine rows of valves. Scale bar = 1 mm. (C) Detail of the junction of the conus arteriosus (ca) and bulbus arteriosus (ba) of *P. senegalus*, labeled with antibodies against tropomyosin (myocardium—red) and smooth muscle actin (green). DAPI labels the nuclei. Smooth muscle cells can be seen extending down into the subendocardium of the conus. A coronary vessel (cv) is also labeled by anti-smooth muscle actin. Scale bar = 100 µm. (panels B and C adapted from Grimes et al. 2010).

confirmed in *Polypterus senegalus,* in which the conus is furnished with a complex arrangement of valves at its luminal side with up to nine rows of valves and between three and six valves in each row (Durán et al. 2014). The shape and size of the valves varies depending on the row, but the most distal valves are of a semilunar configuration with leaflets that have an external fibrosa, a cellular-rich spongiosa and a thin internal fibrosa. Other valves have a more cellularized spongiosa that contain some scattered elastic fibers.

The remaining 10 to 15% of the OFT comprises the smooth muscle bulbus arteriosus. It is a relatively short tube with walls that are thinner than those of the conus and that taper towards the cranial boundary of the pericardial cavity where the bulbus connects with the ventral aorta. The bulbus lacks myocardium and valves and is composed of three layers; an inner layer built by flattened endocardial cells; an intermediate layer consisting of several sheets of circumferentially-oriented smooth muscle cells alternating with thicker sheets of elastin and type I and III collagen; and an external epicardial layer covering a thick subepicardium rich in type I collagen. The subepicardium of the bulbus arteriosus is also crossed by coronary arteries.

The coronary supply is cephalad and originates from the subclavian artery. It supplies the compact myocardium of the conus arteriosus and reaches the ventricle.

As described above for the chondrichthyan OFT, there is a region at the junction of the proximal and distal components of the polypterid OFT where the myocardium of the conus arteriosus forms an overlapping junction with smooth muscle cells that are continuous with those in the walls of the bulbus arteriosus. Some of these smooth muscle cells extend caudally at least as far as the base of the most distal valves.

There is a dearth of information regarding cardiac physiology in the cladistians. However, there is increasing evidence, based on the identification of nerves and ganglion cells, and on the localization of neurotransmitters, that the heart of the bichir receives both cholinergic and adrenergic innervation (Zaccone et al. 2009a,b). Immunohistochemical techniques using antibodies against tyrosine hydroxylase, acetyl cholinesterase, substance P, galanin, vasoactive intestinal polypeptide, pituitary adenylate cyclase-activating peptide and neuronal nitric oxide synthase showed that there is a rich population of nerve terminals in the conus arteriosus, although the origin of these fibers is unknown. Similarly, although several of the neurotransmitters identified in the OFT of the bichir heart are known to be implicated in the regulation of the teleost cardiovascular system, the lack of physiological and pharmacological studies in the bichir means that their function in these species is unclear.

Chondrostei—The Sturgeons and Paddlefish

There are 26 species of Acipenseriformes (the sturgeons) and two species of Polyodontidae (the paddlefishes), although one of these—the Chinese paddlefish—is critically endangered and possibly already extinct as none had been seen in the four years since 2003 (Zhang et al. 2009) until there were reports of a single 250 Kg specimen captured by illegal fishing in 2007.

The OFT of the sturgeons has been the subject of several detailed studies that include embryonic and adult forms. In the adult, the myocardial conus arteriosus comprises approximately two-thirds of the length of the OFT and is formed from compact and well vascularized myocardium lying between subendocardial and subepicardial layers. The remaining one-third, the bulbus arteriosus, possesses a smooth muscle and elastic phenotype. In addition to fat tissue, blood vessels, fibroblasts and collagen, a remarkable feature of the subendocardial layer of the sturgeon heart (including that of the myocardial portion of the OFT) is that it contains a great number of nodular bodies with a large vascular space and lined by a continuous endothelium, which give the heart a cauliflower-like, or cobblestone-like, appearance. Because these nodular bodies contain immature forms of blood cells, along with lymphocytes, granulocytes and macrophages, and because they resemble thymus tissue of other species, it is thought that they are part of a highly complex lympho-hemopoietic organ and may be involved in the sturgeon immune response (Icardo et al. 2002a,b).

As exemplified by the Adriatic sturgeon *A. naccarii,* the conus does not arise symmetrically from the acipenseriform ventricle but is displaced to the anatomical right (in other words, viewed from the dorsal aspect—Icardo et al. 2002a,b. See also Fig. 4). Within the lumen of the conus there are three transverse tiers of valves. The distal-most tier, situated close to the ventral aorta, contains four pocket valve cusps, but the most proximal two tiers have valves with a high degree of morphological and numerical variation, such that a longitudinal system is not discernable. Between the distal-most tier and the proximal two tiers is a long segment of the conus that is devoid of valves. Early studies on two species, the Atlantic sturgeon *Acipenser sturio* and the European sturgeon *Huso huso,* alluded to four tiers of valves. The third tier was

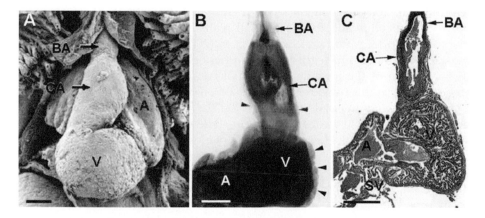

Fig. 4. The Chondrostean OFT. (A) Scanning electron microscope image of the heart of an 18 day post-hatch sturgeon *Acipenser naccarii,* showing the anatomical rightward displacement of the OFT with respect to the ventricular midline. Scale bar = 100 μm (adapted from Icardo et al. 2004). (B) Infra-red optical tomographic image of the heart of a juvenile (14″) *A. stellatus.* The constrictions at the proximal and distal limits of the conus arteriosus can be seen. The asterisk marks the 'tier of the empty space', a region of the conus that is devoid of valves. Arrowheads highlight the cobblestone-like lympho-hemopoietic nodular bodies on the surface of the ventricle and conus. The dark feature in the lumen of the bulbus arteriosus is blood trapped between the endocardial cushions at the time of fixation. Scale bar = 1 mm. (C) Elastic-stained sagittal section through the heart of a juvenile American paddlefish *Polyodon spathula,* showing the presence of elastic proteins extending from the bulbus arteriosus deep into the lumen of the conus arteriosus. Two rows of valves can be seen and the distance between them indicates that paddlefish have the same 'two and three' valvular arrangement as the sturgeons. Scale bar = 1 mm. BA = bulbus arteriosus; CA = conus arteriosus; V = ventricle; A = atrium; SV = sinus venosus.

described as the 'tier of the empty space' because the valves were actually missing or nothing more than rudimentary. It therefore appears that this 'two and three' arrangement of the OFT valves is common among sturgeons (Icardo et al. 2002a). In most vertebrates, as described above for the chondrichthyans, there are four longitudinal expansions of the fibrous subendothelium of the conus from which the cusps of the valves arise. However, in the sturgeon, while during development the invasion of mesenchymal cells form thickenings of the subendothelium in those regions of the conus from where valves will arise (Icardo et al. 2004), a longitudinal pattern is difficult to discern (Icardo et al. 2002a). The diameter of the sturgeon conus is not uniform along its length. It appears to balloon out along the middle section such that there are constrictions at the proximal and distal ends where the valves are located (Fig. 4). The morphology of the proximal cusps makes it unlikely that they function particularly well to prevent backflow; thus, these constrictions may serve an important functional role.

It has been reported that the entire outflow tract of sturgeons is myocardial in early development, but a distal smooth muscle component becomes obvious at later stages and then persists into adulthood (Guerrero et al. 2004; Icardo et al. 2004). As in the chondrichthyans, this structure, interposed between the conus and the pericardial wall, is histologically distinct from the ventral aorta. Although Guerrero et al. (2004) suggested that the smooth muscle component may arise from a phenotypic transition, whereby the distal portion of the initially myocardial conus becomes transformed into smooth muscle, this hypothesis was based on analysis of different fish fixed at various developmental stages, and on the observation that the smooth muscle component was absent in early-stage embryos but present at later stages of development. However, because of more detailed recent observations of other fish species such as the zebrafish (e.g., Grimes et al. 2006; de Pater et al. 2009; Hami et al. 2011; Zhou et al. 2011; Guner-Ataman et al. 2013; Nevis et al. 2013) and because of what is known about OFT cardiogenesis in many other vertebrate models, it appears much more likely that the smooth muscle component is added to the outflow tract secondarily, relatively late in development, and that the entire OFT grows into the pericardial cavity by an accretion of cells from what is now widely known as the secondary heart field (see Waldo et al. 2005; Soufan et al. 2006; van den Berg et al. 2009).

In common with all species described thus far, a myocardial-smooth muscle overlap at the junction of the conus arteriosus and bulbus arteriosus has been described for the OFT of the American paddlefish *Polyodon spathula*, and for four species of sturgeon—*Acipenser stellatus, A. baeri, A. naccarii* and *Scaphirhynchus platorynchus* (Grimes et al. 2010).

A single coronary artery passes along the length of the OFT on the ventral surface and supplies the compact myocardium of the conus arteriosus and the ventricle. In several species, the spongiosa of the ventricle may also receive this coronary blood supply, similar to that observed in chondrichthyans (Romensky 1978). However, the morphology of the coronary supply may be different depending upon species. In the Siberian sturgeon (*Acipenser naccarii*), the supply is cranial (Icardo et al. 2002), whereas in the paddlefish (*Polyodon spathula*) and the shovelnose sturgeon (*Scaphirhynchus platorynchus*) it is caudal and only a rudimentary cephalad coronary vessel becomes "lost on the ventral aorta before reaching the heart" in the paddlefish (Danforth 1916). Moreover, the origin of the caudal coronary supply is different in these species, arising from the subclavian artery in the shovelnose but from the fourth efferent branchial artery in the paddlefish (Danforth 1916).

Similar to the situation in most chondrichthyans, although numerous unmyelinated nerve fibers invade into the subepicardium of the sturgeon heart, there is no parasympathetic innervation and no obvious nerve fibers described in the OFT.

Lepisosteiformes—The Gars

The seven extant species of gar are descendants of a group of fishes that dominated marine and freshwater systems for more than 200 million years, from the mid-Devonian to the Mesozoic eras. They have some remarkable adaptive traits, not the least of which is the ability to breathe bimodally by gulping air and passing it to a highly vascularized gas bladder (Potter 1927). In fact, similar to the bichirs, in poorly oxygenated water gars have an absolute requirement for atmospheric air. The cardiovascular system of these fishes has undoubtedly evolved to accommodate this requirement, but it is unclear how the morphology and function of the OFT has contributed to these changes, or how it responded to adaptations in other cardiopulmonary structures.

Similar to the sturgeons, the conus arteriosus of the gars comprises the majority of the length—around 75%—of the OFT. However, in gross morphological terms, it appears more similar to that described for many chondrichthyan species (see Durán et al. 2008) as it exits the ventricle from a central position and its diameter is much more uniform along the entire length. In *Lepisosteus osseus* (the long nosed gar), the conus supports eight tiers of valves, but these valves are not placed evenly along the length (Fig. 5). There are also abundant elastic fibers and collagen in the subendocardium, along with smooth muscle cells that are continuous with the medial smooth muscle of the distal OFT component, the bulbus arteriosus, and which extend caudally to the level of at least the sixth tier of valves. A large coronary artery and several small blood vessels are associated with the wall of the conus. The distal one quarter of the length of the OFT—the bulbus arteriosus—does not contain myocardium in its wall, but is comprised of smooth muscle, elastic fibers and collagen. It contains no valves.

A cephalad coronary supply arising from the hypobranchial and subclavian arteries extends along the length of the OFT in easily identifiable ventral and less discernable dorsal coronaries (Danforth 1916). It supplies the compact myocardium of the conus arteriosus. Although these vessels reach

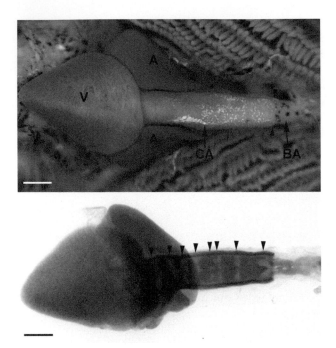

Fig. 5. The OFT of Lepisosteiformes. In the top panel is a ventral view of the heart of *L. osseus*, showing that the conus arteriosus (CA) comprises approximately 75% of the OFT. The single atrium straddles the ventricle and OFT rather like a saddle, so it can be seen on either side of the OFT. Arrowheads highlight the distal limit of the myocardium, where it meets the bulbus arteriosus (BA). V = ventricle; A = atrium (adapted from Zaccone et al. 2011). In the lower panel, an infra-red optical tomographic image of the same heart, fixed and cleared, shows the myocardium and the eight irregularly placed tiers of valves (arrowheads) in the conus arteriosus. Scale bars = 1 mm.

the ventricle (Danforth 1916), the ventricular myocardium lacks a substantial compacta and nothing is known about its vascularization (Farrell et al. 2012).

Immunohistochemical techniques have shown that within the wall of the conus arteriosus of *L. osseus* is a well-developed system of adrenergic, cholinergic, substance P-positive and neuronal nitric oxide synthase-positive nerve terminals (Zaccone et al. 2011, 2012) that co-innervate the myocardium and the aforementioned smooth muscle located in the conal subendocardium, suggesting the presence of both extrinsic and intrinsic nerves. The lack of colocalization of fibers that are immunopositive for choline acetyl transferase and neuronal nitric oxide synthase in these tissues indicates the possibility that the gar OFT is innervated by both vagal motorneurons and by spinal motorneurons, but, similar to the case of the bichirs, the functional significance of this is unknown.

Amiiformes—The Bowfin

The bowfin, *Amia calva*, is the last surviving member of the order Amiiformes. Although the interrelationships of the bowfin, the gars and other teleosts has been controversial, current consensus—based on paleontological and molecular sequence-based evidence—places the bowfin (and its now extinct ancestors) as a sister group to the gars. Together, the bowfin and the gars are the Holostei. The bowfin is a freshwater species indigenous to southeastern Canada and the eastern United States. They thrive in shallow lakes, ponds and the back waters of rivers. Like the gars, the bowfin possesses a highly vascularized swim bladder connected to its gastrointestinal tract, with which it is able to breathe air by gulping at the water's surface. The frequency of air-breathing increases during exercise, but is reduced at rest and at lower temperatures (Johansen et al. 1970; Farmer and Jackson 1998).

In the bowfin, the prominent conus arteriosus persists into adulthood but is considerably smaller than in the chondrichthyans, gars, bichirs and sturgeons, and has three transverse tiers of valves with four or five leaflets in each tier (Sedgwick et al. 1905; Goodrich 1930; Icardo 2006). The leaflets of the most distal tier of valves, although attached to the luminal side of the myocardial conus, extend deep into the smooth muscle component of the OFT, the bulbus arteriosus. The bulbus arteriosus comprises around one half of the length of the OFT and tapers towards the cranial limit of the pericardial cavity where it joins the ventral aorta. At its widest, the diameter of the bulbus is considerably larger than that of the conus. However, its walls are remarkably thin compared to those observed in many apical teleosts, in which the size of the conus has diminished over evolutionary time along with an augmentation of the size of the bulbus and a decrease in the number of valve tiers. Two cephalad coronary arteries—one dorsal and one ventral—run from the first and second gill arches, enter the pericardial cavity and along the length of the bulbus arteriosus, then bifurcate before running along the conus arteriosus towards the ventricle, where it supplies the compact myocardium (Parker and Davis 1899. See also Fig. 6).

Whether the limited conal reduction and bulbal augmentation in *A. calva* is advantageous to bimodal breathing seems unlikely, as there is an extensive conus arteriosus and a relatively small bulbus arteriosus in other bimodal breathers, such as the bichirs, gars, and in the true lungfishes (although the branchial circulation in *Amia* and the bichirs and lungfishes may serve quite dissimilar roles). Moreover, there are examples of bimodal breathers in at least 374 species of fish from 49 families (Graham 1997) that have highly variable OFT and circulatory architecture. However, the arrangement in *Amia* may allow better regulation of blood flow into distinct pathways through the gills, as in the chondrichthyans.

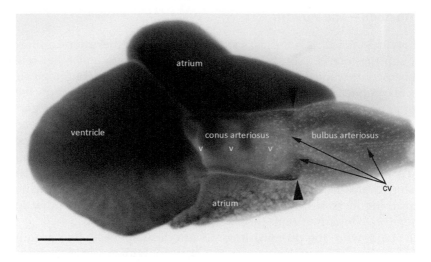

Fig. 6. The OFT of *Amia calva*. An infra-red optical tomographic image of the heart. Similar to the gar, a single atrium straddles the ventricle and OFT, so it can be seen on either side of the OFT. Large arrowheads show the distal limit of the myocardial conus arteriosus and the proximal limit of the bulbus arteriosus. Arrows highlight coronary arteries running along the ventral surface of the OFT, also bordered by pigment cells (the pigment cells are black, but show white due to artifact of the imaging technique). Three tiers of valves can be seen in the conus (v). Scale bar = 1 mm.

Sarcopterygii—The Lobe-Finned Fishes

The phylogenetic relationships of the Sarcopterygii, a group that includes both lobe-finned fishes and land-based tetrapods, has been controversial. While molecular data gives conflicting and inconsistent results, it is now generally accepted that the coelacanths (subclass—Actinistia) form a sister group to the true lungfishes (subclass—Dipnoi) and the land-based tetrapods, who thus share a more recent common ancestor with each other than with the coelacanths. There are only two extant coelacanth species, *Latimeria chalumnae* and *L. menadoensis*, indigenous to the coastlines of the West Indian Ocean and Indonesia, respectively. There are three genera of lungfishes, with four species of the genus Protopterus found in Africa (*Protopterus annectens, P. aethiopicus, P. amphibius* and *P. dolloi*), one species found in South America (*Lepidosiren paradoxa*), and one Australian species (*Neoceratodus forsteri*).

Actinistia—The Coelacanths

Because both species of coelacanth are rare, and both are threatened, the actinistians form the most endangered order of animals in the world. Indeed, the West Indian Ocean coelacanth is critically endangered. This means that there is a dearth of biological information available and the heart of these fishes has been poorly studied. As with the agnathans, it is worth emphasizing that coelacanths are highly specialized and adapted to their particular environment, so it is a mistake to describe them and, more particularly, their heart, as being 'simple'. In "Living Fossil: The Story of the Coelacanth", for example (Thompson 1992), it is stated that "the heart itself is unusually primitive in *Latimeria*, retaining in some respect the features of the embryonic stage of development of the heart in advanced vertebrates". Millot et al. (1978) described the coelacanth heart as having a 'relatively modest size', perhaps representing what Goodrich (1930) proposed as the 'hypothetical primitive state'. However, based on available literature and the drawings of Millot et al. (1978), it would appear that the heart of the adult coelacanth is quite complex, having a compacta, trabeculae, and a coronary circulation in the atrium and ventricle and also in the conus arteriosus. It would therefore likely be classified as a type II heart (Tota 1989).

Images of the heart of the coelacanth are limited to sketches made by Millot et al. (1978) and these have been reproduced in Farrell (2007), in Grimes and Kirby (2009) and in Jones and Braun (2011). The OFT is quite elongate, with a proximal myocardial conus arteriosus comprising about 50% of the total length. The conus arteriosus has a thin myocardial wall and within its lumen are four longitudinal endocardial ridges, arranged dorsally, ventrally, and laterally (left and right). Each ridge supports five or six successive valve leaflets that increase considerably in size from caudal to cranial. Interestingly, as most vertebrates have four longitudinally arranged endocardial ridges, it has been suggested that the 'primitive state' conus arteriosus was originally invested with four longitudinal rows of valve cusps (Goodrich 1930). Throughout evolution, either one of the rows was lost (as is the case in the sturgeons and most chondrichthyans) or, alternatively, one or more was added (as in some sharks). Thus, in adult forms, there are very few species that possess four rows, suggesting that the arrangement in actinistians may well be reflective of the primitive vertebrate state.

The remaining 50% of the OFT comprises a thin-walled bulbus arteriosus that was designated the 'truncus arteriosus' by Millot et al. (1978) and has walls of elastic fibers and smooth muscle. This structure extends to the cranial limit of the pericardial cavity, where it connects to the ventral aorta.

The coronary circulation in coelacanths is pectoral, in that it approaches the heart caudally and is derived from paired efferent branchial arteries, the left and right subclavians. A similar cephalad supply is found in skates, rays and some chondrosteans (Farrell and Jones 1992; Farrell 2007). However, the primary coronary supply in most fishes is cephalad with an occasional additional pectoral supply.

Nothing is known about the innervation of the coelacanth OFT.

Dipnoi—True Lungfishes

There have been several excellent works describing the OFT of the lungfishes, in which there is an apparent gradation of cardiac specialization relative to the individual species' dependence upon aerial oxygen, with *L. paradoxa* appearing to be the most specialized and *N. forsteri* the least (Johansen and Lenfant 1968; Farmer 1999). Experiments with contrast media have demonstrated that specializations in the atrium and ventricle of the lungfishes act to maintain two separate flows of blood through the heart, passing lung-oxygenated blood from the pulmonary vein into the left side of the OFT, and deoxygenated blood from the systemic circulation to the right. As is the case in the sturgeons, the myocardial conus arteriosus of the lungfishes does not arise symmetrically from the ventricle, but is displaced to the anatomic right of the ventral midline, as reported by Icardo et al. (2005). While on the other hand, Burggren and Johansen (1986) reported that the OFT is displaced to the left of the ventricular midline, an image in that work shows that the authors likely viewed the heart from a ventral aspect and thus confused left with right.

The lungfish conus arteriosus is a remarkable structure formed of well-vascularized, circumferentially arranged compact myocardium that bends first to the left and then cranially through two approximate right angles. There is also an approximate 270° counterclockwise rotation about its longitudinal axis. It is not clear whether this twist actually occurs within the myocardial wall itself or whether it results from the spiraling-in of the endothelial septa and spiral valves at the luminal side of the conus. However, the proximal region of the conus is partially septated by a fold of tissue that arises from a ventral row of fused valves and which extends along the entire length of the conus, spiraling through 270°. In *L. paradoxa* and *Protopterus* species, the folds also fuse at the distal end with a similar but smaller fold of tissue arising from the opposite lumenal wall. In the conus arteriosus of *N. forsteri*, this spiral valve is less derived, having several tiers of small valves that prevent reflux of blood during late systole, and large semilunar valves instead of the fused cushions at the distal end. Pocket valves found at the proximal end of the conus arteriosus of other lungfish appear to be vestigial (Johansen et al. 1968).

The position of the two folds of the spiral valve, with their free edges almost in opposition, and their eventual fusion in all lungfish except *L. forsteri*, effectively divides much of the length of the conus and creates a horizontal septum that separates the lumen into dorsal and ventral channels. Oxygenated blood from the left side of the heart flows primarily through the ventral channel towards the two anterior gill arches, while deoxygenated blood flows through the dorsal channel to the two posterior gill arches.

There are conflicting reports about the myocardial extent of the OFT and some contradictory nomenclature. Although the conus arteriosus has been described as possessing a thick myocardial wall for only one-third of its length, with the remaining two-thirds being arterial-like, it should be emphasized that there is in fact a thin myocardial sheath covering almost the entire length of the outflow tract. There is an abundance of smooth muscle cells, collagen and elastic fibers lying within this myocardial sheath that extends down to the level of the valve (Icardo et al. 2005) and it is this phenomenon that has led to some confusion over nomenclature. Additionally, as now reported for ALL fishes, there is a region of the distal OFT that does not have a myocardial coat and may be considered a bulbus arteriosus (Fig. 7). However, similar to the polypterids and the gars, this region is quite small in the lungfishes and may be less than 10% of the entire OFT length.

The myocardium of the conus arteriosus is supplied by a cephalad coronary artery supply, which also extends to the ventricle (Farrell 2007).

Innervation of the lungfish heart is vagal and inhibitory (autonomic innervation is absent). There is an abundance of chromaffin cells that line the inner walls of inflow structures of the heart and these may provide an alternative cardioexcitatory adrenergic control.

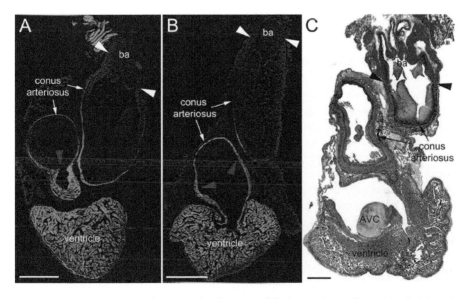

Fig. 7. The OFT of Dipnoi. (A) Longitudinal section of the heart of the African (slender) lungfish *Protopterus dolloi*, labeled with CH1 (myocardium - green) and SM22-alpha (smooth muscle—red). Cell nuclei are labeled with DAPI (blue). Cranial is to the top. The smooth muscle of the outflow tract (OFT) extends to the base of the endocardial valves (red arrowhead). The myocardial collar extends almost as far as the pericardial wall (white arrowheads). Note however that a small portion of the distal OFT contains no myocardium. Thus, the intrapericardial OFT of the lungfishes follows the 'principle of universality' and has two components, a myocardial conus arteriosus and a smooth muscle bulbus arteriosus (ba). (B) Similarly labeled section of the heart of the South American lungfish *Lepidosiren paradoxa*, showing the same general morphology. (C) Elastic trichrome stain of the Australian lungfish *Neoceratodus forsteri*, showing similar general OFT morphology and that elastic proteins are coincident with the OFT smooth muscle cells almost to the proximal limit of the valves. Black arrowheads in C show the distal intrapericardial limit of the conal myocardium. AVP, atrioventricular plug. Scale bars = 1 mm (adapted from Grimes et al. 2010).

Summary

It has been proposed—and indeed, it seems likely—that extant chondrichthyans possess a heart that is representative of the primitive gnathostomate state, and that all other forms are adaptations and specializations derived from that blueprint (see Durán et al. 2008). Thus, following the principle of universality (Opitz and Clark 2000), it is now quite clear that in gross morphological terms the cardiac OFT of most (if not all) fishes is comprised of two main components—a proximal myocardial component known as the conus arteriosus and a smooth muscle and elastic component known as the bulbus arteriosus. Considerable variability exists in the relative contribution these two basic components make to the entire OFT, such that a larger proportion is myocardial in the chondrichthyans and in basal actinopterygians, whereas in many of the apical actinopterygians (i.e., most teleosts), the larger proportion of the OFT possesses a smooth muscle and elastic character. In addition to these two basic structures, a third OFT component can be distinguished using ultrastructural or histological techniques. This region forms an overlapping seam at the junction between myocardial and smooth muscle components, such that smooth muscle cells continuous with the bulbus arteriosus can be seen extending down within the sub-endocardial wall at the luminal side of the myocardial conus arteriosus. The myocardium of this third OFT region forms a collar around a ring of smooth muscle in the vicinity of, or just efferent to the valves. In the lungfishes and the gars, the smooth muscle appears to extend deep into the conus, whereas in some other species the overlap is far less considerable. However, it would appear to be a common feature in the ancient fishes and may exist in all vertebrates (see Grimes et al. 2010). The agnathans appear to be a specific exception to this general rule, certainly in adult form, but this may be the result of degenerative evolutionary processes—the gradual disappearance of the myocardial component and the retention of an almost vestigial smooth muscle component. Future investigations into the details of cardiogenesis in these animals may suggest hitherto unrecognized similarities with other, more apical species.

References

Boas, J.E.V. 1880. Über Herz und Arterienbogen bei *Ceratodus* und *Protopterus*. Morphologisches Jahrbuch 6: 321–354.

Burggren, W.W. and K. Johansen. 1986. Circulation and respiration in lungfishes (Dipnoi). J. Morphol. Suppl. 1: 217–236.

Chapman, C.B., D. Jensen and K. Wildenthal. 1963. On circulatory mechanisms in the Pacific hagfish. Circ. Res. 12: 427–440.

Dahl, E., B. Ehinger, B. Falck, C. Von Mecklenburg, H. Myhrberg and E. Rosengren. 1971. On the monoamine-storing cells in the heart of *Lampetra fluviatilis* and *L. planeri* (Cyclostomatal). Gen. Comp. Endocrinol. 17: 241–246.

Danforth, C.H. 1916. The relation of the coronary and hepatic arteries in the common ganoids. Am. J. Anat. 9: 391–400.

Davie, P.S. and A.P. Farrell. 1991. The coronary and luminal circulations of the myocardium of fishes. Can. J. Zool. 69: 1993–2001.

De Andres, A.V., R. Munoz-Chapuli, V. Sans-Coma and L. Garcia-Garrido. 1990. Anatomical studies of the coronary system in elasmobranchs: I. Coronary arteries in lamnoid sharks. Am. J. Anat. 187: 303–310.

De Andres, A.V., R. Munoz-Chapuli, V. Sans-Coma and L. Garcia-Garrido. 1992. Anatomical studies of the coronary system in elasmobranchs: II. Coronary arteries in hexanchoid, squaloid, and carcharhinoid sharks. Anat. Rec. 233: 429–439.

de Pater, E., L. Clijsters, S.R. Marques, Y.F. Lin, Z.V. Garavito-Aguilar, D. Yelon and J. Bakkers. 2009. Distinct phases of cardiomyocyte differentiation regulate growth of the zebrafish heart. Development 136: 1633–1641.

Delarbre, C., C. Gallut, V. Barriel, P. Janvier and G. Gachelin. 2002. Complete mitochondrial DNA of the hagfish, *Eptatretus burgeri:* the comparative analysis of mitochondrial DNA sequences strongly supports the cyclostome monophyly. Mol. Phylogenet. Evol. 22: 184–192.

Durán, A.C., B. Fernandez, A.C. Grimes, C. Rodriguez, J.M. Arque and V. Sans-Coma. 2008. Chondrichthyans have a bulbus arteriosus at the arterial pole of the heart: morphological and evolutionary implications. J. Anat. 213: 597–606.

Durán, A.C., I. Reyes-Moya, B. Fernández, C. Rodríguez, V. Sans-Coma and A.C. Grimes. 2014. The anatomical components of the cardiac outflow tract of the gray bichir, *Polypterus senegalus:* their evolutionary significance. Zoology (Jena) 117: 370–6.

Eschmeyer, W.N. and J.D. Fong. 2014. Catalogue of Fish Species: species by family/subfamily. Retrieved 02/14/2014, 2014, from http://research.calacademy.org/research/ichthyology/catalog/SpeciesByFamily.asp.

Farmer, C.G. 1999. Evolution of the vertebrate cardio-pulmonary system. Annu. Rev. Physiol. 61: 573–592.

Farmer, C.G. and D.C. Jackson. 1998. Air-breathing during activity in the fishes *Amia calva* and *Lepisosteus oculatus.* J .Exp. Biol. 201: 943–948.

Farrell, A.P. 2007. Cardiovascular systems in primitive fishes. pp. 53–120. In: D.J. McKenzie, A.P. Farrell and C.J. Brauner (eds.). Fish Physiology: Primitive Fishes. Academic Press, San Diego.

Farrell, A.P. and D.R. Jones. 1992. The heart. pp. 1–88. In: W.S. Hoar, D.J. Randall and A.P. Farrell (eds.). Fish Physiology, XII: The Cardiovascular System. Academic Press, San Diego.

Farrell, A.P., N.D. Farrell, H. Jourdan and G.K. Cox. 2012. A perspective on the evolution of the coronary circulation in fishes and the transition to terrestrial life. pp. 75–102. In: D. Sedmera and T. Wang (eds.). Ontogeny and Phylogeny of the Vertebrate Heart. Springer, New York.

Foxon, G.E. 1950. A description of the coronary arteries in dipnoan fishes and some remarks on their importance from the evolutionary standpoint. J. Anat. 84: 121–131.

Gegenbaur, C. 1891. Über den Conus arteriosus der Fische. Morphologisches Jahrbuch 17: 596–610.

Gegenbaur, C. 1901. Vergleichende Anatomie der Wirbelthiere mit Berücksichtigung der Wirbellosen Leipzig, Wilhelm Engelmann.

Goodrich, E.S. 1930. Studies on the Structure and Development of Vertebrates. Macmillan and Co., London.

Graham, J.B. 1997. Air-Breathing Fishes: Evolution, Diversity, and Adaptation. Academic Press, San Diego.

Grimes, A.C. and M.L. Kirby. 2009. The outflow tract of the heart in fishes: anatomy, genes and evolution. J. Fish Biol. 74: 983–1036.

Grimes, A.C., H.A. Stadt, I.T. Shepherd and M.L. Kirby. 2006. Solving an enigma: arterial pole development in the zebrafish heart. Dev. Biol. 290: 265–276.

Grimes, A.C., A.C. Duran, V. Sans-Coma, D. Hami, M.M. Santoro and M. Torres. 2010. Phylogeny informs ontogeny: a proposed common theme in the arterial pole of the vertebrate heart. Evol. Dev. 12: 552–567.

Guerrero, A., J.M. Icardo, A.C. Duran, A. Gallego, A. Domezain, E. Colvee and V. Sans-Coma. 2004. Differentiation of the cardiac outflow tract components in alevins of the sturgeon *Acipenser naccarii* (Osteichthyes, Acipenseriformes): implications for heart evolution. J. Morphol. 260: 172–183.

Guner-Ataman, B., N. Paffett-Lugassy, M.S. Adams, K.R. Nevis, L. Jahangiri, P. Obregon, K. Kikuchi, K.D. Poss, C.E. Burns and C.G. Burns. 2013. Zebrafish second heart field development relies on progenitor specification in anterior lateral plate mesoderm and nkx2.5 function. Development 140: 1353–1363.

Hami, D., A.C. Grimes, H.J. Tsai and M.L. Kirby. 2011. Zebrafish cardiac development requires a conserved secondary heart field. Development 138: 2389–2398.

Heimberg, A.M., R. Cowper-Sallari, M. Semon, P.C.J. Donoghue and K.J. Peterson. 2010. microRNAs reveal the interrelationships of hagfish, lampreys, and gnathostomes and the nature of the ancestral vertebrate. Proc. Natl. Acad. Sci. USA 107: 19379–19383.

Helfman, G., B.B. Collette, D.E. Facey and B.W. Bowen. 1989. The Diversity of Fishes: Biology, Evolution, and Ecology. Wiley-Blackwell, Chichester.

Hill, R., G.A. Wyse and M. Anderson. 2008. Animal Physiology. Sinauer Associates, Inc.

Hirsch, E.F., M. Jellinek and T. Cooper. 1964. Innervation of the systemic heart of the California hagfish. Circ. Res. 14: 212–217.

Holmes, E.B. 1975. Reconsideration of phylogeny of tetrapod heart. J. Morphol. 147: 209–228.

Icardo, J.M. 2006. Conus arteriosus of the teleost heart: dismissed, but not missed. Anat. Rec. 288A: 900–908.

Icardo, J.M., E. Colvee, M.C. Cerra and B. Tota. 2002a. Structure of the conus arteriosus of the sturgeon (Acipenser naccarii) heart. I: the conus valves and the subendocardium. Anat. Rec. 267: 17–27.

Icardo, J.M., E. Colvee, M.C. Cerra and B. Tota. 2002b. The structure of the conus arteriosus of the sturgeon (Acipenser naccarii) heart: II. The myocardium, the subepicardium, and the conus-aorta transition. Anat. Rec. 268: 388–398.

Icardo, J.M., J.L. Schib, J.L. Ojeda, A.C. Duran, A. Guerrero, E. Colvee, D. Amelio and V. Sans-Coma. 2003. The conus valves of the adult gilthead seabream (*Sparus auratus*). J. Anat. 202: 537–550.

Icardo, J.M., A. Guerrero, A.C. Duran, A. Domezain, E. Colvee and V. Sans-Coma. 2004. The development of the sturgeon heart. Anat. Embryol. (Berl) 208: 439–449.

Icardo, J.M., E. Brunelli, I. Perrotta, E. Colvee, W.P. Wong and Y.K. Ip. 2005. Ventricle and outflow tract of the African lungfish *Protopterus dolloi*. J. Morphol. 265: 43–51.

Janvier, P. 1996. The dawn of the vertebrates: characters versus common ascent in the rise of current vertebrate phylogenies. Palaeontology 39: 259–287.

Janvier, P. 2011. Living primitive fishes and fishes from deep time. pp. 1–51. *In*: A.P. Farrell (ed.). Fish Physiology: Primitive Fishes. Academic Press, San Diego.

Jensen, D. 1965. The aneural heart of the hagfish. Ann. N. Y. Acad. Sci. 127: 443–458.

Johansen, K. and C. Lenfant. 1968. Respiration in the African lungfish *Protopterus aethiopicus*. II. Control of breathing. J. Exp. Biol. 49: 453–468.

Johansen, K., D.L. Franklin and R.I. Vancitte. 1966. Aortic blood flow in free-swimming elasmobranchs. Comp. Biochem. Physiol. 19: 151–160.

Johansen, K., D. Hanson and C. Lenfant. 1970. Respiration in a primitive air breather, *Amia calva*. Respir. Physiol. 9: 162–174.

Jones, D.R. and M.H. Braun. 2011. Design and physiology of the heart: the outflow tract from the heart. pp. 1015–1029. *In*: A.P. Farrell (ed.). Encyclopedia of Fish Physiology—From Genome to Environment. Academic Press, San Diego.

Kapoor, B.G. and B. Khanna. 2004. Ichthyology Handbook. Springer, New York.

Kardong, K. 2006. Vertebrates: Comparative Anatomy, Function, Evolution. McGraw Hill, Boston.

Kikugawa, K., K. Katoh, S. Kuraku, H. Sakurai, O. Ishida, N. Iwabe and T. Miyata. 2004. Basal jawed vertebrate phylogeny inferred from multiple nuclear DNA-coded genes. BMC Biol. 2: 3.

Kuraku, S., A. Meyer and S. Kuratani. 2009. Timing of genome duplications relative to the origin of the vertebrates: did cyclostomes diverge before or after? Mol. Biol. Evol. 26: 47–59.

Lauder, G.V. and K.F. Liem. 1983. The evolution and interrelationships of the actinopterygian fishes. Bull. Mus. Comp. Zool. 150: 95–197.

Meyer, A. and M. Schartl. 1999. Gene and genome duplications in vertebrates: the one-to-four (-to-eight in fish) rule and the evolution of novel gene functions. Curr. Opin. Cell Biol. 11: 699–704.

Millot, J., J. Anthony and D. Robineau. 1978. Anatomie de *Latimeria chalumnae*. Appareil digestif, Appareil respiratoire, Appareil urogenital, Glandes endocrines, Appareil circulatoire, Teguments, Ecailles, Conclusions generales. Paris, Centre national de la recherche scientifique.

Moorman, A.F., V.M. Christoffels, R.H. Anderson and M.J.B. van den Hoff. 2007. The heart-forming fields: one or multiple? Philos. Trans. R. Soc. London B 362: 1257–1265.

Müller, J. 1846. Über den Bau und die Grenzen der Ganoiden und über das natürliche System der Fische. Abhandlungen der Deutschen Akademie der Wissenschaften zu Berlin 1844: 117–216.

Nakao, T., S. Suzuki and M. Saito. 1981. An electron microscopic study of the cardiac innervation in larval lamprey. Anat. Rec. 199: 555–563.

Nelson, J.S. 2006. Fishes of the World. Wiley, New York.

Nevis, K., P. Obregon, C. Walsh, B. Guner-Ataman, C.G. Burns and C.E. Burns. 2013. Tbx1 is required for second heart field proliferation in zebrafish. Dev. Dyn. 242: 550–559.

Noack, K., R. Zardoya and A. Meyer. 1996. The complete mitochondrial DNA sequence of the bichir (*Polypterus ornatipinnis*), a basal ray-finned fish: ancient establishment of the consensus vertebrate gene order. Genetics 144: 1165–1180.

Opitz, J.M. and E.B. Clark. 2000. Heart development: an introduction. Am. J. Med. Genet. 97: 238–247.

Parker, G.H. and F.K. Davis. 1899. The blood vessels of the heart in *Carcharias, Raja*, and *Amia*. Proc. Boston. Soc. Nat. Hist. 29: 1–356.

Parsons, C.W. 1929. Memoirs: the conus arteriosus in fishes. Q. J. Microsc. Sci. s2: 145–176.

Patterson, C. and D.E. Rosen. 1977. Review of ichthyodectiform and other Mesozoic teleost fishes and the theory and practice of classifying fossils. Bull. Am. Mus. Nat. Hist. 158: 81–172.

Potter, G.E. 1927. Respiratory function of the swim bladder in *Lepidosteus*. J. Exp. Zool. 49: 45–67.

Randall, D.J. and P.S. Davie. 1980. The hearts of urochordates and cephalochordates. pp. 41–59. *In*: G.H. Bourne (ed.). Hearts and Heart-Like Organs. Academic Press, New York.

Rodríguez, C., V. Sans-Coma, A.C. Grimes, B. Fernández, J.M. Arqué and A.C. Durán. 2013. Embryonic development of the bulbus arteriosus of the primitive heart of jawed vertebrates. Zoologischer Anzeiger—A Journal of Comparative Zoology 252: 359–366.

Romer, A.S. and T.S. Parsons. 1985. The Vertebrate Body. Holt Rinehart & Winston, New York.

Sans-Coma, V. 2007. New insight into the morphology of the cardiac outflow tract in chondrichthyans and actinopterygians: implications for heart evolution. 8th International Congress of Vertebrate Morphology, Paris.

Satchell, G.H. 1976. The circulatory system of air-breathing fish. pp. 105–123. *In*: G.M. Hughes (ed.). Respiration of Amphibious Vertebrates. Academic Press, London.

Satchell, G.H. 1991. Physiology and Form of Fish Circulation. Cambridge University Press, Cambridge.

Satchell, G.H. and M.P. Jones. 1967. Function of conus arteriosus in Port Jackson shark *Heterodontus portusjacksoni*. J. Exp. Biol. 46: 373–382.

Schmidt-Nielsen, K. 2006. Animal Physiology: Adaptation and Environment. Cambridge University Press, Cambridge.

Sedgwick, A., J.J. Lister and A.E. Shipley. 1905. A Student's Text book of Zoology: Amphioxus, Vertebrata. Swan Sonnenschein and Company, London.

Senior, H.D. 1907a. The conus arteriosus in *Tarpon atlanticus* (Cuvier & Valenciennes). Biol. Bull. 12: 146–151.

Senior, H.D. 1907b. Note on the conus arteriosus of *Megalops cyprinoides* (Broussonet). Biol. Bull. 12: 378–379.

Smith, W.C. 1918. On the process of disappearance of the conus arteriosus in teleosts. Anat. Rec. 15: 65–67.

Soufan, A.T., G. van den Berg, J.M. Ruijter, P.A. de Boer, M.J. van den Hoff and A.F. Moorman. 2006. Regionalized sequence of myocardial cell growth and proliferation characterizes early chamber formation. Circ. Res. 99: 545–552.

Takeuchi, M., M. Okabe and S. Aizawa. 2009. The genus *Polypterus* (bichirs): a fish group diverged at the stem of ray-finned fishes (Actinopterygii). Cold Spring Harb Protoc. 2009: pdb emo117.

Thompson, K.S. 1992. Living Fossil: The Story of the Coelacanth. W. Norton & Company, New York.

Tota, B. 1989. Myoarchitecture and vascularization of the elasmobranch heart ventricle. J. Exp. Zool. S2: 122–135.

Tota, B., V. Cimini, G. Salvatore and G. Zummo. 1983. Comparative study of the arterial and lacunary systems of the ventricular myocardium of elasmobranch and teleost fishes. Am. J. Anat. 167: 15–32.

van den Berg, G., R. Abu-Issa, B.A. de Boer, M.R. Hutson, P.A. de Boer, A.T. Soufan, J.M. Ruijter, M.L. Kirby, M.J. van den Hoff and A.F. Moorman. 2009. A caudal proliferating growth center contributes to both poles of the forming heart tube. Circ. Res. 104: 179–188.

Waldo, K.L., M.R. Hutson, C.C. Ward, M. Zdanowicz, H.A. Stadt, D. Kumiski, R. Abu-Issa and M.L. Kirby. 2005. Secondary heart field contributes myocardium and smooth muscle to the arterial pole of the developing heart. Dev. Biol. 281: 78–90.

Wright, G.M. 1984. Structure of the Conus Arteriosus and Ventral Aorta in the Sea Lamprey, *Petromyzon marinus*, and the Atlantic Hagfish, *Myxine glutinosa*—Microfibrils, a Major Component. Canadian Journal of Zoology-Revue Canadienne De Zoologie 62: 2445–2456.

Yamauchi, A. 1980. Fine structure of the fish heart. pp. 119–148. *In*: G.H. Bourne (ed.). Hearts and Heart-Like Organs. Vol. 1: Comparative Anatomy and Development. Academic Press, New York.

Zaccone, D., A.C. Grimes, A. Sfacteria, M. Jaroszewska, G. Caristina, M. Manganaro, A.P. Farrell, G. Zaccone, K. Dabrowski and F. Marino. 2011. Complex innervation patterns of the conus arteriosus in the heart of the longnose gar, *Lepisosteus osseus*. Acta Histochem. 113: 578–584.

Zaccone, G., A. Mauceri, M. Maisano and S. Fasulo. 2009a. Innervation of lung and heart in the ray- finned fish, bichirs. Acta Histochem. 111: 217–229.

Zaccone, G., A. Mauceri, M. Maisano, A. Giannetto, V. Parrino and S. Fasulo. 2009b. Distribution and neurotransmitter localization in the heart of the ray-finned fish, bichir (*Polypterus bichir bichir* Geoffroy St. Hilaire 1802). Acta Histochem. 111: 93–103.

Zhang, H., Q.W. Wei, H. Du, L. Shen, Y.H. Li and Y. Zhao. 2009. Is there evidence that the Chinese paddlefish (*Psephurus gladius*) still survives in the upper Yangtze River? Concerns inferred from hydroacoustic and capture surveys, 2006–2008. J. Appl. Ichthyol. 25: 95–99.

Zhou, Y., T.J. Cashman, K.R. Nevis, P. Obregon, S.A. Carney, Y. Liu, A. Gu, C. Mosimann, S. Sondalle, R.E. Peterson, W. Heideman, C.E. Burns and C.G. Burns. 2011. Latent TGF-beta binding protein 3 identifies a second heart field in zebrafish. Nature 474: 645–648.

Zummo, G. and F. Farina 1989. Ultrastructure of the conus arteriosus of *Scyliorhinus stellaris*. J. Exp. Zool. 2: 158–164.

8

Control of Breathing in Primitive Fishes

Michael S. Hedrick[1],* and Stephen L. Katz[2]

Introduction

The 'ancient' fishes represent a diverse group occupying a wide variety of marine and freshwater niches. The respiratory strategies of these fishes are likewise diverse, ranging from water breathing to bimodal breathing to obligate air-breathing. These primitive fish lineages were the first vertebrates to exploit atmospheric respiratory gases, in addition to dissolved gases in their aquatic milieu, well prior to the colonization of terrestrial habitats by amphibians. Indeed, the ancient fish lineages are viewed as the archetypes for the physiological adaptations that then were leveraged in the emergence of amphibious life; likewise, the extant descendants of these primitive fishes are often the subjects of study as physiological models for this same evolutionary transition. While much research has been focused on the consequences of this transition for gas exchange physiology, it cannot be forgotten that there are also biomechanical and physical consequences from transitioning from gill to lung ventilation.

Ventilation, whether gill ventilation or lung ventilation, serves to maintain homeostasis of blood and tissue O_2 and CO_2 (and pH) levels. However, in the aquatic environment, buoyancy and maneuverability are an important aspect of the behavior and natural history of these animals and, if air-breathing occurs,

[1] Department of Biological Sciences, California State University, East Bay, Hayward, CA 94542 USA.

[2] School of the Environment, Washington State University, Pullman, WA 99164 USA.
   E-mail: steve.katz@wsu.edu

* Corresponding author: michael.hedrick@csueastbay.edu

changing lung volumes will change buoyancy. Control of breathing, therefore, requires that animals be capable of sensing and responding to changes in external and internal partial pressures of respiratory gases (O_2 and CO_2) via chemoreceptors as well as sensing changes in the volume of the 'lung' via mechanoreceptors. These sensory feedback mechanisms are integrated into Central Nervous System (CNS) respiratory centers and alter motor output from the CNS to muscles that regulate gill ventilation or drive observable air-breathing behaviors.

The two major clades of Osteichthyes, the Actinopterygii (ray-finned fishes) and Sarcopterygii (lobe-finned fishes), diverged sometime in the late Silurian (438–408 million years ago), and it is generally conceded that air-breathing evolved in both major lineages prior to the fish-tetrapod transition in the Devonian approximately 385 to 360 million years ago (Long 1995; Graham 1997; Clack 2009, 2012). Air-breathing species persist among the Actinopterygii and Sarcopterygii and the extant primitive forms in these groups, therefore, exhibit primitive characteristics that may provide clues to the evolution of air-breathing. Among the extant actinopterygians, the most primitive forms are air-breathing, including bichirs (*Polypterus*), gars (*Lepisosteus*) and the bowfin (*Amia*). Among the sarcopterygians, the only remaining extant air-breathing species are the dipnoan lungfishes (*Protopterus, Lepidosiren* and *Neoceratodus*). The control of air-breathing in these primitive fishes therefore provides a convenient backdrop for discussing the evolution of air-breathing and the possible factors that gave rise to aerial ventilation and the transition to terrestriality in vertebrates (Clack 2012).

Although air-breathing in fishes occurs widely, and has evolved independently many times (Graham 1997), the use of a 'lung' for aerial ventilation is the common feature that links all of the primitive fishes with respect to air-breathing structures. This is in contrast to the many different teleost groups that have secondarily evolved air-breathing and have adapted a variety of structures used for gas exchange (Graham 1997). Although the driving forces that led to the evolution of air-breathing are unknown, the prevailing view is that hypoxic environments, particularly in the middle to late Devonian, was a significant factor in the rise of air-breathing forms in the stem tetrapodomorph fish that represent the primitive condition for tetrapod ancestors (Graham 1997; Clack 2007, 2012).

Our primary focus in this chapter will be on those primitive forms capable of air-breathing since we are interested in the archetypes that reflect designs on the major lineages of Osteichthyes. Therefore, our focus is on a few species of primitive fishes that provide insight into the evolution of air-breathing in those lineages. Although air-breathing has occurred in a large number of diverse groups of fishes (Graham 1997), in most cases these animals are modern, derived teleost fishes that are secondarily adapted to air-breathing and do not represent stem lineages within the Osteichthyes.

Several excellent reviews on the control of breathing in the primitive fishes are available (Graham 1994; Smatresk 1994; Graham and Lee 2004; Long and

Gordon 2004; Coolidge et al. 2007), thus it is not our intention to review this literature in detail. Instead, we will briefly review the various mechanisms for air-breathing, the chemoreceptor and mechanoreceptor inputs that control air-breathing in these fishes, and then devote the bulk of the chapter on a select group of primitive air-breathing fishes that provide insight into the evolution of air-breathing. In particular, we will focus on empirical data from the bowfin (*Amia calva*) a primitive actinopterygian fish, to point out how this fish uses separate and distinct feedback inputs to control air-breathing. Data from previous studies suggest that *Amia* use blood chemoreceptor and lung mechanoreceptor afferent inputs to regulate two distinct air-breathing mechanisms; one mechanism for gas exchange and the other mechanism to regulate buoyancy (Hedrick and Jones 1993, 1999). We test this hypothesis with a computer model using empirical data from *Amia* and from other sources to illustrate the importance of these separate inputs for controlling air-breathing in primitive fishes and to point to future directions for study within the primitive air-breathing fishes.

Mechanisms of Air-Breathing in Ancient Fishes

Although details vary, all primitive fishes (and amphibians) use a positive-pressure buccal force pump to ventilate the lungs. A variety of techniques, including electromyography, X-ray cine film and pressure measurements, have established the use of the buccal force pump as a common feature of the air-breathing mechanism in actinopterygean and sarcopterygean fishes (McMahon 1969; Liem 1988, 1989). One exception seems to be a recoil aspiratory-like mechanism in the polypterid fishes *Polypterus* and *Erpetoichthys* (Brainerd et al. 1989). This aspiratory mechanism is a unique mechanical feature of the ganoid scales and body wall and is not related to the aspiratory mechanism for lung ventilation in tetrapods.

The buccal force pump lung ventilation mechanism has been described as a two-stroke buccal pump, characteristic of the sarcopterygian fishes (and amphibians) or a four-stroke buccal pump mechanism, characteristic of the actinopterygian fishes (Brainerd et al. 1993). The two-stroke mechanism, shared by the Dipnoi and Amphibia, is characterized by the buccal cavity making two-movements, one expansion and one compression, during the ventilatory cycle (Bishop and Foxon 1968; McMahon 1969; de Jongh and Gans 1969). In the two-stroke mechanism, buccal expansion is used to fill the buccal cavity while expiration occurs during this phase by opening the glottis and allowing lung air to escape through the mouth (Dipnoi) or nares (Amphibia); the inspired air is then forced into the lung through the open glottis by buccal compression. Therefore, both inspiration and expiration occur during the two-stroke mechanism. By contrast, the four-stroke buccal pump mechanism is characterized by four movements (two expansion, two compression) of the buccal cavity during an air-breathing cycle. In this scenario, air is transferred from the lung to the buccal cavity during buccal expansion, exhaled by buccal compression through the mouth (e.g., *Amia*, Deyst and Liem 1985) or opercular cavity (e.g., *Lepisosteus*, Rahn et al. 1971 and *Polypterus*, Brainerd et al. 1989) fresh air is inhaled through the mouth with a second buccal expansion, and finally transferred to the lung with a second buccal compression. An important exception to this scheme, which we view as highly significant (see below), is that *Amia* use both a two-stroke mechanism and a four-stroke mechanism, depending on the apparent 'needs' of the animal (Hedrick and Jones 1993, 1999). The four-stroke mechanism in *Amia*, originally described by Wilder (1877), was subsequently confirmed by later studies (Johansen et al. 1970; Randall et al. 1981; Deyst and Liem 1985; Liem 1988, 1989). The type of two-stroke mechanism used by *Amia* was first described by Hedrick and Jones (1993) which prompted a classification scheme to distinguish these two breath types in this species. The four-stroke mechanism for *Amia* was named a 'type I' breath and the two-stroke mechanism was named a 'type II' breath (Fig. 1). The type I breath therefore consists of an exhalation followed by inhalation during one breath cycle (i.e., a four-stroke mechanism), whereas the type II breath consists of an inhalation only, with no associated exhalation (Hedrick and Jones 1993). It is unknown whether any of the other primitive actinopterygians use a two-stroke mechanism like the one described for *Amia*, but some teleostean air-breathers use a two-stroke mechanism (*Hoplerythrinus* and *Gymnotus*, Graham 1997); likewise, not all amphibians use a two-stroke lung ventilation pattern (*Amphiuma*, Brainerd and Ditelberg 1993 and *Xenopus*, Brett and Shelton 1979). Thus, the two-stroke, four-stroke distinction may have little usefulness as a character state for determining evolutionary relationships but, rather, may reflect the plasticity and variation of centrally-generated motor patterns among the Osteichthyes.

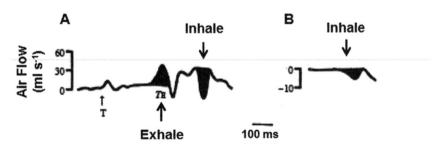

Fig. 1. Records of air flow (ml s^{-1}) for *Amia* air-breathing in normoxic conditions. (A) Type I air-breath illustrating both exhalation and inhalation. An artifact was created during the transfer (T) of gas from the lung to the buccal cavity prior to exhalation. The expiratory time interval (T$_E$) is also shown. (B) Type II air-breath for the same animal illustrating inhalation only for this breath type (see Text). From Hedrick and Jones (1993).

Lungs as Gas Exchangers and Buoyancy Organs

A common feature that links all of the primitive fishes is the presence of a 'lung' that is used for gas exchange and buoyancy regulation in the aquatic environment. This is in contrast to the various gas exchangers that evolved secondarily in the teleost lineages. The term 'lung' is not consistently used to describe the air-breathing organ found in all the primitive fishes. Graham (1997) uses the term 'gas bladder' and suggests there are fundamental differences between a 'lung' and a 'gas bladder' including differences in embryonic origin, location of the glottis and differences in the pulmonary circulation. The question of homology between lungs and gas bladders has been extensively debated since Owen (1846) first described the homology between these two structures. However, a recent study using microCT technology has shown that pulmonary arteries are retained in *Amia* which supports the homology between lungs and gas bladders due to a shared vascular supply (Longo et al. 2013). In light of these recent findings, we will use the term 'lung' to describe the structure used in both lineages of Osteichthyes. It is generally agreed that lungs evolved first in a common ancestor of the Actinopterygii and Sarcopterygii (Clack 2012). Over the course of evolution and divergence of these two clades, the gas exchange function of the lung was intensified in the Sarcopterygii, leading to terrestriality in the first amphibians, but the buoyancy function of the lung was intensified in the Actinopterygii leading to the teleosts with a gas bladder that functions solely as a buoyancy organ (Liem 1988, 1989).

Although this evolutionary scenario may be accurate, most work has not addressed the physical and physiological problem of using a single organ (lung) for both gas exchange and buoyancy. This is a basic problem that must be overcome in all the primitive fishes, and *Amia* has evolved a respiratory strategy to overcome this problem that we outline below. The fundamental problem is that there is a functional conflict between a gas exchanger and a buoyancy (hydrostatic) organ (Gee and Graham 1978). Any air-filled cavity used as a gas exchanger will automatically change the density and, therefore, buoyancy of the animal (Alexander 1966). To be effective as a gas exchanger, a lung must permit oxygen diffusion through its thin, vascularized membrane. Oxygen diffusion from lung to blood is not replaced by equal volumes of CO_2 (Johansen 1970), therefore, the volume of the lung decreases and the animal becomes negatively buoyant. In order to remain neutrally buoyant, a fish must replace the lost volume either by secretion of gas into the lung or, if secretory mechanisms are too slow or not present, they must gulp air (Alexander 1966). Because efficient hydrostatic organs have low gas permeabilities, thus they are not useful as gas exchangers. The swim bladders of teleosts are usually lined with guanine crystals that prevent significant amounts of oxygen diffusion (Fange 1976). Secretory mechanisms, which regulate buoyancy in physoclistous fishes, are usually too slow or absent in physostomous fishes (Jones and Marshall 1953) which is characteristic of all primitive fishes that breathe air. In most physostomous fishes, air-breathing is essential for buoyancy regulation (Jones and Marshall 1953). The structural and functional conflict for the gas bladder or lung of primitive fishes is not well appreciated and has not received much attention in the literature.

Central Chemoreceptors in the Control of Air-Breathing

Respiratory rhythm generation of gill ventilatory activity is driven by brainstem neural circuits that activate respiratory motoneurons that, in turn, drive ventilatory musculature of the branchial and opercular cavities. Respiratory rhythm generating neurons are located in the brainstem of all vertebrates and provide rhythmic neural output in the absence of peripheral sensory input (Shelton et al. 1986). The first unequivocal support for this was found with an *in vitro* brainstem preparation from goldfish that generated spontaneous respiratory related neural activity (Adrian and Buytendijk 1931). Rovainen (1974) also used an *in vitro* brainstem preparation from lamprey to demonstrate that respiratory rhythm generation in this species may be driven by pacemaker-like neural activity because respiratory activity continued with blockade of synaptic inhibition, a typical requirement of network-driven rhythmic activity.

There has been very little work examining the potential rhythm-generating respiratory networks in any primitive fishes. Although gill ventilation appears to be driven continuously by brainstem neural circuits, as in goldfish, air-breathing behavior has been viewed as representing an 'on-demand' response to sensory input from peripheral chemoreceptors or mechanoreceptors (Shelton and Croghan 1988). A single study has examined the potential role for an air-breathing central pattern generator and its regulation by CO_2 chemoreceptors (Wilson et al. 2000). Their study found that isolated brainstem preparations from the gar *Lepisosteus* were capable of spontaneous motor outputs that occasionally resembled air-breathing responses in intact fish. However, the isolated brainstem also produced multiple fictive air-breaths that did not resemble the natural air-breathing behavior of gar. These fictive air-breathing motor outputs increased in frequency with exposure to hypercarbia, suggesting a role for central chemoreceptors in this species. This result contrasts with previous work in *Lepisosteus* exposed to hypercarbia which failed to elicit any increase in air-breathing (Smatresk and Cameron 1982b). In *Amia*, perfusion of a hypercarbic or acidic mock cerebrospinal fluid (CSF) surrounding the brain also failed to elicit air-breathing responses (Hedrick et al. 1991). Similar studies in the lungfish *Lepidosiren*, also using mock CSF perfusion, showed that these animals have central CO_2/pH receptors that regulate lung ventilation (Sanchez et al. 2001; Amin-Naves et al. 2007). These data indicate that central CO_2/pH sensitivity is associated with air-breathing in sarcopterygian fishes, but the results in actinopterygian fishes are equivocal. These data also indicate that lungfish and tetrapods share in common the central regulation of lung ventilation via brain CO_2/pH receptors; however, because many details are lacking in the location and mechanisms of the receptors in these two groups, and given the uncertain relationship of the lungfishes with the tetrapods (Takezaki et al. 2004), it is not clear if the regulation of ventilation by central chemoreceptors is homologous or the result of homoplasy.

Peripheral Receptors in the Control of Air-Breathing

Chemoreceptors

Control of breathing, whether gill ventilation or air-breathing, is influenced by feedback from peripheral and/or central nervous system receptors that respond to changes in PO_2, PCO_2 and/or pH. The major contrast between fishes and terrestrial vertebrates is that control of ventilation is dominated by peripheral chemoreceptors located on the gill arches, primarily in response to changes in PO_2, whereas ventilation in terrestrial vertebrates is primarily driven by central chemoreceptors in response to changes in PCO_2/pH (Jones and Milsom 1982).

Among the primitive air-breathing fishes, two respiratory strategies emerge: facultative air-breathing species with functional gills and lungs that switch between the two respiratory gas exchangers depending on the needs of the animal (e.g., *Lepisosteus*, *Amia* and *Neoceratodus*), and obligate air-breathing species that have morphologically reduced gills and acquire O_2 primarily through lung ventilation (e.g., *Protopterus* and *Lepidosiren*). In *Protopterus*, for example, approximately 90% of total O_2 uptake occurs by lung ventilation and about 10% of total O_2 uptake across the gills and skin (Lenfant and Johansen 1968; McMahon 1970). The distinction between facultative and obligate air-breathing is not always so clear cut, however, because some so-called facultative air-breathers, such as *Lepisosteus*, may be obligate air-breathers at higher temperatures or in severe aquatic hypoxia (Rahn et al. 1971; Smatresk and Cameron 1982a), while some so-called obligate air-breathers respond to aquatic hypoxia (Jesse et al. 1968).

In general, facultative air-breathing species, such as *Lepisosteus* and *Neoceratodus*, reduce gill ventilation and increase aerial ventilation in reponses to aquatic hypoxia (Johansen et al. 1967; Smatresk and Cameron 1982a; Fritsche et al. 1993), however, there are conflicting data on whether *Amia* reduce gill ventilation in response to aquatic hypoxia (Johansen et al. 1970; Randall et al. 1981; McKenzie et al. 1991). Some obligate air-breathing species have no gill ventilatory response to aquatic hypoxia (Johansen and Lenfant 1967, 1968), but some species do increase gill ventilation with aquatic hypoxia (Jesse et al. 1968). All facultative and obligate air-breathers studied increase aerial ventilation in response to aerial hypoxia or nitrogen injected into the lung (Burggren and Johansen 1968; Johansen and Lenfant 1968; Hedrick and Jones 1993; Seifert and Chapman 2006). These experiments suggest that many of the primitive air-breathing fishes have both internally-oriented and/or externally-oriented chemoreceptors that respond to hypoxia (Smatresk et al. 1986).

Denervation of cranial nerves innervating the gills of facultative and obligate air-breathing fishes has been used to test whether the O_2-chemoreceptors mediating the ventilatory reflexes are located on gill tissues. In *Amia*, cutting the gill arch and pseudobranch nerve supply from cranial nerves VII, IX and X abolished the short term (< 15 minutes) response to air-breathing hypoxia (McKenzie et al. 1991), but does not abolish air-breathing responses to hypoxia over longer time domains (Hedrick and Jones 1999). The study by Hedrick and Jones (1999) was noteworthy by showing that normal air breathing mechanics—the capture and transfer of inhaled gas—was compromised by gill arch denervation. This suggests that some aspect of afferent activity carried in gill arch nerves, most likely gill arch mechanoreceptors, is important for normal air-breathing mechanics. Branchial denervation in *Protopterus* also abolished the gill ventilatory response to hypoxia, but enhanced the air-breathing response to hypoxia (Lahiri et al. 1970). Collectively, these studies point to O_2-sensitive chemoreceptors located on the branchial arches mediating gill and/or air-breathing responses to either aquatic or aerial hypoxia. Although the specific chemoreceptors mediating these reflexes have not been conclusively identified, work with water-breathing teleosts implicates the neuroepithelial cells (NECs) as the putative O_2-sensitive chemoreceptors in gill tissue (Porteus et al. 2012, 2013). Some of the primitive air-breathing fishes, such as *Protopterus*, *Polypterus* and *Amia* have been shown to have NECs in the gills and/or lungs (Zaccone et al. 1989, 2007, 2008, 2012; Adriaensen et al. 1990; Goniakowska-Witalinska et al. 1995), although the roles of these receptors and their innervation in the control of breathing are unclear. More work is needed in this area to determine the structure and function of putative chemoreceptors in primitive fishes.

Mechanoreceptors

The lungs of ancient air-breathing fishes are known to have mechanoreceptors that sense changes in volume of these organs. These mechanoreceptors are generally characterized as Slowly-Adapting Receptors (SARs) or Rapidly-Adapting Receptors (RARs) based on their afferent firing patterns in response to inflation or deflation of these organs. In general, the afferent firing frequency of single unit neurons within the vagus nerve increases upon inflation and decreases upon deflation. For example, *Amia*, *Lepisosteus* and *Protopterus* have SARs that increase firing frequency of vagus nerve single units in response to lung inflation, and decrease firing frequency in response to lung deflation (DeLaney et al. 1983; Milsom and Jones 1985; Smatresk and Azizi 1987). *Protopterus* and *Lepisosteus* also have RARs that transiently increase firing frequency in response to lung inflation (DeLaney et al. 1983; Smatresk and Azizi 1987), but these receptors are not located in *Amia* (Milsom and Jones 1985). In addition, the SARs of *Protopterus* and *Lepisosteus* were also found to be inhibited by CO_2, but SARs in *Amia* did not display any CO_2 sensitivity. In this respect, the lung stretch receptors of *Protopterus* and *Lepisosteus*, but not *Amia*, display characteristics similar to the lung receptors found in terrestrial vertebrates.

One would expect *a priori* that lung stretch receptors play a role in inhibiting (inflation) or promoting (deflation) air-breathing activity in air-breathing fishes. A major question that arises is how the input from these receptors is integrated into the overall air-breathing behavior in these species, but there is very little known about CNS integration of any peripheral sensory input that affects air-breathing behavior. Lung deflation stimulates, and lung inflation inhibits, air-breathing responses in *Amia* (Johansen et al. 1970; Hedrick and Jones 1999), *Lepisosteus* (Smatresk and Cameron 1982c) and *Protopterus* (Pack et al. 1992). The reduction in lung breath duration in response to lung inflation in decerebrate *Protopterus* was abolished after vagotomy indicating that the responses were carried in vagal afferents from lung mechanoreceptors (Pack et al. 1992). In *Amia*, gas bladder deflation in conscious fish caused an increase in air-breathing frequency and frequency was significantly correlated with the volume of gas removed from the lung (Hedrick and Jones 1999); moreover, nearly every breath taken by the fish was a type II breath (inhalation only) which supports the hypothesis that type II breaths regulate buoyancy in *Amia* (see below). The correlation between frequency of type II breaths and volume of gas removed suggests that type II breaths are regulated primarily by stretch receptors in the lung (Hedrick and Jones 1999). In the same study, lung inflation above an apparent volume threshold of 3 ml kg^{-1} prevented an air breath attempt if the animal was approaching the water surface. Because the responses are on a short time scale, the interpretation is that volume-related sensory information plays a significant role in shaping air-breathing behavior.

Evolutionary Implications of Buoyancy-Related Breathing in *Amia*

The most parsimonious interpretation for the presence of type II breaths in *Amia* (Fig. 1) is that this evolved as a mechanism to replace gas volume in the lung lost through diffusion. This implies a buoyancy-regulating function for type II breaths in this species. It also questions the primary driving force for the evolution of air-breathing. Although hypoxia is generally considered the primary driving force for the evolution of air-breathing (Graham 1997; Clack 2009, 2012), a strong argument can be made for buoyancy as a primary driving force. It has been suggested that air-breathing motor patterns evolved from reorganization of pre-existing neuromuscular patterns for aquatic (gill) ventilation, feeding or coughing (McMahon 1969; Smatresk 1990). Primitive fishes evolved feeding motor patterns much earlier than air-breathing, so the neuromuscular motor patterns would have been in place before the evolution of air-breathing. Lauder (1980) suggested that the neuromuscular motor pattern for the feeding mechanism in *Amia* represents the plesiomorphic condition for the Teleostomi. The feeding pattern in *Amia* involves a suction mechanism in which negative pressures are generated by lowering the buccal floor (Lauder 1980). This pattern superficially resembles a Type II air-breath in which a single aspiratory mechanism creates a negative pressure in the buccal cavity to inhale air. The resemblance between the two patterns suggests that aspiration of air or prey into the buccal cavity may have evolved first. This would argue that air bladder volume regulation, rather than gas exchange, may have been the primary selection pressure for air-breathing in *Amia*.

A second line of argument for buoyancy as a primary driving force for the evolution of air-breathing is based on physics. The argument is a simple one: any gas taken into the body of an aquatic animal necessarily affects buoyancy, regardless of whether the gas is used for a respiratory function. In order for inhaled gas to assume a respiratory function, it must be in contact with a respiratory structure that is perfused with blood. It is more parsimonious to argue that air-gulping at the air-water interface would affect buoyancy first before any gas exchange function could be used. As *Amia* uses a distinct and possibly, 'simpler' mechanism to regulate gas bladder volume, hence buoyancy, this questions if the inhalation (type II) breath may be a more primitive air-breathing mechanism.

It has also been argued that surface (water) breathing behavioral mechanisms pre-dated the origin of air-breathing (Perry et al. 2001). An hypothesized sequence of events in the evolution of air-breathing organs is the origin of a central oscillator which drives an air-breathing motor pattern; the use of the posterior pharynx and an arterial branch of the sixth embryonic branchial arch; specialization of dorsal (Actinopterygii) or ventral (Sarcopterygii including Polypteriformes) pharyngeal regions for gas exchange; and finally differentiation of the two- and four-cycle respiratory mechanisms (Perry et al. 2001). Because *Amia* use both two- (Type II breaths) and four-stroke (Type I breaths) for buoyancy and gas-exchange, respectively, this would further question if a two-stroke mechanism (air-gulping) evolved first. Many non-air-breathing fish gulp or skim air from the oxygen-rich air-water interface (e.g., tambaqui, *Colossoma macropomum*) suggesting that this behavior likely evolved prior to gas bladders or lungs.

The earliest Paleozoic fishes were heavy-bodied, with dermal armoring (Carroll 1988; Long 1995) and, without a lung, would have been negatively buoyant. Gulping air would have the advantage of allowing these animals to exploit the three-dimensional water column thus providing niche expansion for greater food sources or avoiding predators. Thus, these animals would have been pre-adapted to use the inhaled air as a source of oxygen as perfused respiratory structures evolved.

Modeling Air-Breathing Behavior in *Amia*

The functional conflict between gas exchange and buoyancy, outlined above, and the observation that *Amia* has evolved different air-breathing mechanisms to overcome these conflicting demands (Jones and Hedrick 1993, 1999), motivated the development of a computer simulation to test the hypothesis that different breathing behaviors are deployed and regulated separately to meet separate physiological needs. The model, described below, uses empirically-derived observations from the literature for *Amia* (Johansen et al. 1970; Randall et al. 1981; Hedrick 1991; Hedrick and Jones 1993, 1999; Hedrick et al. 1994) or from the gar *Lepisosteus* (Rahn et al. 1971) where data for *Amia* are unavailable.

At present, there appears to be only one comparable model that has simulated the ventilatory pattern of intermittently breathing vertebrates (Shelton and Croghan 1988). Their model used data from the air-breathing electric eel (*Electrophorus electricus*), and an aquatic anuran, *Xenopus laevis*, to simulate intermittent breathing patterns in response to changes in lung, blood and tissue oxygen stores. Both models rely on monitoring a decline in a physiological variable (i.e., blood/lung PO_2 and lung volume) until a threshold is crossed, at which point there is a behavioral resetting. These models are both of a general type known as 'integrate and fire' models (Glass and Mackey 1988). A unique feature of the *Amia* model is the addition of gas bladder volume-related air breaths (type II breaths) suggested by the empirical data. Consequently, there are two independent inputs for triggering air breaths. A schematic diagram and flow chart of the essential features of the model is depicted in Fig. 2.

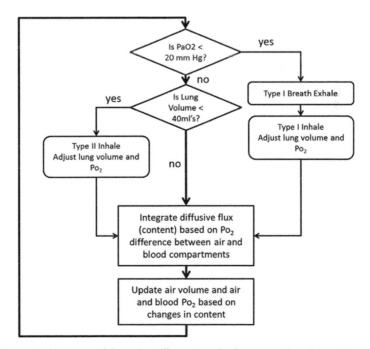

Fig. 2. Schematic diagram and flow chart illustrating the features of the air-breathing model for *Amia* based upon specific thresholds for blood PO_2 and lung volume.

In the *Amia* model, air breaths, either Type I or Type II, were modeled as discrete events triggered by independent thresholds from blood or lung receptor feedback (Fig. 3). It was assumed that following either Type I or Type II breath, O_2 diffused into the blood flowing past the lung (Q_L). Oxygen flux after an air breath was modeled in two discrete steps: (1) diffusion from lung to blood in pulmonary capillaries; (2) convective mixing of O_2 enriched blood leaving the lung and systemic venous return before arrival at the heart. Respiratory gas exchange ratios (R) in the air-breathing organs of air-breathing fishes are low owing to preferential loss of CO_2 from the gills to the water (Shelton et al. 1986). Thus, O_2 flux into the blood is accompanied by a reduction in lung volume and PO_2. Oxygen diffusion from the lung to blood was calculated as the product of the PO_2 gradient (ΔPO_2) and the lung diffusion capacity ($D_L O_2$). Values for $D_L O_2$ have not been determined for *Amia*, so this parameter was estimated from empirical measurements of $D_L O_2$ for the bullfrog (0.03 ml min^{-1} mmHg^{-1}; Glass et al. 1981). Lung surface area in *Lepisosteus* is approximately three-four times lower than in anuran amphibians (Rahn et al. 1971) so $D_L O_2$ was scaled proportionately lower for *Amia* (0.0085 · 0.03 ml min^{-1} mmHg^{-1}).

Overview of air-breathing model

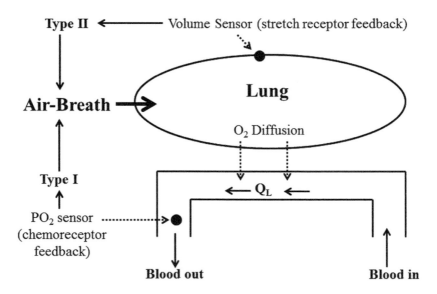

Fig. 3. Overview of the model for air-breathing in *Amia*. Type I and type II air breaths are shown in relation to the lung, a volume sensor for detecting changes in lung volume, lung blood flow (Q_L) and a blood PO_2 sensor (see text for details).

The convective step was simplified and modeled as a single vessel containing blood flowing past the lung (Q_L; Fig. 3). Oxygen flux from the lung was added to blood flow iteratively in discrete steps. The iteration process resulted in the integration of the O_2 flux or blood flow rates over each discrete time interval. Because the volume (ml O_2 or ml blood) is the integral of O_2 flux or blood flow rate, volumes of O_2 or blood were calculated at each iteration step. Thus, the convective step was modeled as the summation of blood O_2 content (Co_2) until the threshold was reached. Blood was therefore treated as a pool rather than blood flow rate in the strict sense. Changing blood flow rate would have the effect of changing the size of the blood pool.

Blood O_2 content at any level of blood PO_2 was calculated by interpolating previously published O_2 dissociation curves for *Amia* (Johansen et al. 1970). Blood flow toward the lung ('blood in'; Fig. 3) is analogous to systemic venous blood returning from the body and was kept constant in the model. Blood flow leaving the lung ('blood out'; Fig. 3) is a mixture of systemic venous blood and O_2-saturated blood diffusing from the lung. The amount of oxygenation of systemic venous blood would reflect the degree of extraction from respiring tissues, thus a constant value in the model is analogous to a constant metabolic rate. The issue of changing metabolic rate is not incorporated into the model since only the lung and its immediate environment is modeled. The two blood pools, systemic venous blood and blood leaving the lung, are mixed before entering the heart in *Amia*, prior to entering the ventral aorta and gill vasculature (Randall et al. 1981). Strictly, the PO_2 sensor placed in this location is analogous to a chemoreceptor sensing venous PO_2 between the lung and gill vasculature (Fig. 3). There is experimental evidence for a venous O_2 receptor in water-breathing fish (Barrett and Taylor 1984), and a venous O_2 receptor has been used in a model of ventilatory control in water-breathing fish (Taylor et al. 1968). Although prior work on modeling fluctuating O_2 content leaving the gill capillaries suggest that the location of a sensor is unimportant for detecting changes in PO_2, an arterial PO_2 sensor would actually perceive a stronger signal than a venous sensor (Katz 1996) for a given variance in venous PO_2. A threshold value of 20 mmHg was used for the PO_2 sensor, which represents an O_2 saturation of 60% measured for ventral aortic blood in *Amia* (Johansen et al. 1970; Randall et al. 1981).

Lung volume was set initially at 40 ml which is approximately 8% of body mass for a 500 g fish (Hedrick and Jones 1999). Type I breath volumes were initially set at 25 ml kg^{-1} (12.5 ml), based on the mean expired volume, assuming 100% transfer efficiency of inspired gas to the gas bladder (see below). A threshold of 1.5 ml (3 ml kg^{-1}; Hedrick and Jones 1999) less than the initial gas bladder volume triggered a simulated type II breath of 1.5 ml to replace the lost volume.

A key feature of the model is the incorporation of a variable error term associated with both breath types. This error term was necessary to account for the observation that inhaled volume of either breath type is not 100% efficient; that is, there is normally some loss of inspired gas during the transfer phase from the buccal cavity to the lung (i.e., inhaled gas bubbles of various size are seen to leave the opercular cavity during the transfer of gas from the buccal cavity to the lung). The error term was varied in the model as a percentage error around the mean breath volume for Type I and Type II breaths. For example, a 15% error of the mean Type II breath (1.5 ml) was calculated by allowing the program to randomly choose a value from the approximated Gaussian distribution of values ranging from 1.50 ± 0.225 ml (i.e., 1.275 to 1.725 ml).

Model Results and Comparison with Empirical Data

In the absence of any error associated with the simulated breaths (i.e., 100% efficiency), there was a regular alternation of Type I and Type II air breaths (Fig. 4). Type I air breaths occurred every 21 minutes after a type II air breath,

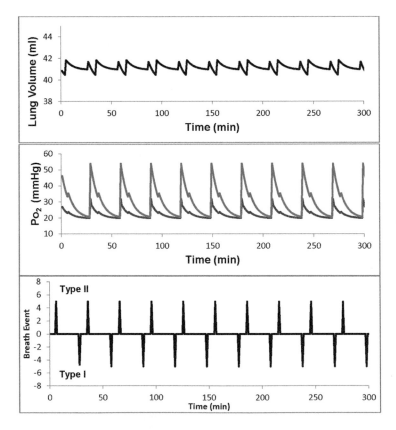

Fig. 4. Model results with no error in type I or type II air-breaths showing lung volume (top), PO$_2$ (middle) in lung (blue) and blood (red) and the occurrence of type I and type II breaths (bottom). With no error in either breath type, there is a regular alternation of type I and type II breaths over time.

and type II air breaths followed 9 minutes after each type I air breath. Efferent blood PO_2 increased to a maximum of 32 mmHg after each type I air breath and declined to the threshold value of 20 mmHg every 30 minutes which triggered another type I air breath (Fig. 4). This pattern is very similar to changes in PO_2 in the lung of *Polypterus* measured over several air-breathing cycles that have a distinct 'threshold' appearance (Abdel Magid et al. 1970). It is apparent that type II air breaths in the model had very little effect on PO_2 over time except to slightly lengthen the time to reach the blood PO_2 threshold. This is not surprising since each simulated type II air breath with an inspired volume of 1.5 ml of air contains about 0.3 ml O_2 and does not appreciably affect the lung to blood PO_2 gradient. With fixed breath volumes, the model produced a constant inter-breath interval for both breath types which is similar to the results of Shelton and Croghan (1988) for air-gulping in *Electrophorus*.

The addition of small amounts of random error (± 15% of mean breath volume) in both breath types had a marked effect on the simulated breathing pattern (Fig. 5). The timing between type I and type II air breaths with variable

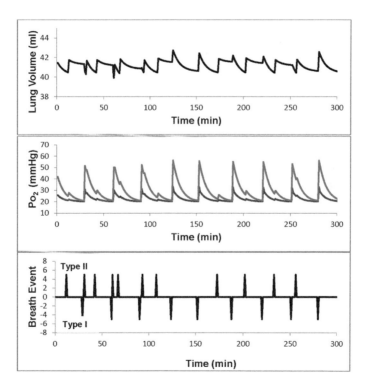

Fig. 5. Model results with ± 15% error in both type I and type II breaths with the same parameters shown as in Fig. 4. The addition of error produces greater fluctuations in lung volume (top), variable peaks of PO_2 in lung and blood (middle) and a more irregular appearance in the occurrence of air-breaths (bottom panel).

volumes was no longer fixed, which gave the overall breathing pattern a highly 'irregular' appearance. This produced a simulated breathing pattern that qualitatively resembled the breathing pattern for *Amia* in normoxic conditions (see Hedrick and Jones 1993). Both the model and empirical data showed an alternation between type I and type II air breaths, with type II air breaths often occurring at short time intervals after type I air breaths, or several type II air breaths following type I air breaths. This can be explained in the model by the transfer efficiency of type I air breaths that were much less than 100% which did not meet the threshold for lung volume; therefore, type II air breaths were triggered to compensate for the reduced volume. If type I air breaths were at the high end of the distribution for breath volume, then lung volume was also above threshold and the PO_2 threshold was reached before the volume threshold. In this case, there was no type II air breath between successive type I air breaths (Fig. 5).

The simulated pattern produced by the model that resembles empirical data can be partially explained by examining the effect on blood PO_2 over time. Variable amounts of inhaled air contain a constant fraction, but different amounts of O_2, which changed the gas bladder to blood PO_2 gradient with each breath. This created variable decay times for blood PO_2 in which one threshold or another (PO_2 vs. volume) was reached at different times depending on the blood PO_2 immediately following an air breath. Thus, even small variations in breath volume had profound effects on the qualitative pattern of breathing in the model. The error in breath volumes that changed the simulated breath interval from constant to highly variable is analogous to incorporating feedback delays in models of human respiratory control (Glass and Mackey 1988).

Despite the qualitatively irregular appearance of model results with the addition of an error term, time series analysis revealed that the underlying rhythmicity is preserved (Fig. 6). Analysis of 10 random data sets from the model, all with ± 15% error in both breath types, was associated with a mean interval of 30 minutes between successive type I air breaths (Fig. 6). Empirical data from undisturbed *Amia* also revealed a periodicity in the breathing pattern with type I breaths occurring at 30 minutes intervals in normoxia (Fig. 7; Hedrick et al. 1994). The agreement between model and empirical data with time series analysis supports the model's foundation hypothesis. Because breathing frequencies in the model and empirical data can be attributed to the interval between type I breaths, this is consistent with the hypothesis that the underlying periodicity during air-breathing is dependent on periodic feedback from O_2 chemoreceptors in contact with blood, rather than arising from the central nervous system (see Wilson et al. 2000).

Gas bladder volume-related breaths are an important element of the breathing pattern in *Amia* since they contribute to the appearance of irregular breathing and tend to mask the normal periodicity of type I breaths. Type II breaths appear to be responsible for making breath by breath adjustments of gas bladder volume on a finer temporal scale than type O_2-related breaths.

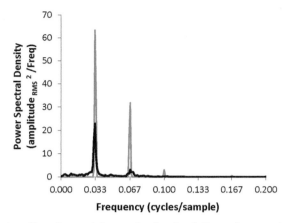

Fig. 6. Fourier spectra of breath event time series for simulations of normoxic conditions with no error in the ventilation events (gray), and with ± 15% error in both type I and type II breaths (black). The spectrum for the simulations with error is an average of 10 spectra, each of a single simulated time series of 1000 minutes sampled once per minute.

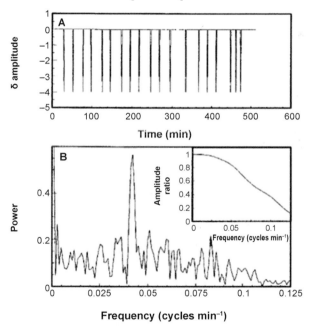

Fig. 7. Time series of one 8 hour record of an *Amia* in normoxia showing the occurrence of type I breaths only illustrated as discrete events (δ amplitude) and the power spectrum of that time series. The peak in the power spectrum reflects the periodic nature of the type I breaths. There is a dominant peak at 0.042 cycles min^{-1} indicating that a type I air breath occurs approximately every 24 minutes. Note that this spectrum reflects a single 400 minute time series rather than a spectrum average and so reflects more spectral energy across a range of harmonics. In spite of this, there is strong similarity between the model time series output (Fig. 6) and the air-breathing fish observations. See Hedrick et al. (1994) for details of techniques used to accommodate spectral analysis of point processes.

The model also accurately predicted the qualitative changes in breathing pattern when aerial hyperoxia (100% O_2) or hypoxia (8% O_2) was used in place of air (21% O_2) in simulated breaths. In aerial hyperoxia, each type II breath replaced the O_2 lost to the blood one-for-one, thus blood PO_2 remained above threshold and type I breaths were never triggered. The higher PO_2 gradient from gas bladder to blood resulted in a greater O_2 flux, causing more rapid reductions in volume that were corrected by more frequent type II breaths. The model response to aerial hyperoxia with type II breaths only is what occurred when live *Amia* were exposed to aerial hyperoxia (Hedrick and Jones 1993). With simulated aerial hypoxia the reverse occurred: the reduction in gas bladder PO_2 caused the blood PO_2 threshold to be reached more frequently than the gas bladder volume threshold, and the model produced a respiratory pattern dominated by more frequent type I breaths. These model responses also mirror observations of live *Amia* forced to breathe hypoxic gas (Hedrick and Jones 1993).

The similarity between model results and empirical data from undisturbed *Amia* under different conditions support the model's premise: ventilatory control in these fish is based upon regulating both lung volume and blood oxygen. If lung volume regulation is a critical factor in determining overall breathing patterns, it suggests that ventilatory regulation of buoyancy is an important feature of the physiology and ecology of this representative primitive fish. Given the diversity of circulatory and neural anatomy seen in other primitive air-breathing fish (Graham 1997), it is unclear to what extent the breathing patterns of other ancient air-breathing fishes are determined by the need to regulate buoyancy. What can be stated is that in these extant fish, thought to be representative of the primitive state that lead to terrestriality, the necessary control mechanisms to regulate buoyancy were already in place.

Acknowledgements

The authors wish to thank Drs. Peter Scheid and Johannes Piiper for helpful discussions and input at key times in the development of the modeling approach.

References

Abdel Magid, A.M., Z. Vokac and N. El Din Ahmed. 1970. Respiratory function of the swim bladder of the primitive fish *Polypterus senegalus*. J. Exp. Biol. 52: 27–37.

Adriaensen, D., D.W. Scheuermann, J.-P. Timmermans and M.H.A. De Groodt-Lassel. 1990. Neuroepithelial endocrine cells in the lung of the lungfish *Protopterus aethiopicus*. An electron- and fluorescence-microscopial investigation. Acta Anat. 139: 70–77.

Adrian, E.D. and F.J.J. Buytendijk. 1931. Potential changes in the isolated brain stem of the goldfish. J. Physiol. 71: 121–135.

Alexander, R.M. 1966. Physical aspects of swim bladder function. Biol. Rev. 41: 141–176.

Amin-Naves, J., H. Giusti, A. Hoffmann and M.L. Glass. 2007. Central ventilatory control in the South American lungfish, *Lepidosiren paradoxa*: contributions of pH and CO_2. J. Comp. Physiol. B. 177: 529–534.

Barrett, D.J. and E.W. Taylor. 1984. Changes in heart rate during progressive hypoxia in the dogfish, *Scyliorhinus canicula* L.: evidence for a venous oxygen receptor. Comp. Biochem. Physiol. 78A: 697–703.

Bishop, I.R. and G.E.H. Foxon. 1968. The mechanism of breathing in the South American lungfish, *Lepidosiren paradoxa*: A radiological study. J. Zool. (Lond.). 154: 263–271.

Brainerd, E.L., K.F. Liem and C.T. Samper. 1989. Air ventilation by recoil aspiration in Polypterid fishes. Science 246: 1593–1595.

Brainerd, E.L., J.S. Ditelberg and D.M. Bramble. 1993. Lung ventilation in salamanders and the evolution of vertebrate air-breathing mechanisms. Biol. J. Linn. Soc. 49: 163–183.

Brett, S.S. and G. Shelton. 1979. Ventilatory mechanisms of the amphibian, *Xenopus laevis*; the role of the buccal force pump. J. Exp. Biol. 80: 251–269.

Burggren, W.W. and K. Johansen. 1986. Circulation and respiration in lungfishes (Dipnoi). J. Morph. Suppl. 1: 217–236.

Carroll, R.L. 1988. Vertebrate Paleontology and Evolution. W.H. Freeman and Co. New York, NY.

Clack, J.A. 2007. Devonian climate change, breathing, and the origin of the tetrapod stem group. Int. Comp. Biol. 47: 510–523.

Clack, J.A. 2009. The fish-tetrapod transition: new fossils and interpretations. Evol. Edu. Outreach 2: 213–223.

Clack, J.A. 2012. Gaining Ground: The Origin and Evolution of Tetrapods. Indiana University Press, Bloomington, IN.

Coolidge, E., M.S. Hedrick and W.K. Milsom. 2007. Ventilatory systems. pp. 181–211. In: D.J. McKenzie, A.P. Farrell and C.J. Brauner (eds.). Primitive Fishes. Fish Physiology, Vol. 26. Academic Press, San Diego, CA.

deJongh, H.J. and C. Gans. 1969. On the mechanism of respiration in the bullfrog *Rana catesbeiana*: A reassessment. J. Morph. 127: 259–290.

DeLaney, R.G., P. Laurent, R. Galante, A.I. Pack and A.P. Fishman. 1983. Pulmonary mechanoreceptors in the dipnoi lungfish *Protopterus* and *Lepidosiren*. Am. J. Physiol. 244: R418–R428.

Deyst, K.A. and K.F. Liem. 1985. The muscular basis of aerial ventilation of the primitive lung of *Amia calva*. Respir. Physiol. 59: 213–223.

Fange, R. 1976. Gas exchange in the swimbladder. pp. 189–211. In: G.M. Hughes (ed.). Respiration of Amphibious Vertebrates. Academic Press, London.

Fritsche, R., M. Axelsson, C.E. Franklin, G.G. Grigg, S. Holmgren and S. Nilsson. 1993. Respiratory and cardiovascular responses to hypoxia in the Australian lungfish. Respir. Physiol. 94: 173–187.

Gee, J.H. and J.B. Graham. 1978. Respiratory and hydrostatic functions of the intestine of the catfishes *Hoplosternum thoracatum* and *Brochis splendens* (Callichthyidae). J. Exp. Biol. 74: 1–16.

Glass, L. and M.C. Mackey. 1988. From Clocks to Chaos. Princeton University Press, Princeton, NJ.

Glass, M.L., W.W. Burggren and K. Johansen. 1981. Pulmonary diffusing capacity of the bullfrog (*Rana catesbeiana*). Acta Physiol. Scand. 113: 485–490.

Goniakowska-Witalinska, L., G. Zaccone, S. Fasulo, A. Mauceri, A. Licata and J. Youson. 1995. Neuroendocrine cells in the gills of the bowfin *Amia calva*. An ultrastructural and immunocytochemical study. Folia Histochem. Cytobiol. 33: 171–177.

Graham, J.B. 1994. An evolutionary perspective for bimodal respiration: A biological synthesis of fish air breathing. Am. Zool. 34: 229–237.

Graham, J.B. 1997. Air-breathing Fishes: Evolution, Diversity and Adaptation. Academic Press, San Diego, CA.

Graham, J.B. and H.J. Lee. 2004. Breathing air in air: in what ways might extant amphibious fish biology relate to prevailing concepts about early tetrapods, the evolution of vertebrate air breathing, and the vertebrate land transition? Physiol. Biochem. Zool. 77: 720–731.

Hedrick, M.S. 1991. Air-breathing in the bowfin (*Amia calva* L.). Ph.D. Thesis, University of British Columbia, Vancouver, British Columbia.

Hedrick, M.S. and D.R. Jones. 1993. The effects of altered aquatic and aerial respiratory gas concentrations on air-breathing patterns in a primitive fish (*Amia calva*). J. Exp. Biol. 181: 81–94.

Hedrick, M.S. and D.R. Jones. 1999. Control of gill ventilation and air-breathing in the bowfin *Amia calva*. J. Exp. Biol. 202: 87–94.

Hedrick, M.S., M.L. Burleson, D.R. Jones and W.K. Milsom. 1991. An examination of central chemosensitivity in an air-breathing fish (*Amia calva*). J. Exp. Biol. 155: 165–174.

Hedrick, M.S., S.L. Katz and D.R. Jones. 1994. Periodic air-breathing behavior in a primitive fish revealed by spectral analysis. J. Exp. Biol. 197: 429–436.

Jesse, M.J., C. Shub and A.P. Fishman. 1967. Lung and gill ventilation of the African lungfish. Respir. Physiol. 3: 267–287.

Johansen, K. 1970. Air breathing fishes. pp. 361–411. *In*: W.S. Hoar and D.J. Randall (eds.). Fish Physiology Vol. IV. Academic Press. New York.

Johansen, K. and C. Lenfant. 1967. Respiratory function in the South American lungfish, *Lepidosiren paradoxa*. J. Exp. Biol. 46: 205–218.

Johansen, K. and C. Lenfant. 1968. Respiration in the African lungfish *Protopterus aethiopicus*. II. Control of breathing. J. Exp. Biol. 49: 453–468.

Johansen, K., C. Lenfant and G.C. Grigg. 1967. Respiratory control in the lungfish, *Neoceratodus forsteri* (Krefft). Comp. Biochem. Physiol. 20: 835–854.

Johansen, K., D. Hanson and C. Lenfant. 1970. Respiration in a primitive air breather *Amia calva*. Respir. Physiol. 9: 162–174.

Jones, D.R. and W.K. Milsom. 1982. Peripheral receptors affecting breathing and cardiovascular function in non-mammalian vertebrates. J. Exp. Biol. 100: 59–91.

Jones, F.R.H. and N.B. Marshall. 1953. The structure and functions of the teleostean swimbladder. Biol. Rev. 28: 16–83.

Katz, S.L. 1996. Ventilatory control in a primitive fish: signal conditioning via non-linear O_2 affinity. Respir. Physiol. 103: 165–175.

Lahiri, S., J.P. Szidon and A.P. Fishman. 1970. Potential respiratory and circulatory adjustments to hypoxia in the African lungfish. Fed. Proc. 29: 1141–1148.

Lauder, G.V. 1980. Evolution of the feeding mechanism in primitive actinopterygian fishes: a functional anatomical analysis of *Polypterus*, *Lepisosteus*, and *Amia*. J. Morph. 163: 283–317.

Lenfant, C. and K. Johansen. 1968. Respiration in the African lungfish *Protopterus aethiopicus*. I. Respiratory properties of blood and normal patterns of breathing and gas exchange. J. Exp. Biol. 49: 437–452.

Liem, K.F. 1988. Form and function of lungs: the evolution of air breathing mechanisms. Am. Zool. 28: 739–759.

Liem, K.F. 1989. Respiratory gas bladders in teleosts: functional conservatism and morphological diversity. Am. Zool. 29: 333–352.

Long, J.A. 1995. The Rise of Fishes. The Johns Hopkins University Press. Baltimore, MD.

Long, J.A. and M.S. Gordon. 2004. The greatest step in vertebrate history: a paleobiological review of the fish-tetrapod transition. Physiol. Biochem. Zool. 77: 700–719.

Longo, S., M. Riccio and A.R. McCune. 2013. Homology of lungs and gas bladders: insights from arterial vasculature. J. Morph. 274: 687–703.

McKenzie, D.J., S. Aota and D.J. Randall. 1991. Ventilatory and cardiovascular responses to blood pH, plasma PCO_2, blood O_2 content, and catecholamines in an air-breathing fish, the bowfin (*Amia calva*). Physiol. Zool. 64: 432–450.

McMahon, B.R. 1969. A functional analysis of the aquatic and aerial respiratory movements of an African lungfish, *Protopterus aethiopicus*, with reference to the evolution of the lung ventilation mechanism in vertebrates. J. Exp. Biol. 51: 407–430.

McMahon, B.R. 1970. The relative efficiency of gaseous exchange across the lungs and gills of an African lungfish *Protopterus aethiopicus*. J. Exp. Biol. 52: 1–15.

Milsom, W.K. and D.R. Jones. 1985. Characteristics of mechanoreceptors in the air-breathing organ of the holostean fish, *Amia calva*. J. Exp. Biol. 117: 389–399.

Owen, R. 1846. Lectures on the Comparative Anatomy and Physiology of Vertebrate Animals. Longman, Brown, Green, and Longmans, London.

Pack, A.I., R.J. Galante and A.P. Fishman. 1992. Role of lung inflation in control of air breath duration in African lungfish (*Protopterus annectens*). Am. J. Physiol. 262: R879–R884.

Perry, S.F., R.J.A. Wilson, C. Straus, M.B. Harris and J.E. Remmers. 2001. Which came first, the lung or the breath? Comp. Biochem. Physiol. 129A: 37–47.

Porteus, C.S., D.L. Brink and W.K. Milsom. 2012. Neurotransmitter profiles in fish gills: putative gill oxygen chemoreceptors. Respir. Physiol. Neurobiol. 184: 316–325.

Porteus, C.S., D.L. Brink, E.H. Coolidge, A.Y. Fong and W.K. Milsom. 2013. Distribution of acetylcholine and catecholamines in fish gills and their potential roles in the hypoxic ventilatory response. Acta Histochem. 115: 158–169.

Rahn, H., K.B. Rahn, B.J. Howell, C. Gans and S.M. Tenney. 1971. Air breathing of the garfish (*Lepisosteus osseus*). Respir. Physiol. 11: 285–307.

Randall, D.J., J.N. Cameron, C. Daxboeck and N. Smatresk. 1981. Aspects of bimodal gas exchange in the bowfin *Amia calva* L. (Actinopterygii: Amiiformes). Respir. Physiol. 43: 339–348.

Rovainen, C.M. 1974. Respiratory motoneurons in lampreys. J. Comp. Physiol. 94: 57–68.

Sanchez, A.P., A. Hoffmann, F.T. Rantin and M.L. Glass. 2001. Relationship between cerebro-spinal fluid pH and pulmonary ventilation of the South American lungfish, *Lepidosiren paradoxa* (Fitz). J. Exp. Zool. 290: 421–425.

Shelton, G. and P.C. Croghan. 1988. Gas exchange and its control in non-steady state systems: the consequences of evolution from water to air breathing in the vertebrates. Can. J. Zool. 66: 109–123.

Shelton, G., D.R. Jones and W.K. Milsom. 1986. Control of breathing in ectothermic vertebrates. *In*: A.P. Fishman, N.S. Cherniak, J.G. Widdicombe and S.R. Geiger (eds.). Handbook of Physiology, sect. 3, The Respiratory System, Vol. II, Control of Breathing, Part 2. Am. Physiol. Soc. Bethesda, MD.

Seifert, A.W. and L.J. Chapman. 2006. Respiratory allocation and standard rate of metabolism in the African lungfish, *Protopterus aethiopicus*. Comp. Biochem. Physiol., Pt. A. 143: 142–148.

Smatresk, N.J. 1990. Chemoreceptor modulation of endogenous respiratory rhythms in vertebrates. Am. J. Physiol. 259: R887–R897.

Smatresk, N.J. 1994. Respiratory control in the transition from water to air breathing in vertebrates. Am. Zool. 34: 264–279.

Smatresk, N.J. and J.N. Cameron. 1982a. Respiration and acid-base physiology of the spotted gar, a bimodal breather. I. Normal values and the response to severe hypoxia. J. Exp. Biol. 96: 263–280.

Smatresk, N.J. and J.N. Cameron. 1982b. Respiration and acid-base physiology of the spotted gar, a bimodal breather. II. Responses to temperature change and hypercapnia. J. Exp. Biol. 96: 281–294.

Smatresk, N.J. and J.N. Cameron. 1982c. Respiration and acid-base physiology of the spotted gar, a bimodal breather. III. Response to a transfer from freshwater to 50% sea water, and the control of ventilation. J. Exp. Biol. 96: 295–306.

Smatresk, N.J. and S.Q. Azizi. 1987. Characteristics of lung mechanoreceptors in spotted gar, *Lepisosteus oculatus*. Am. J. Physiol. 252: R1066–R1072.

Smatresk, N.J., M.L. Burleson and S.Q. Azizi. 1986. Chemoreflexive responses to hypoxia and NaCN in longnose gar: evidence for two chemoreceptor loci. Am. J. Physiol. 251: R116–R125.

Takezaki, N., F. Figueroa, Z. Zaleska-Rutczynska, N. Takahata and J. Klein. 2004. The phylogenetic relationship tetrapod, coelacanth, and lungfish revealed by the sequences of forty-four nuclear genes. Mol. Biol. Evol. 21: 1512–1524.

Taylor, W., A.H. Houston and J.D. Horgan. 1968. Development of a computer model simulating some aspects of the cardiovascular-respiratory dynamics of the salmonid fish. J. Exp. Biol. 49: 477–493.

Wilder, B.G. 1877. On the respiration of *Amia*. Proc. Am. Acad. Adv. Sci. 26: 306–313.

Wilson, R.J.A., M.B. Harris, J.E. Remmers and S.F. Perry. 2000. Evolution of air-breathing and central CO_2/H^+ respiratory chemosensitivity: new insights from an old fish? J. Exp. Biol. 203: 3505–3512.

Zaccone, D., M. Sengar, E.R. Lauriano, S. Pergolizzi, F. Macri, L. Salpietro, A. Favaloro, L. Satora, K. Dabrowski and G. Zaccone. 2012. Morphology and innervation of the teleost swim bladder and their functional evolution in non-teleostean lineages. Acta Histochem. 114: 763–772.

Zaccone, G., L. Goniakowska-Witalinska, J.M. Lauweryns, S. Fasulo and G. Taglaferro. 1989. Fine structure and serotonin immunohistochemistry of the neuroendocrine cells in the lungs of *Polypterus delhezi* and *P. ornatipinnis*. Basic Appl. Histochem. 33: 277–287.

Zaccone, G., A. Mauceri, M. Maisano, A. Giannetto, V. Parrino and S. Fasulo. 2007. Innervation and neurotransmitter localization in the lung of the Nile bichir *Polypterus bichir bichir*. Anat. Rec. 290: 1166–1177.

Zaccone, G., A. Mauceri, M. Maisano, A. Giannetto, V. Parrino and S. Fasulo. 2008. Neurotransmitter localization in the neuroepithelial cells and unipolar neurons of the respiratory tract in the bichir, *Polypterus bichir bichir* G. ST-HIL. Acta Histochem. 110: 143–150.

9

The Lung-Swimbladder Issue: A Simple Case of Homology— Or Not?

*Markus Lambertz** and *Steven F. Perry*

Introduction

The origin of evolutionary novelties is often difficult to determine, and the situation becomes complex when related taxa within a single evolutionary lineage show similar traits that intuitively and subjectively might be homologized *a priori*. Many classical anatomists, embryologists and zoologists, such as Karl Ernst von Baer (1792–1876), Martin Heinrich Rathke (1793–1860), Johann Friedrich Theodor 'Fritz' Müller (1822–1897) and of course Ernst Heinrich Philipp August Haeckel (1834–1919) to name a few, were well aware of the importance of developmental studies for comparative analyses. It was, however, specifically the advances in molecular genetics during the late 20th century that eventually (and fortunately) revived the ontogenetic approach under the heading of Evo-Devo (e.g., Laubichler and Maienschein 2007). This molecular biological input doubtlessly represents a very powerful suite of tools for the study of evolutionary questions (see also Shubin et al. 2009). Concerning the discussion of homology of a given structure, however, if taken alone, molecular data may result in overlooking the evolutionary role of the entire organism as a functional unit within its natural environment.

One classic example of doubtful homology and uncertain evolutionary origin that, in our opinion, remains insufficiently resolved concerns the lungs

Institut für Zoologie, Rheinische Friedrich-Wilhelms-Universität Bonn, Poppelsdorfer Schloss, 53115 Bonn, Germany.
* Corresponding author: lambertz@uni-bonn.de

and swimbladders of vertebrates (Perry 2007; Hsia et al. 2013; Cass et al. 2013; Longo et al. 2013). The question concerning lung-swimbladder homology can be reduced to: did one of them transform into the other? The present chapter critically addresses conceptual aspects of lung-swimbladder homology and aims at providing an integrative evolutionary scenario to explain the diversity of derivatives of the posterior pharynx among osteognathostomes (osteichthyians).

Structural Diversity of the Derivatives of the Posterior Pharynx

A tremendous anatomical diversity of lungs and swimbladders exists among vertebrates (Fig. 1). It ranges from simple, sac-like protrusions of the posterior pharynx to highly hierarchically structured and extremely complex organs.

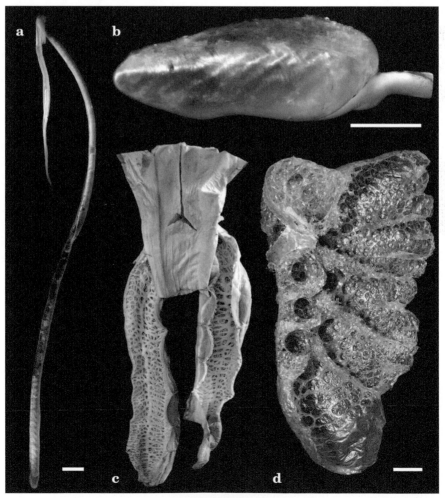

Fig. 1. contd....

Since recent detailed reviews on the ontogeny and anatomy of lungs and swimbladders can be found elsewhere (Maina 2002; Perry 2007; Hsia et al. 2013), we provide here only a generalized overview.

Vertebrate lungs can be defined as paired organs that embryologically derive from the ventral portion of the posterior pharynx (Moser 1902; Marcus 1937; Witschi 1956). According to this definition, lungs are found in the Polypteriformes (*Polypterus* spp., *Erpetoichthys calabaricus*), Dipnoi (*Lepidosiren paradoxa, Neoceratodus forsteri, Protopterus* spp.), Lissamphibia (frogs and toads, salamanders and newts, and caecilians), and Amniota (mammals, turtles, crocodiles, birds, tuatara, lizards and snakes). Although several of these taxa (e.g., *N. forsteri*, some lizards and the majority of snakes) only have a single lung, this is only true for the adult forms. Embryologically, their lungs also begin as paired structures, one of which either remains in an early developmental stage or may become entirely reduced during ontogeny (e.g., Neumayer 1930; Lambertz et al. 2015).

The curious, fat-filled and unpaired organ of the Coelacanthiformes (*Latimeria chalumnae, L. menadoensis*) should be considered autapomorphic and it remains unclear if or how it is related to lungs. In addition to its ventral origin with respect to the alimentary canal, there is also a small diverticulum located at its anterior portion that may be considered a rudimentary second 'fat-lung' (see Millot et al. 1978 for details). Thus, in spite of the apomorphic fat filling, the entire organ might be interpreted as a lung (ventral, paired, posterior pharynx), but embryological data, which so far are lacking, are ultimately needed to address this question appropriately.

A swimbladder, on the other hand, can be defined as an unpaired organ that embryologically derives from the dorsal portion of the posterior pharynx (e.g., Moser 1904; Wassnezow 1928; Neumayer 1930; Wassnetzov 1932; Liu 1993; Poulain and Ober 2011). Swimbladders are found in all non-polypteriform Actinopterygii (ray-finned fishes): the Actinopteri (Fig. 2). Exceptions to this generalization occur among certain benthic, deep-sea or highly pelagic teleosts, which have completely reduced their swimbladder (see McCune and Carlson 2004 for summary).

All lungs (with the exception of the fat-filled organ of coelacanths, if considered to be a lung) have some respiratory function, while few swimbladders do (see Graham 1997). In such respiratory (including pulmonoid) swimbladders, the tissue for gas exchange (parenchyma) frequently extends into the lumen in a very similar way than it does in lungs. These similarities

---

Contd.

Fig. 1. Overview of the diversity of derivatives of the posterior pharynx in osteognathostomes. (a) paired lungs of *Erpetoichthys calabaricus* (Polypteriformes) in dorsal view, cranial is towards the top. (b) unpaired swimbladder of *Acipenser ruthenus* (Acipenseridae) in right lateral view, cranial is towards the right. (c) paired lungs of *Andrias japonicus* (Lissamphibia: Caudata) in dorsal view (lungs opened), cranial is towards the top. From Lambertz et al. (2015). (d) left lung of *Trachemys scripta* (Amniota: Testudines) in ventral view (opened), cranial is towards the top. Scale bars equal 1 cm.

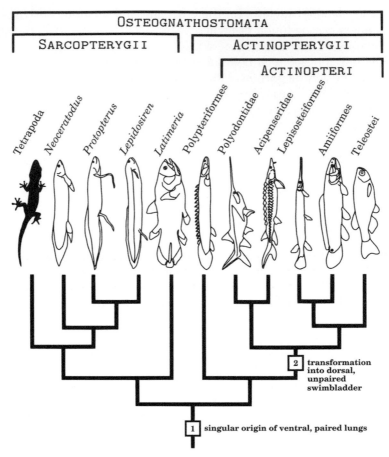

Fig. 2. Traditional view of the lung-swimbladder relationship among the Osteognathostomata. Lungs originated first (1) and later (2) were transformed into a swimbladder. Redrawn and modified after Cass et al. (2013), with kind permission of John Wiley & Sons, Inc.

even apply to the histological and ultrastructural levels (e.g., Icardo et al. 2015). Additional important functions for swimbladders include buoyancy control and communication (sound production as well as sound perception) (e.g., Alexander 1966). In many adult teleosts, the swimbladder even loses its embryological connection to the pharynx (*ductus pneumaticus*): the so-called physoclist state.

The Parsimonious Point of View

Emil Hans Willi Hennig's (1913–1976) phylogenetic systematics revolutionized the scientific classification of organisms (Hennig 1950; Schmitt 2013). His conceptual approach became especially attractive and was further refined once computer-aided tools became available that, according to the parsimony principle, calculate the 'cheapest' solution to explain how characters and character states evolved (see, e.g., Wägele 2001 for details). Based on this logic, the transformation of paired ventral lungs into an unpaired dorsal swimbladder usually is considered to require a single evolutionary step (Fig. 2). It is the favored hypothesis in the phylogenetic literature, and consequently has been adopted by textbook authors (e.g., Mickoleit 2004; Kardong 2012). Strictly speaking, however, it involves two evolutionary steps: the translocation from ventral to dorsal, and the complete reduction of one of the two *Anlagen*.

Vascular Supply and its Bearing on Lung-Swimbladder Homology

The vascularization of lungs and swimbladders recently has been extensively studied, and excellently described and illustrated by Longo et al. (2013). Blood vessels often serve as a conservative element that helps to establish homology (i.e., criterion of specific quality *sensu* Remane 1952). However, we disagree with one of the main conclusions of Longo et al. (2013). We do not concur that vascular supply of lungs and swimbladders by homologous blood vessels necessarily means that the organs are homologous, i.e., that one was derived from the other. This homologous arterial supply, in our view, could be the result of the fact that lungs and swimbladders are both derivatives of the posterior pharynx (compare above), which is supplied by branches of the sixth (= last) aortic arch (= fourth branchial arch in fishes). It simply represents the closest 'available' artery not only for the pharynx, but also for its derivatives: lungs and swimbladders.

Some Conceptual Aspects of Homology

There is a vast amount of literature dealing with the homology concept, its theoretical foundations and its application to biological systems (e.g., Owen 1843, 1848; Remane 1952; Osche 1973; Hall 1994; Wagner 2007; McCune and Schimenti 2012; Wagner 2014). In the present chapter, it is our specific aim to point out different hierarchical levels on which homology can be approached and to stress that the combination of all of these is essential before one can make a definitive statement about the evolutionary relationship between lungs and swimbladders among vertebrates.

We distinguish three principal hierarchical levels at which homology can be approached (Fig. 3). The ground level concerns the genetic code and the congruence (or difference) in the sequence of base pairs. The next level deals with the spatiotemporal interaction of genetic processes and how they eventually generate specific structures. The ultimate level concerns the organism as a whole, which is constantly interacting with its environment.

Basic questions at the sequence level for instance can be: is gene x, which is present in organism A also present in organism B? Are there gene duplications?

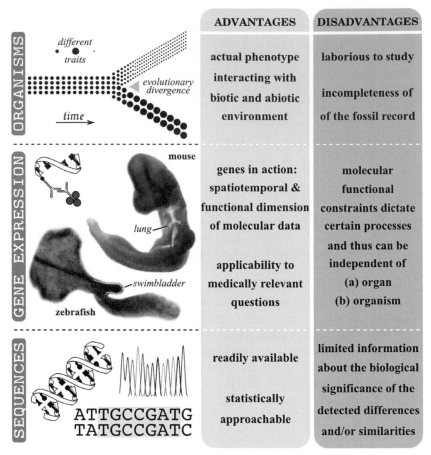

Fig. 3. Three hierarchical levels on which homology can be assessed. *In situ* hybridizations are adapted from Poulain and Ober (2011) (doi: 10.1242/dev.055921) and Sagai et al. (2009) (doi: 10.1242/dev.032714), with kind permission of The Company of Biologists Ltd.

What are the detailed differences in base-pair sequences? Recent technological advancements have made sequence data very accessible and it is relatively easy to acquire large data sets that can be statistically analyzed. However, one drawback of this approach is that it, on its own, rarely provides any direct information about the biological significance of the detected differences, even if the function(s) of the gene is (are) known.

The next hierarchical level, namely gene expression, provides information on the exact location and timing of gene activity, and its modulatory interplay. This can reveal the mechanisms behind the formation of a given phenotype and it consequently is much more meaningful for approaching the homology of morphological structures than knowing sequences alone. But exactly which genes should one look at? From a cell biological point of view, one faces certain functional problems. It is for instance to be expected that certain genes, especially highly conserved regulatory ones, are recruited during a large variety of ontogenetic stages, as well as during the formation of different organs and tissues. Shared expression patterns in such fundamental genes therefore do not necessarily shed an unequivocal light on the homology of a given structure. They are likely to be involved in the development of different organs and even in different (distantly related) organisms, simply because of similar 'products' (e.g., tubular outgrowth of epithelial tissue), which may require equivalent developmental pathways. We consider this to constitute a molecular functional constraint. What is critical for the assessment of homology is the specific congruence of the sequence of molecular processes that generate the identity of a given structure (compare Wagner 2014). 'What' regulates 'what', 'when', 'where' and 'how?'

The organismic and ultimate level is more than simply the sum of the previous two. One is facing an entire functional organism that is interacting with its biotic and abiotic environment. The structure in question represents only one integral part of the whole. At this level one is furthermore not limited to extant organisms (or model species), but can include fossils.

Scenario(s) for the Evolution of Lungs and Swimbladders

How can one explain the different structures that derive from the posterior pharynx in vertebrates? In order to have (apomorphic) swimbladders that are homologous to (plesiomorphic) lungs, one has to postulate a transformation of the ventral, paired organ into a dorsal, unpaired one. There are several possibilities how this could have taken place. The most intuitive ones are that either the posterior pharyngeal region experienced torsion and the ventral lungs became dorsal, or that the ventral lungs migrated dorsad. In either of these cases there must have been an additional modification: reduction of the second pulmonary *Anlage*. However, embryological data support neither the torsion nor the migration hypothesis. Swimbladders directly develop from the dorsal portion of the posterior pharynx (compare above). Several instances are known in which one lung becomes partially or even completely reduced and the remaining one lies dorsally. However, embryologically they always have a paired, ventral origin (compare above). It should be pointed out also that paired lungs often eventually occupy a dorsal position within the animal (e.g., in turtles), but that this is a secondary phenomenon independent of their ventral origin.

An alternative to the torsion or migratory hypotheses would be a 'switch' in the regulatory network that initiates the development of the lungs. The genetic cascade that starts pharyngeal outgrowth on the ventral side could have been modified to simply start on the dorsal side. Such a switch could have taken place from one generation to the next, as for instance in the case of the *situs inversus* phenomenon (e.g., Yokoyama et al. 1993; Bartolini et al. 2002). If evidence for such a switch eventually becomes available, lungs and swimbladders indeed would be homologous, or strictly speaking: ventral lungs transformed into a dorsal swimbladder. However, also in this scenario one would still have to postulate a secondary reduction of one of the pulmonary *Anlagen*.

A fundamentally different, yet equally plausible scenario is that lungs and swimbladders are derived separately from a 'respiratory pharynx' (Perry et al. 2001; Perry 2007; Hsia et al. 2013). The respiratory pharynx hypothesis assumes that multiple protrusions of the posterior pharynx, derivatives of former gill pouches, were present at the base of the Osteognathostomata (Fig. 4). Embryological data for instance by Wassnetzov (1932) indicate that

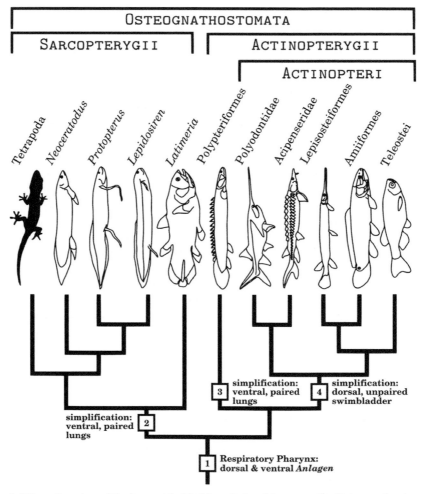

Fig. 4. Alternative view of the lung-swimbladder relationship among the Osteognathostomata. A respiratory pharynx with multiple dorsal and ventral *Anlagen* evolved first (1). Lungs originated twice (2 and 3) through a simplification, whereby only the ventral *Anlagen* were retained. A swimbladder evolved independently (4) through simplification and the retention of a dorsal *Anlage*. Diagram redrawn and modified after Cass et al. (2013), with kind permission of John Wiley & Sons, Inc.

separate ventral and dorsal *Anlagen* indeed are simultaneously present in fishes, amphibians and reptiles, although no case is known where lungs and swimbladders develop simultaneously. It is consequently assumed that sarcopterygians and polypteriforms (most likely independently) simplified this anatomical configuration and restricted it to the ventral part. The Actinopteri, on the other hand, are assumed to have followed a different path: they retained an unpaired dorsal element. This would explain why lungs directly develop on the ventral side and swimbladders on the dorsal side.

Conclusions

The issue of lung-swimbladder homology is not as simple as it might appear. The parsimony-based argument, in our view, is not compelling. The evolution of a swimbladder requires two steps if lungs were plesiomorphically present: the translocation from ventral to dorsal, and the reduction of one lung. The respiratory pharynx hypothesis, on the other hand, requires only one additional step. Although numerically less parsimonious, the latter hypothesis requires exclusively reductions/simplifications, which frequently occur in repetitive structures (e.g., body segmentation, number of gills, excretory organs, etc.).

We are convinced that the lung-swimbladder issue can only be resolved through a detailed understanding of the genetic mechanisms that underlie the formation of the two organs. The regulatory elements that initiate the development are of critical importance, as they determine whether the organ develops on the ventral or on the dorsal side. Genes that are active and shared during the later, visible growth of lungs and swimbladders are probably less important for assessing the homology. It is not surprising that several genes involved in epithelial tube formation are the same in closely related taxa. Even in distantly related taxa such as mammals and insects, the same molecular mechanisms are involved in pulmonary airway branching and tracheae formation, respectively (e.g., Behr 2010). Recent studies such as those of Zheng et al. (2011) and Cass et al. (2013) are crucial in that they provide important data on the transcriptome and even on parts of the genetic regulation during lung and swimbladder development, respectively. While such studies represent important steps in the right direction, we strongly believe that more research is needed in order to reveal the complete developmental pathways.

What ultimately is needed to conclusively answer the question of lung-swimbladder homology is the identification of the entire genetic cascade that leads to both organs: a proper mechanistic ontogenetic framework that includes data from sarcopterygians and polypteriforms and actinopterians. Only then can a meaningful comparison be drawn, which eventually might indicate whether the lungs 'switched' towards a dorsal swimbladder, or whether both organs are independent derivatives of the posterior pharynx.

Acknowledgments

We wish to thank Giacomo Zaccone for the invitation to contribute this chapter and for his and the other editors' patience until it eventually was written. The *Institut für Zoologie* (Bonn) in general and Michael H. Hofmann in particular are thanked for their ongoing support.

References

Alexander, R. McN. 1966. Physical aspects of swimbladder function. Biol. Rev. 41: 141–176.
Bartolini, L., J.-L. Blouin, Y. Pan, C. Gehrig, A.K. Maiti, N. Scamuffa, C. Rossier, M. Jorissen, M. Armengot, M. Meeks, H.M. Mitchison, E.M.K. Chung, C.D. Delozier-Blanchet, W.J. Craigen and S.E. Antonarakis. 2002. Mutations in the *DNAH11* (axonemal heavy chain dynein type 11) gene cause one form of *situs inversus* totalis and most likely primary ciliary dyskinesia. Proc. Natl. Acad. Sci. USA 99: 10282–10286.
Behr, M. 2010. Molecular aspects of respiratory and vascular tube development. Respir. Physiol. Neurobiol. 173(Suppl.): S33–S36.
Cass, A.N., M.D. Servetnick and A.R. McCune. 2013. Expression of a lung developmental cassette in the adult and developing zebrafish swimbladder. Evol. Dev. 15: 119–132.
Graham, J.B. 1997. Air-Breathing Fishes—Evolution, Diversity, and Adaptation. Academic Press, San Diego.
Hall, B.K. (ed.). 1994. Homology—The Hierarchical Basis of Comparative Biology. Academic Press, San Diego.
Hennig, W. 1950. Grundzüge einer Theorie der phylogenetischen Systematik. Deutscher Zentralverlag, Berlin.
Hsia, C.C.W., A. Schmitz, M. Lambertz, S.F. Perry and J.N. Maina. 2013. Evolution of air breathing: oxygen homeostasis and the transitions from water to land and sky. Compr. Physiol. 3: 849–915.
Icardo, J.M., E. Colvee, E.R. Lauriano, G. Capillo, M.C. Guerrera and G. Zaccone. 2015. The structure of the gas bladder in the spotted gar, *Lepisosteus oculatus*. J. Morph. 276: 90–101.
Kardong, K.V. 2012. Vertebrates—Comparative Anatomy, Function, Evolution. 6th ed. McGraw-Hill, New York.
Lambertz, M., K. Grommes, T. Kohlsdorf and S.F. Perry. 2015. Lungs of the first amniotes: why simple if they can be complex? Biol. Lett. 11: 20140848.
Laubichler, M.D. and J. Maienschein (eds.). 2007. From Embryology to Evo-Devo: A History of Developmental Evolution. MIT Press, Cambridge and London.
Liu, W.S. 1993. Development of the respiratory swimbladder of *Pangasius sutchi*. J. Fish Biol. 42: 159–167.
Longo, S., M. Riccio and A.R. McCune. 2013. Homology of lungs and gas bladders: insights from arterial vasculature. J. Morphol. 274: 687–703.
Maina, J.N. 2002. Functional Morphology of the Vertebrate Respiratory Systems. Science Publishers, Enfield.
Marcus, H. 1937. Lungen. pp. 909–988. *In*: L. Bolk, E. Göppert, E. Kallius and W. Lubosch (eds.). Handbuch der vergleichenden Anatomie der Wirbeltiere—Dritter Band. Urban & Schwarzenberg, Berlin und Wien.
McCune, A.R. and R.L. Carlson. 2004. Twenty ways to loose your bladder: common natural mutants in zebrafish and widespread convergence of swim bladder loss among teleost fishes. Evol. Dev. 6: 246–259.
McCune, A.R. and J.C. Schimenti. 2012. Using genetic networks and homology to understand the evolution of phenotypic traits. Curr. Genomics 13: 74–84.
Moser, F. 1902. Beiträge zur vergleichenden Entwicklungsgeschichte der Wirbeltierlunge (Amphibien, Reptilien, Vögel, Säuger). Arch. mikrosk. Anat. 60: 587–668 + Pl. XXX–XXXIII.
Moser, F. 1904. Beiträge zur vergleichenden Entwicklungsgeschichte der Schwimmblase. Arch. Mikrosk. Anat. 63: 532–574 + Pl. XXII–XXV.

Mickoleit, G. 2004. Phylogenetische Systematik der Wirbeltiere. Verlag Dr. Friedrich Pfeil, München.

Millot, J., J. Anthony and D. Robineau. 1978. Anatomie de *Latimeria chalumnae*. Tome III. CNRS, Paris.

Neumayer, L. 1930. Die Entwicklung des Darms von Acipenser. Acta Zool. 11: 39–150 + Pl. I–V.

Osche, G. 1973. Das Homologisieren als eine grundlegende Methode der Phylogenetik. Aufsätze u. Red. Senckenb. Naturf. Ges. 24: 155–165.

Owen, R. 1843. Lectures on the Comparative Anatomy and Physiology of the Invertebrate Animals, delivered at the Royal College of Surgeons in 1843. Longman, Brown, Green and Longmanns, London.

Owen, R. 1848. On the Archetype and Homologies of the Vertebrate Skeleton. Richard and John E. Taylor, London.

Perry, S.F. 2007. Swimbladder-lung homology in basal osteichthyes revisited. pp. 41–54. *In*: M.N. Fernandes, F.T. Rantin, M.L. Glass and B.G. Kapoor (eds.). Fish Respiration and Environment. Science Publishers, Enfield.

Perry, S.F., R.J.A. Wilson, C. Straus, M.B. Harris and J.E. Remmers. 2001. Which came first, the lung or the breath? Comp. Biochem. Physiol. A 129: 37–47.

Poulain, M. and F.A. Ober. 2011. Interplay between Wnt2 and Wnt2bb controls multiple steps of early foregut-derived organ development. Development 138: 3557–3568.

Remane, A. 1952. Die Grundlagen des natürlichen Systems, der vergleichenden Anatomie und der Phylogenetik—Theoretische Morphologie und Systematik I. Akademische Verlagsgesellschaft Geest & Portig K.-G., Leipzig.

Sagai, T., T. Amano, M. Tamura, Y. Mizushina, K. Sumiyama and T. Shiroishi. 2009. A cluster of three long-range enhancers directs regional *Shh* expression in the epithelial linings. Development 136: 1665–1674.

Schmitt, M. 2013. From Taxonomy to Phylogenetics—Life and Work of Willi Hennig. Brill, Leiden.

Shubin, N., C. Tabin and S. Carroll. 2009. Deep homology and the origins of evolutionary novelty. Nature 457: 818–823.

Wagner, G.P. 2007. The developmental genetics of homology. Nat. Rev. Genet. 8: 473–479.

Wagner, G.P. 2014. Homology, Genes, and Evolutionary Innovation. Princeton University Press, Princeton and Oxford.

Wägele, J.-W. 2001. Grundlagen der Phylogenetischen Systematik, 2nd edn. Verlag Dr. Friedrich Pfeil, München.

Wassnetzov, W. 1932. Über die Morphologie der Schwimmblase. Zool. Jahrb. Abt. Anat. Ontog. Tiere 56: 1–36.

Wassnezow, W. 1928. Zur Frage über die Morphologie der Schwimmblase—Vorläufige Mitteilung. Anat. Anz. 66: 161–166.

Witschi, E. 1956. Development of Vertebrates. W.B. Saunders Company, Philadelphia—London.

Yokoyama, T., N.G. Copeland, N.A. Jenkins, C.A. Montgomery, F.F. Elder and P.A. Overbeck. 1993. Reversal of left-right asymmetry: a *situs inversus* mutation. Science 260: 679–682.

Zheng, W., Z. Wang, J.E. Collins, R.M. Andrews, D. Stemple and Z. Gong. 2011. Comparative transcriptome analyses indicate molecular homology of zebrafish swimbladder and mammalian lung. PLoS ONE 6: e24019.

10

The Gut and Associated Organs in the African Lungfish *Protopterus annectens*

José M. Icardo

Introduction

Lungfish (Dipnoi) are air breathing fish (Graham 1997) that thrive in freshwater in tropical areas of Africa (*Protopterus*), South America (*Lepidosiren*) and Australia (*Neoceratodus*). They live in low-flowing waters, often poor in oxygen, feeding on a wide variety of small preys such as worms, molluscs, crustaceans, and other fish. They also feed on plants and dead bodies, and are considered to be opportunistic omnivores (Almeida-Val et al. 2011; Kind 2011; Mlewa et al. 2011). The study of the anatomy and physiology of the feeding apparatus of lungfish has mostly been focussed on the comparative analysis of several structures such as skull, jaws, masticator musculature and tooth plates (see Bemis et al. 1986). Much less information is available on the organs in charge of the digestive processes.

Early studies (Parker 1892; Purkerson et al. 1975; Rafn and Wingstrand 1981) have described the gastrointestinal tract of the lungfish as a longitudinally organized organ which includes a very short oesophagus, an elongated stomach, a pyloric region, a spiral valve and a cloaca. This general organization can be applied to all lungfishes (see Icardo et al. 2010), although

Department of Anatomy and Cell Biology, University of Cantabria, 39011-Santander, Spain.
  E-mail: icardojm@unican.es

the nature of some of the chambers may have been erroneously interpreted in the past (see below).

On the other hand, lungfish have the ability to adapt to extreme climate changes. This is especially true for the individuals of the African genus, *Protopterus*. They are able to excavate a nest in the mud and to survive within a mucous cocoon through the tropical drought until the beginning of the next rainy season. This process, named aestivation, lasts several months and implies the depression of the general metabolism and the depression or suppression of most of the body functions such as those related to the heart and kidney (Burggren and Johansen 1986; Fishman et al. 1986). It also includes the absence of food intake and, hence, drastic downregulation of the intestinal function. Curiously, the general organization of the gut is not modified during the aestivation process (Icardo et al. 2010). In this regard, it must be underscored that the gut is very sensitive to diet modifications, being able to undergo remarkable changes in structure and function in response to food restriction (Secor 2005). Downregulation of intestinal performance, loss of intestinal mass, reduction in microvillous height and cell death are among the commonest modifications found after fasting episodes (Starck and Beese 2002; Secor 2008; German et al. 2010). Despite this wealth of information, the study of the possible structural modifications undergone by the lungfish gut during aestivation has not been performed until very recently (Icardo et al. 2012a). This has been made possible by the development of a method that allows selected specimens to aestivate individually in a controlled environment (Chew et al. 2004; Ip et al. 2005).

It should also be noted that the gastrointestinal tract of the lungfish is a composite formed by several organs. In addition to the gut, it contains the spleen, the pancreas, and a lympho-granulocytic tissue that follows the coils of the spiral valve. Small amounts of this tissue are also observed in the cloaca. We (Icardo et al. 2012b, 2013) and others (Rafn and Wingstrand 1981; Hassanpour and Joss 2011) have studied the structural characteristics of these organs and tissues and noted several discrepancies. Different or contradictory observations may arise from the study of aestivating fish instead of fish living in freshwater, or from erroneous species identification. Early observations could also be handicapped by technical limitations. This work is based on our most recent studies and reviews the anatomical and structural organization of the gastrointestinal tract of the African lungfish *Protopterus annectens*. The existence of common and uncommon morphological traits between lungfishes is analyzed whenever possible.

General Morphology

From the pharynx to the cloaca the alimentary canal of *P. annectens* shows a longitudinal, straight organization (Fig. 1a). The entire system is wrapped by the peritoneal serosa. Dissection of the serosa reveals the different segments of

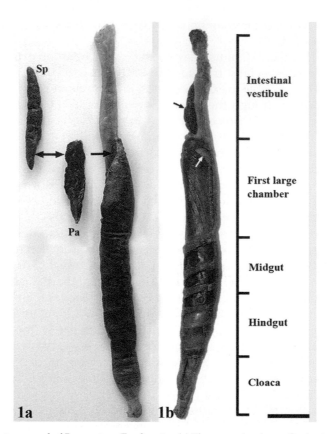

Fig. 1. Alimentary canal of *P. annectens*. Freshwater. (a) The serosa has been eliminated. The spleen (Sp) and the pancreas (Pa) have been dissected. Arrow indicates the shallow area occupied by the pancreas. Two-headed arrow indicates the area of spleen-pancreas overlapping. (b) The spiral valve has been opened through the external wall. It is formed by six cones piled one on top of the next. The coiling starts at the cranial pyloric border. White arrow indicates pyloric leaflets. The first chamber is very large and shows oblique mucosal ridges. The rest of the intestine shows a smooth surface. The spleen (black arrow) follows the right side of the intestinal vestibule. Magnification bar: 1 cm (a: From Icardo et al. 2010, Anat. Rec. 293: 1146–1154, with modifications. b: From Icardo et al. 2011, J. Morphol. 272: 769–779, with modifications).

the gut and the presence of two associated organs, the spleen and the pancreas (Fig. 1a). These two organs are attached to the gut by connective sheaths and perforating vessels, but they can be dissected easily. Opening of the gut reveals its internal organization (Fig. 1b). Below the pharynx, a very short oesophagus, an elongated sac (the intestinal vestibule), the pyloric region, a prominent spiral valve and a cloaca constitute the main components of the gut. The straight organization along the cranio-caudal axis and the presence of a spiral valve are primitive features shared by all lungfishes (Parker 1892; Rafn and Wingstrand 1981; Icardo et al. 2010). Straight organization of the gut is also present in other primitive fish such as lampreys and chimaeras whereas a spiral valve can be observed in chimaeras, sturgeons and in several elasmobranches (Kardong 2006; Ostos-Garrido et al. 2009).

The Intestinal Vestibule

Although the oesophagus constitutes the first portion of the gut, it appears to be a mere segmental transition between the pharynx and the intestinal vestibule. It is very short and poorly delimited (Hassanpour and Joss 2011), being marked in *Protopterus* by the presence of discrete luminal protrusions (Icardo et al. 2011).

The intestinal vestibule is an elongated, thin-walled sac that extends between the oesophagus and the pyloric region. This chamber has classically been considered the stomach of the lungfish (Parker 1892; Rafn and Wingstrand 1981). However, it lacks any of the morphological characteristics normally associated with the stomach such as gastric pits, gastric glands and well-organized tissue layers. Therefore, lungfishes can be considered stomach-less (Icardo et al. 2011).

The internal surface of the vestibule is marked by the presence of longitudinal folds. Thus, the lumen appears irregular in cross-sections (Fig. 2a). The wall consists of a mucosa surrounded by a layer of dense connective tissue. The mucosal epithelium is stratified, being formed by cuboidal cells that appear organized in, at least, two cell layers (Fig. 2a). Many cells show large, distended cytoplasms filled with vacuoles that often appear discharging into the lumen (Fig. 2b). All the epithelial cells show periodic acid of Schiff (PAS)-positive cytoplasm. The cell apical surface bulges into the lumen, giving a cobblestone appearance to the epithelium (Fig. 2c). Numerous microridges, apical junctional complexes, surface pits and areas of cell extrusion are often observed (Fig. 2c). The epithelium also contains a few basal mitoses, wandering leukocytes and capillary vessels restricted to the basal part of the epithelium (Fig. 2b). In addition, scattered ciliated cells appear in the epithelium near the pyloric area. The dense connective tissue layer that surrounds the epithelium contains numerous smooth muscle cells, many of them oriented circumferentially (Fig. 2b). In addition, bundles of longitudinally-oriented cells appear at the wall periphery. However, coherent muscle layers are not observed. Also, a distinct *lamina propria* is absent (see Icardo et al. 2011).

The mucosal epithelium of the intestinal vestibule shows most of the characteristics of a transitional epithelium, including a variable number of cell layers. The number of layers that can be observed depends on the degree of wall contraction before fixation (Fawcett 1994). From the functional point of view, the most plausible hypothesis is that the intestinal vestibule may simply serve to store food, or to slow food transit (Icardo et al. 2010). Indirect evidence of the lack of digestive functions comes from the observations made

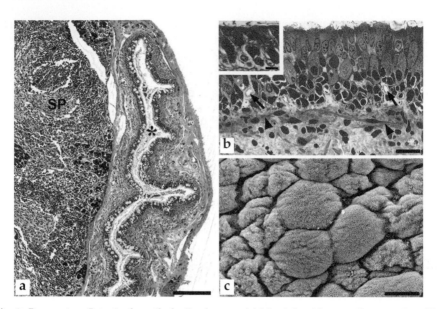

Fig. 2. *P. annectens*. Intestinal vestibule. Freshwater. (a) Martin's trichrome. Cross-section. The lumen profile is irregular due to the presence of longitudinal folds. The mucosa contains numerous cells filled with vacuoles and is surrounded by a thick layer of connective tissue. Mucus and cell debris (asterisk) occupy the lumen. The spleen (SP) is located to the right of the vestibule. (b) Toluidine blue. Note several cell layers and the presence of large cell cytoplasms filled with vacuoles. Arrows indicate intraepithelial capillaries. Arrowheads indicate bundles of smooth muscle cells. Inset: PAS-staining of the epithelium. (c) SEM. The bulging surface of the epithelial cells shows microridges. Note cobblestone appearance of the epithelium. Scale bars: a, 500 μm; b, 40 μm; inset of b, 20 μm; c, 5 μm.

in aestivating fish. Should the intestinal vestibule be directly implicated in digestion, important structural modifications would be expected during aestivation. However, except for some focal, transient bursts of secretory activity, the morphology of the epithelium after six months of aestivation is quite similar to that of aquatic specimens (Icardo et al. 2012a). This includes the maintenance of the PAS-positive staining of the epithelial cells (Icardo et al. 2012a).

The Spiral Valve

The spiralling portion of the gut extends between the pyloric region and the cloaca. The pylorus shows an oblique aperture guarded by a valve formed by two, right and left, thick flap-like leaflets (Fig. 1b). The surface of the leaflets is covered by cells with microridges. Numerous ciliated cells are also observed. The morphology of the pyloric region appears to be similar between the different species of lungfish (Rafn and Wingstrand 1981; Icardo et al. 2010).

In *P. annectens*, spiralling of the gut starts from the dorsally, below the pyloric aperture (Fig. 1b). The beginning of coiling is marked externally by the presence of a shallow area. This area is occupied by the pancreas (Fig. 1a). The gut spirals downwards to form six coils that attach to the outer wall. Each coil is shaped like a cone with the cones piled one on the other (Icardo et al. 2010). However, coiling of the gut is not uniform. Spiralling creates first a very large chamber, the size of the successive coils being much smaller. The last coil ends into the cloaca (Fig. 1b).

The morphology of the spiral valve appears to be similar in all lungfish species. In *N. forsteri*, however, coiling of the intestine has been reported to start above the pylorus (Rafn and Wingstrand 1981), in what has been dubbed the pre-pyloric intestine (Hassanpour and Joss 2009; Hassanpour and Joss 2011). This area coincides with the intestinal vestibule of *P. annectens*, that only shows a longitudinal depression or fold to accommodate the spleen. Furthermore, this fold may just be a deformation produced during fixation. On the other hand, some confusion arises from the fact that both *P. annectens* and *N. forsteri* have, below the pylorus, a large first chamber that occupies about one-third of the total length of the spiral valve. In the two cases, this is the 'first spiral' (Hassanpour and Joss 2011). Of note, the spiral valve of *N. forsteri* shows nine full coils (Rafn and Wingstrand 1981; Hassanpour and Joss 2009; Hassanpour and Joss 2011), while only six have been described in *P. annectens* (Parker 1892; Icardo et al. 2010, 2011).

From the structural viewpoint there are two different areas in the spiral valve. The inner wall of the first large chamber is characterized by the presence of oblique ridges (Icardo et al. 2010) or rugae (Hassanpour and Joss 2011) (Figs. 1b, 3a). This contrasts with the smooth appearance of the rest of the spiral valve surface (Fig. 1b). The presence of ridges has been misinterpreted in the past since, when sectioned, they resemble intestinal villi (Fig. 3b). This has led to some confusion related to the anatomical and functional identification of the gut chambers (Purkerson et al. 1975; Rafn and Wingstrand 1981). The ridges are covered by a pseudostratified columnar epithelium which contains three main cell types: enterocytes, goblet cells and ciliated cells (Fig. 3c). The enterocytes show a distinct brush-border, the goblet cells show loosely arranged microvilli, and the ciliated cells are characterized by the presence of prominent ciliary tufts. Goblet cells are frequently seen discharging their content into the lumen (Fig. 3d), and, both, enterocytes and goblet cells show a supranuclear cytoplasm filled with vacuoles. In addition to the three cell types mentioned, the ridge epithelium also contains melanocytes, wandering leukocytes and endothelial cells pertaining to intraepithelial capillaries (Fig. 3d). Under the ridge epithelium the *lamina propria* supports the delicate vasculature that forms the ridge core. The *lamina propria* also contains subepithelial fibroblasts, granulocytes and macrophages (Fig. 3d), and a single cell layer formed by smooth muscle cells oriented longitudinally and encased within fine collagen sheaths (Fig. 3b) (for details, see Icardo et al. 2011).

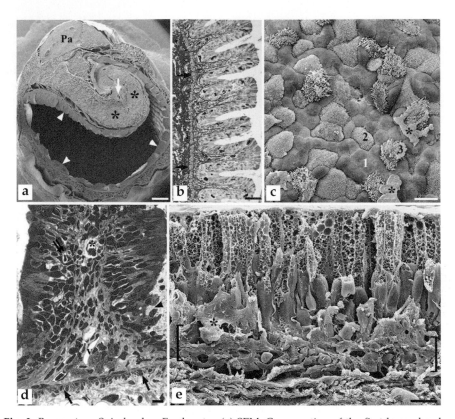

Fig. 3. *P. annectens.* Spiral valve. Freshwater. (a) SEM. Cross-section of the first large chamber showing dilated lumen and the presence of ridges (arrowheads). The mucosa is surrounded by a reticular tissue housing lymph-like nodes (asterisks). The mesenteric artery (arrow) runs along the inner border of the spiral valve. The section includes the pancreas (Pa). (b) Ridge area. Sirius red. A collagenous *lamina propria* is common to all the ridges. Smooth muscle cells form a single cell layer under the epithelium (arrows). Black dots correspond to melanocytes. (c) Ridge area. SEM. The apical ridge surface is a mosaic formed by enterocytes (1), goblet cells (2) and ciliated cells (3). Variable amounts of mucus are also present. (d) Ridge area. Toluidine blue. Goblet cells in the epithelium (clear cells) show active secretion (arrowhead). The ridge core contains vessels (v) surrounded by macrophages and lymphatic micropumps (asterisk). Arrows: smooth muscle cell layer (compare with b). Double arrow: intraepithelial capillary. (e) Smooth portion. SEM. Columnar cells show supranuclear cytoplasms filled with vacuoles. Brackets indicate subvascular plexus. Asterisk: large melanocyte. Scale bars: a, 500 μm; b, 100 μm; c, 10 μm; d, 20 μm; e, 20 μm (a: From Icardo et al. 2010, Anat. Rec. 293: 1146–1154, with modifications).

The epithelium of the smooth portion of the spiral valve contains the same six cell types observed in the ridge area. In this portion, enterocytes and goblet cells also display supranuclear cytoplasms filled with vacuoles. The vacuoles are very prominent at the transition zone between the ridge area and the smooth portion (Fig. 3e), decrease in number and size in the second and third coil, and are greatly reduced in the last coil. A similar decrease is also observed in the transverse plane, from the outer to the inner portion of the spiral valve. Curiously, the distribution of the ciliated cells follows a similar pattern. The mucosa of the smooth portion of the spiral valve is underlined by a vascular plexus (Fig. 3e). The *lamina propria* underlying the vascular plexus contains collagen and fibroblasts, abundant vascular spaces, and a layer of muscle cells arranged circularly. This area also contains granulocytes and macrophages. Of note, the different muscle cell layers associated with the mucosa of the spiral valve are irregular and discontinuous. However, a relatively thick layer of circularly-oriented muscle cells runs in the external side of the spiral valve wall. This is the single muscle layer common to all the intestine coils (Icardo et al. 2011).

In general, brush-border enterocytes are endowed with secretory and absorptive properties, the mucus secreted by the goblet cells hydrates and protects the epithelium from abrasion, and ciliated cells redistribute the mucous secretion and facilitate fluid movements (Ross et al. 2003). While there is not direct evidence of the role played by these cell types in the lungfish gut, they likely play a similar role than in other vertebrates. The presence of ciliated cells should be underscored since it constitutes a primitive feature shared by several reptilian (Giraud et al. 1978) and fish (Radaelli et al. 2000) species. Nonetheless, the distribution of the ciliated cells may vary between lungfish species since they are mostly absent in the ridge area of *N. forsteri* (Rafn and Wingstrand 1981; pers. observations). Wandering leukocytes are common in the epithelial intestine. Their presence is related to immunological surveillance. On the other hand, intraepithelial capillaries were previously detected in the lungfish gut (Purkerson et al. 1975; Rafn and Wingstrand 1981). They are found through the entire gastrointestinal tract in *P. annectens*, but appear to be restricted to the midgut area in *N. forsteri* (Rafn and Wingstrand 1981).

In all the specimens studied the gut lumen contains large amounts of mucus. Melanocyte cell bodies and numerous cell debris and cell nuclei appear embedded in the mucus. Thus, it is evident that several cell types traverse the epithelium and reach the gut lumen. Epithelial cells also appear to desquamate into the lumen. However, the mechanisms of epithelial cell replacement remain obscure. The number of mitotic figures in the mucosa is very low, and a distinct germinal zone could not be identified (Icardo et al. 2011). Of note, the mucosa of both the intestinal vestibule and the spiral valve shows numerous cells with basal nucleus and abundant heterochromatin. Cells with similar characteristics appear to constitute a reserve of stem cells in the mammalian intestine (Fawcett 1994). Whether basal cells may constitute a similar reserve in the lungfish gut is currently unknown.

The presence of granulocytes in the *lamina propria* is another feature typical of the lungfish intestine. In the mammalian intestine, enteroendocrine cells and Paneth cells are granulocytes that secrete hormones involved in the regulation of intestinal physiology (Ross et al. 2003). Identification of the granule content is necessary before a similar role could be ascribed to the lungfish granulocytes.

The Spiral Valve during Aestivation

The spiral valve undergoes important structural modifications during aestivation. However, these modifications do not affect the general organization of the tissues. The gut lumen may appear collapsed, and the ridges of the first large chamber appear slightly deformed. However, gross modifications such as loss of the intestine mass, gut shortening, and reduction in the surface area of villi are not observed (Icardo et al. 2011). These modifications accompany fasting in a wide range of vertebrate groups, from fish to mammals (Dunel-Erb et al. 2001; Hume et al. 2002; Starck and Beese 2002; German et al. 2010). Despite the absence of changes in gross anatomy, the structure of the epithelium of the spiral valve undergoes severe modifications during aestivation. These modifications affect more to the anterior than to the posterior levels of the spiral valve, and to the external than to the internal areas. These difference may be related to the existence of differences in the digestive capabilities across the mucosa. The areas more actively involved in digestion appear to be the areas more affected during aestivation. In addition, modifications are time dependent since they are different at four and at six months of aestivation.

At four months of aestivation the epithelium of the spiral valve appears mostly disintegrated. The regular, columnar appearance is lost, the height of the epithelium is reduced, and the epithelium appears composed by one or two cell layers (Fig. 4a). In other areas, however, the epithelium may be completely lost (see Icardo et al. 2011). Enterocytes, goblet cells and ciliated cells could not be recognized (Fig. 4a). Instead, the epithelium is formed by cells that show rounded or oval nucleus, heterochromatin clumps and pale cytoplasm. Numerous dead cells can be observed in the epithelium and in the *lamina propria*, and the number of dark pigment cells appears to have increased with respect to the aquatic condition. The gut lumen is filled with large amounts of mucus, melanocytes, cell debris and dead cells. Strikingly, the epithelium appears stratified at six months of aestivation (Fig. 4b), showing a homogeneous height, a clear basal boundary, and a more uniform apical surface. A few number of mast cells filled with granules can also be observed. The number of dead cells in the epithelium and in the *lamina propria* is reduced when compared to four months of aestivation. Despite this, the structure of the epithelium is very different from that observed in aquatic animals (Fig. 4c). Under the scanning microscope the epithelial surface appears formed by overlapping flattened cells that adopt, in many areas, a roof-tile configuration (Fig. 5a). Cells with microvilli are not apparent. Ciliated cells are seldom observed and show loss of cilia. Fractures of the epithelium reveal the presence of cell columns extended across the entire epithelial thickness (Fig. 5b). These cells are joined to each other by slender cell processes and appear to be migrating from the basal to the apical zones of the epithelium. The modifications of the epithelium affect both to the ridge area and to the

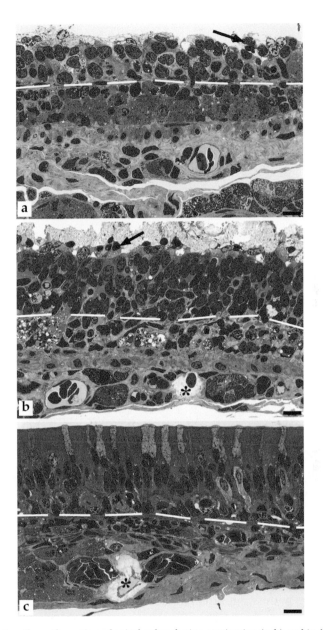

Fig. 4. *P. annectens*. Smooth portion of spiral valve during aestivation (a, b) and in freshwater (c). Toluidine blue. The white broken lines mark the basal limit of the mucosa. Note the differences in thickness. A vascular plexus underlies the mucosa. Arrows in a and b indicate necrotic cells close to the epithelial surface. Asterisks in b and c indicate lymphatic micropumps. (a) 4-months aestivation. The normal epithelial organization is lost. (b) 6-months aestivation. The epithelium becomes pseudostratified. (c) Freshwater. The epithelium is formed by columnar cells. Goblet cells show clear cytoplasm. Scale bars: a–c, 20 μm.

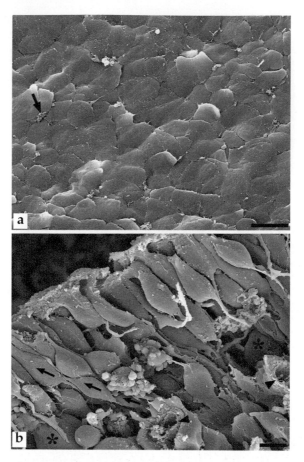

Fig. 5. *P. annectens*. Spiral valve. 6-months aestivation. SEM. (a) Epithelial surface. Note smooth apical cell surfaces and roof-tile overlapping (compare to Fig. 3c). A ciliary tuft (arrow) appears under the flattened cell surface. (b) Columns of connected cells (arrows) extend across the epithelium (compare to Fig. 3e). Cells with surface blebbing correspond to granulocytes. Arrowheads: intraepithelial capillaries. Asterisks, dark pigment cells. Scale bars: a, 20 μm; b, 10 μm blebbing.

smooth portion of the spiral valve. In general, however, the epithelial changes are more severe in the ridges than in the smooth portion, and at the crest than at the lateral parts of the ridges.

The morphologic observations indicate that the normal constitutive cells are substituted by a cell population that lacks the phenotypic characteristics found in aquatic conditions. In addition, the presence of cells in the gut lumen, and the formation of cell chains across the epithelium, indicates that cell migration and desquamation may be a continuous process. Thus, the entire epithelium appears to be renewed during aestivation. The origin of the new cells remains obscure. The intestinal mucosa of lungfishes lacks a distinct germinal zone (Icardo et al. 2011). This absence becomes more evident during aestivation, when the epithelium is completely disorganized. In addition, epithelial mitoses are scarce in any condition. In other vertebrates, epithelial cell turnover is maintained by proliferation of undifferentiated cells situated in the deepest areas of the intestinal crypts (Fawcett 1994). The high rate of mitosis, together with the remarkable plasticity of the intestinal tissues, allows for fast and reversible changes of the epithelium. In the lungfish, plasticity of the intestine is also remarkable. However, the origin of the cells that populate the mucosa remains speculative (see Icardo et al. 2012a).

Functional studies such as the identification of brush border membrane enzyme activities or the study of the expression of key transporters have not been performed in lungfishes. However, the severity of the structural changes and the disappearance of the normal constitutive cells indicate that the digestive and absorptive functions are lost during aestivation. This can indirectly be demonstrated by different procedures. For instance, in aquatic conditions, enterocytes and goblet cells are PAS-positive (Fig. 6a). PAS-positive

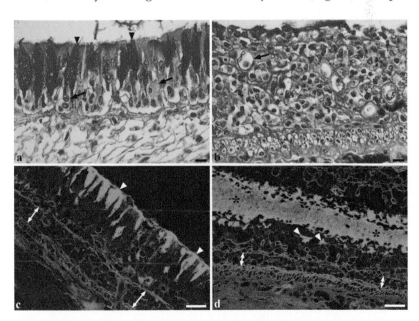

Fig. 6. *P. annectens*. Spiral valve. (a) Freshwater. PAS-staining. Columnar enterocytes and goblet cells (arrowheads) are PAS-positive. Arrows: intraepithelial vessels. (b) 6-months aestivation. PAS-staining. Columnar cells are no longer observed. Most cell cytoplasms are PAS-negative. Arrow: intraepithelial vessel. (c) Freshwater. Wheat germ agglutinin (WGA) lectin. Goblet cells (arrowheads) are intensely stained by this lectin. Double-headed arrow: Vascular plexus under the epithelium. (d) 6-months aestivation WGA staining. Only a few superficial cells are WGA-positive (arrowheads). Note the decrease in thickness of both the epithelium and the subepithelial vascular plexus (double-headed arrow). Most of the positivity in this panel corresponds to mucus (asterisks) in the gut lumen. Scale bars: a–d, 20 µm.

staining is related to the presence of mucosubstances that may be implicated in defence and absorptive functions. PAS-positive staining of the epithelium is lost during aestivation (Fig. 6b). Only the mucus contained in the gut lumen retains some positivity. This contrasts with the maintenance of the positive staining in the intestinal vestibule (see above). On the other hand, the presence of sugar residues has been used to investigate cell function in numerous systems. The presence of these residues can be traced by fluorescent lectins that serve as markers of several physiological and pathological states. For instance, the agglutinin of the wheat germ (WGA), which localizes N-acetyl-D-glucosamine residues, marks intensely the goblet cells in the intestine of the aquatic lungfish (Fig. 6c). However, epithelial cells become negative to WGA during aestivation (Fig. 6d). While the significance of the modifications in the lectin-binding patterns are still obscure, these techniques may be of help to bring light into the functional characteristics of the intestinal mucosa of the lungfish.

As an additional observation, nematodes are frequently observed accommodated between the ridges of the first large chamber in freshwater fish (Rafn and Wingstrand 1981; Icardo et al. 2010). However, parasites are never observed in aestivating animals. Unidentified parasites can be observed in other organs such as the spleen (Hassanpour and Joss 2011) and the kidney. Small parasites also disappear during aestivation (unpubl. observations). Thus, parasite cleaning may be an additional benefit of aestivation.

The Cloaca

The spiral valve opens into the cloaca, which constitutes the end of the gut. The cloaca is a short chamber which shows a thick wall and an irregular contour in cross-sections. The inner side of the cloaca is lined by a pseudostratified epithelium containing cells with microridges and goblet cells. As it occurs in the intestinal vestibule, the epithelium shows a cobblestone appearance and the number of cell layers that can be observed varies in relation to the degree of wall contraction. In the cloaca, a large proportion of the wall thickness is due to the presence of well-organized muscle components (see Icardo et al. 2011). This is a distinct feature since the *muscularis mucosa* is discontinuous or absent in many areas of the gut wall.

The mucosal lining of the cloaca reunites, like the epithelium of the intestinal vestibule (see above), most of the characteristics of a transitional epithelium. Similar to what occurs in the intestinal vestibule, few changes are observed in the cloaca after six months of aestivation. Loss of surface microridges, attenuation of the cobblestone appearance and some cytoplasm overlapping are the main modifications observed in this chamber during aestivation (Icardo et al. 2012a).

Recovery after Aestivation

The study of the effects of aestivation also includes a number of observations made at several time intervals during the period of arousal (Icardo et al. 2011). When the period of aestivation is considered to be finished, the surface of the lungfish is humidified, the animal wakes up, and the cocoon cracks and tears apart. Release of the animals from the mucus envelop is accelerated by manual elimination of parts of the cocoon. Then, the fish are submersed in a tank filled with fresh water. Initially, the fish show little or no activity. Most movements are clumsy, and many of them are directed toward the liquid surface, to gulp air. Normal activity is recovered about two weeks after arousal. During this period the animals swallow water but they refuse to eat at least until the beginning of the second week. The progressive acquisition of the body activities indicates that normal functioning of the different organs and systems requires some time for adaptation. Adaptation times appear to depend on the characteristics of each system. In the case of the alimentary canal, the first signs of structural recovery appear by the end of the first week. In the intestinal vestibule and in the cloaca the cells recover the surface microridges and the epithelium shows a regular appearance. Six days after arousal the ridges in the spiral valve are well aligned and the epithelium appears formed by columnar cells that adopt a pseudostratified organization. Many of these cells develop supranuclear vesicles and become PAS-positive. While this indicates some functional recovery, cell phenotypes are not clearly established. It is not until two weeks after arousal that enterocytes, goblet cells and ciliated cells can be recognized. However, brush border enterocytes have small apical surfaces, goblet cells present a reduced number of microvilli, and many ciliated cells show just a few cilia of variable height (Icardo et al. 2011). This means that recovery of the structural features typical of freshwater conditions is progressive and continues beyond the period studied (two weeks). The feeding behaviour during this time is probably related to the lack of functional and structural maturation of the intestinal mucosa.

Organs Associated to the Gut

The lungfish gut is a composite that includes the spleen, the pancreas and a reticular tissue that has remained ill-defined until very recently (Icardo et al. 2014). In the genus *Protopterus*, the spleen is a brownish, compact, elongated organ that runs along the right side of the intestinal vestibule (Fig. 1). The caudal end of the spleen reaches the midlevel of the pyloric aperture. The existence of a rod-shaped spleen is a common feature in lungfish (Parker 1892; Rafn and Wingstrand 1981). In addition, a second spleen has been described in the Australian lungfish *Neoceratodus forsteri* (Rafn and Wingstrand 1981; Hassanpour and Joss 2011). This caudal spleen runs along the inner border of the spiral valve forming a kind of long intestinal axis (Hassanpour and Joss 2009, 2011). In *Protopterus*, however, the place of the caudal spleen is occupied by a series of nodes, initially described as lymphatics (Parker 1892; Coujard and Coujard-Champy 1947), which run along the mesenteric artery (Icardo et al. 2010). The existence of two spleens in *N. forsteri* could be a major anatomical difference between lungfish species. However, the morphological situation is not entirely clear and further studies may be necessary to clear this point.

Histological studies carried out in both *N. forsteri* (Rafn and Wingstrand 1981; Hassanpour and Joss 2009, 2011) and *P. annectens* (Icardo et al. 2012b) have shown that the spleen is formed by a cortical tissue and by a parenchyma which contains both white and red pulp. The spleen cortex shows irregular thickness and was previously described as a kind of lymphoid tissue (Rafn and Wingstrand 1981; Hassanpour and Joss 2009). In freshwater *Protopterus*, however, the cortex barely contains lymphocytes. Instead, it is a complex reticular tissue which contains two types of granulocytes, developing and mature plasma cells and melanomacrophage centres. The cortical tissue may simply provide an adequate environment for homing and differentiation of cells of the white series (Icardo et al. 2012b). Systematic analyses of the structure of the spleen in *P. annectens* have detected the presence of a subcapsular sinus and have provided evidence to support the involvement of the red pulp in destruction of effete erythrocytes and in plasma cell differentiation. Additionally, the white pulp appears to be related to the production of immune responses (for details, see Icardo et al. 2012b).

Throughout the years, the lungfish spleen has been attributed to important haemopoietic functions such as the production of erythrocytes, lymphocytes, granulocytes and platelets (Jordan and Speidel 1931; Dustin 1934; Rafn and Wingstrand 1981; Hassanpour and Joss 2009). In *Protopterus*, however, the lack of blood islands and the virtual absence of mitosis appear to discard haemopoiesis. Rather, the spleen could be a site colonized by early erythrocytes and by immature forms of the white series. These cells would reach the red pulp to complete their differentiative processes (Icardo et al. 2012a). On the whole, the *Protopterus* spleen reunites both primitive and advanced features and shares structural characteristics with other secondary lymphoid organs such as the mammalian haemal and haemolymph nodes.

The pancreas is another organ associated to the lungfish gut. In *Protopterus*, the pancreas is a dark, elongated organ which appears embedded in the deep furrow that marks the beginning of the intestine coiling (Fig. 1) (Icardo et al. 2010). Thus, a large part of the lungfish pancreas is situated in a dorsal, post-pyloric position. The pancreas maintains a close relationship with the spleen since the caudal end of the spleen usually overlaps the cranial end of the pancreas (Fig. 7a). The post-pyloric position and the close association with the spleen are maintained in all lungfishes (Parker 1892; Rafn and Wingstrand 1981; Hassanpour and Joss 2011). Differences in the external surface include colouration since the pancreas in *N. forsteri* is light grey while it appears much darker in *Lepidosiren* and *Protopterus* (Fig. 1a) (Rafn and Wingstrand 1981). This may depend on the presence of melanocytes. These cells are very numerous

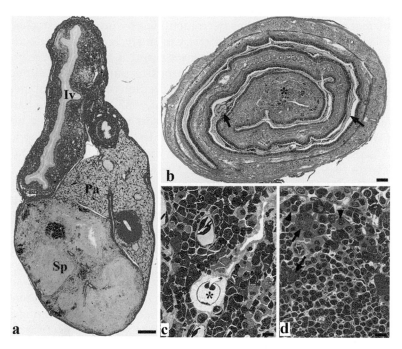

Fig. 7. *P. annectens.* (a) Prepyloric portion of the gut. Sirius red. The spleen (Sp) and pancreas (Pa) overlap in front of the intestinal vestibule (Iv). A large vessel connects the spleen and the pancreas. Black dots in the pancreas and vestibule correspond to melanocytes. Black dots in the spleen correspond to macrophages and melanomacrophage centres. (b) Midgut. Martin's trichrome. The coils of the mucosa are followed by the reticular tissue. The gut lumen contains large amounts of mucus (arrows). Asterisk indicates lymph-like nodes. (c) Reticular tissue. Toluidine blue. Light- and dark blue-cells correspond to two different types of granulocytes. Plasma cells stain purple. Note numerous blood vessels. Asterisk, lymphatic micropump. (d) Node. Toluidine blue. Lymphocytes predominate in the parenchyma. Macrophages (arrows) and plasma cells (arrowheads) co-localize very often. Scale bars: a–b, 500 μm; c–d, 20 μm.

in the *Protopterus* pancreas (Fig. 7a). On the other hand, the spleen and the pancreas are attached to each other and to the surrounding structures by sheaths of connective tissue and by numerous blood vessels of small calibre that have to be cut to dissect the two organs. Arterial supply for the two organs entered the area of spleen/pancreas overlapping (Icardo et al. 2012b).

In *N. forsteri*, the exocrine component of the pancreas consists of tubular acini formed by glandular cells that contain supranuclear zymogen granules and appear arranged around a narrow duct. The endocrine component consists of islets of Langerhans dispersed among the acini. The islets are formed by small cell masses permeated by a rich capillary plexus and surrounded by a delicate layer of connective tissue that separates the cell patches from the surrounding acini (Rafn and Wingstrand 1981). The production of the four main hormones typical of the vertebrate pancreas (glucagon, insulin, somatostatin and pancreatic polypeptides) has been identified by immunohistochemistry in *P. aethiopicus* (Scheuermann et al. 1991). Curiously, only three endocrine cell types have been identified in both *P. aethiopicus* and *N. forsteri*. In *P. aethiopicus*, the glucagon (type-A) cells may also produce pancreatic polypeptides (Scheuermann et al. 1991). However, the production of specific polypeptides has not been detected in *N. forsteri* (Hansen et al. 1987). On the other hand, the distribution of the endocrine component appears to differ between lungfish species. While in *Neoceratodus* it is very similar to that found in tetrapods, the islets of Langerhans appear concentrated in the caudal part of the pancreas in *Lepidosiren* and *Protopterus* (Joss 2011).

The wall of the spiral valve is associated with an additional type of tissue (Fig. 7b). The basal side of the mucosa is followed by a well-vascularized reticular tissue. This tissue was initially described as adenoid (Parker 1892) or lymphoid (Rafn and Wingstrand 1981) in nature, appeared to form part of the intestine mucosa and submucosa, and was more developed in *Protopterus* and *Lepidosiren* than in *Neoceratodus* (Rafn and Wingstrand 1981). However, recent evidence indicates that several of the early observations have to be revisited. In *N. forsteri*, the submucosa is mostly formed by a band of adipose tissue. This band follows all the spiralling, being sandwiched between adjacent gut coils (Hassanpour and Joss 2011). In addition, discrete aggregations of lymphoid tissue associate to the submucosa throughout the spiral valve. It has been suggested that these aggregations may be equivalent to the Peyer's patches found in other vertebrates (Hassanpour and Joss 2011).

The situation is very different in *Protopterus*. In *P. annectens*, a distinct reticular tissue follows the gut coils under the mucosa. It also occupies most of the space situated along the inner border of the spiral valve. The reticular tissue is attached to the mucosa by a system of connective tissue sheaths (Icardo et al. 2010). Because of the attachments, the tissue could be considered as a highly specialized submucosa or, at least, as part of the submucosa (Icardo et al. 2014). Of note, the presence of a system of connective sheaths that holds the components of the alimentary canal together has only been described in *P. annectens* (Icardo et al. 2010, 2011). However, this system is also present in *N. forsteri* (pers. observations).

The structure and composition of the reticular tissue has recently been described (Icardo et al. 2014). The reticular spaces house two types of granulocytes, developing and mature plasma cells, a few lymphocytes and melanomacrophage centres (Fig. 7c). The granulocytes show empty granules and other structural signs of secretory activity. This composition is similar to that observed in the spleen cortex (Icardo et al. 2012b) and in the surface of the most anterior portion of the kidney (unpubl. observations). In *P. annectens*, the two types of granulocytes have tentatively been identified as eosinophils and mast cells (Icardo et al. 2012b, 2013). However, granulocyte identification in fish is controversial, especially if it is solely based on structural criteria. Comparison with data obtained from other lungfish species only adds confusion to the subject. Several types of eosinophils have been described in

P. aethiopicus (DeLaney et al. 1976) whereas *L. paradoxa* and *N. forsteri* have three and four granulocyte types, respectively (Hine et al. 1990; Bielek and Strauss 1993). Furthermore, the ultrastructure of the cytoplasmic inclusions of *P. annectens* does not coincide with any of the previous descriptions (Hine et al. 1990; Bielek and Strauss 1993). Clearly, proper granulocyte identification depends on the identification of the granule components. Nonetheless, the enormous amount of granulocytes contained in the reticular tissue should be underscored. This is even more remarkable if the extension of the reticular tissue is taken into account. On the whole, the reticular tissue may function as a glandular structure controlling the activity of the gut and, perhaps, of other body organs. The fact that this tissue is absent in *N. forsteri* suggests that its presence may be related to metabolic modifications and adjustments during the freshwater/aestivation cycles of body activity.

In addition to the reticular tissue, numerous small nodes of lymphatic nature appear to be located along the inner side of the spiral valve in *Protopterus* (Icardo et al. 2010, 2011). The nodes appear surrounded by the reticular tissue (Fig. 7b) and are attached to it by connective sheaths. Taken together, the two components constitute a lympho-granulocytic system that resembles the Leydig organ described in the oesophagus of several elasmobranches (Zapata and Cooper 1990). Despite this close association the nodes are independent structures, can be dissected, and their structure is different from that of the reticular tissue (Fig. 7d). Indeed, the nodes share structural characteristics with secondary lymphatic organs of other vertebrates (Icardo et al. 2013). The nodes contain a cortex whose composition is similar to that of the spleen cortex, and a parenchyma that contains diffuse lymphoid tissue, a high number of mitoses and plasma cell clusters (Icardo et al. 2013). The structural data indicate that the nodes are involved in the production of lymphocytes, in erythrocyte destruction and in iron metabolism. They also appear to be sites for development and maturation of plasma cells.

All the available information strongly indicates that the spleen and the nodes are involved in the production of immune responses. Most likely, these activities are modified during aestivation. The spleen from aestivating animals shows lymphocyte and monocyte infiltration, especially in the cortex area. The fact that the lymphocytes do not organize into distinct structures suggests that they simply are sequestered from the circulation. In addition, the red pulp is invaded by granulocytes, and numerous dead cells can be observed (Icardo et al. 2012b). The nodes during aestivation also show granulocyte invasion and cell death. By contrast, the parenchyma shows lymphocyte depletion and suppression of mitoses (Icardo et al. 2014). While some of the data are compatible with a decrease of the immune functions during aestivation, the real significance of all these changes is unclear. The lack of studies directed to analyze in specific ways the production of immune responses makes these conclusions speculative at this time.

Finally, the submucosa of the entire gut shows an extensive lymphatic system formed by lymphatic micropumps and associated microvessels.

In histological sections, lymphatic micropumps are spherical structures consisting of a central vessel surrounded by a clear matrix (Fig. 7c). The central vessel is covered by a smooth muscle cell, and the clear space is formed by a system of extracellular fibrils that appear oriented in a radial direction and anchor the central vessel to the surrounding tissues (Icardo et al. 2011, 2013). The lymphatic micropumps were initially described in the gills and fins of *Protopterus* and *Lepidosiren* (Vogel and Matheus 1998). They are also present in the gut, in the reticular tissue (Icardo et al. 2014), in the cortex of the spleen and nodes (Icardo et al. 2012b, 2014), and in the wall of other organs such as the lung (unpubl. observations). The central vessel appears to absorb the interstitial fluid, releasing it into neighbouring veins. The connection between the micropumps and the drainage vessels may be direct, or it may be made through short interposed vessels that lack pericytes and a basement membrane. Thus, the excess of interstitial fluid may be evacuated in the absence of a real circulatory system. Lymphatic micropumps appear to be widely distributed in both *Protopterus* and *Lepidosiren*. Whether they are also present in *Neoceratodus* is currently unknown.

Acknowledgements

The author was supported by grant CGL2008-04559/BOS from the "Ministerio de Educación y Ciencia", Spain.

References

Almeida-Val, V.M.F., S.R. Nozawa, N.P. Lopes, P.H.R. Aride, L.S. Mesquita-Saad, M.N. Paula-Silva, R.T. Honda, M.S. Ferreira-Nozawa and A.L. Val. 2011. Biology of the South American lungfish, *Lepidosiren paradoxa*. pp. 129–147. *In*: J.M. Jorgensen and J. Joss (eds.). The Biology of Lungfishes. Science Publishers, Enfield.

Bemis, W.E., W.W. Burggren and N.E. Kemp (eds.). 1986. The biology and evolution of lungfishes. J. Morphol. Suppl. 1.

Bielek, E. and B. Strauss. 1993. Ultrastructure of the granulocytes of the South American lungfish, *Lepidosiren paradoxa*: morphogenesis and comparison to other leukocytes. J. Morphol. 218: 29–41.

Burggren, W.W. and K. Johansen. 1986. Circulation and respiration in lungfishes (Dipnoi). J. Morphol. Suppl. 1: 217–236.

Chew, S.F., N.K.Y. Chan, A.M. Loong, K.C. Hiong, W.L. Tam and Y.K. Ip. 2004. Nitrogen metabolism in the African lungfish, *Protopterus dolloi*, aestivating in a mucus cocoon on land. J. Exp. Biol. 207: 777–786.

Coujard, R. and C. Coujard-Champy. 1947. Recherches sur l'epithelium intestinale du Protoptere et sur l'evolution des enterocytes chez les vertebrates. Arch. d'Anat. d'Hist. et d'Emb. 30: 69–97.

DeLaney, R.G., C. Shub and A.P. Fishman. 1976. Hematologic observations on the aquatic and estivating African lungfish, *Protopterus aethiopicus*. Copeia 1976(3): 423–434.

Dunel-Erb, S., C. Chevalier, P. Laurent, A. Bach, F. Decrock and Y. Le Maho. 2001. Restoration of the jejunal mucosa in rats refed after prolonged fasting. Comp. Biochem. Physiol. A 129: 933–947.

Dustin, P. 1934. Recherches sur les organes hématopoietiques du *Protopterus dolloi*. Arch. Biol. (Paris) 45: 1–26.

Fawcett, D.W. 1994. *In*: W. Bloom and D.W. Fawcett (eds.). A Textbook of Histology, 12th edition. Chapman & Hall, New York.

Fishman, A.P., A.I. Pack, R.G. Delaney and R.J. Galante. 1986. Aestivation in *Protopterus*. J. Morphol. Suppl. 1: 237–248.

German, D.P., D.T. Neuberger, M.N. Callahan, N.R. Lizardo and D.H. Evans. 2010. Feast to famine: the effects of food quality and quantity on the gut structure and function of a detritivorous catfish (Teleostei: Loricariidae). Comp. Biochem. Physiol. A 155: 281–293.

Giraud, A.S., C.R. Hunter and D.J.B. St. John. 1978. Epithelial surfaces of the upper gastrointestinal tract of the blue-tongued lizard, *Tiliqua scincoides*: a scanning electron microscopic study. Aust. J. Zool. 26: 241–247.

Graham, J.B. 1997. Air-breathing Fishes. Evolution, Diversity and Adaptation. Academic Press, San Diego.

Hansen, G.N., B.L. Hansen and P.N. Jorgensen. 1987. Insulin-, glucagons- and somatostatin-like immunoreactivity in the endocrine pancreas of the lungfish, *Neoceratodus forsteri*. Cell Tissue Res. 248: 181–185.

Hassanpour, M. and J. Joss. 2009. Anatomy and histology of the spiral valve intestine in juvenile Australian lungfish, *Neoceratodus forsteri*. Open Zool. J. 2: 62–85.

Hassanpour, M. and J. Joss. 2011. The lungfish digestive system. pp. 341–357. *In*: J.M. Jorgensen and J. Joss (eds.). The Biology of Lungfishes. Science Publishers, Enfield.

Hine, P.M., R.J.G. Lester and J.M. Wain. 1990. Observations on the blood of the Australian lungfish, *Neoceratodus forsteri* klefft. I. Ultrastructure of granulocytes, monocytes and thrombocytes. Aust. J. Zool. 38: 131–144.

Hume, I.D., C. Beiglböck, T. Ruf, F. Frey-Roos, U. Bruns and W. Arnold. 2002. Seasonal changes in morphology and function of the gastrointestinal tract of free-living alpine marmots (*Marmota marmota*). J. Comp. Physiol. B. 172: 197–207.

Icardo, J.M., W.P. Wong, E. Colvee, A.M. Loong and Y.K. Ip. 2010. The anatomy of the gastrointestinal tract of the African lungfish *Protopterus annectens*. Anat. Rec. 293: 1146–1154.

Icardo, J.M., W.P. Wong, E. Colvee, A.M. Loong and Y.K. Ip. 2011. The gut of the juvenile African lungfish *Protopterus annectens*. A light- and scanning electron-microscope study. J. Morphol. 272: 769–779.

Icardo, J.M., W.P. Wong, E. Colvee, A.M. Loong and Y.K. Ip. 2012a. The alimentary canal of the African lungfish *Protopterus annectens* during aestivation and after arousal. Anat. Rec. 295: 60–72.

Icardo, J.M., W.P. Wong, E. Colvee, A.M. Loong and Y.K. Ip. 2012b. The spleen of the African lungfish *Protopterus annectens*: freshwater and aestivation. Cell Tissue Res. 350: 143–156.

Icardo, J.M., W.P. Wong, E. Colvee, A.M. Loong, A.G. Zapata and Y.K. Ip. 2014. Lympho-granulocytic tissue associated with the wall of the spiral valve in the African lungfish *Protopterus annectens*. Cell Tissue Res. 355: 397–407.

Ip, Y.K., P.J. Yeo, A.M. Loong, K.C. Hiong, W.P. Wong and S.F. Chew. 2005. The interplay of increased urea synthesis and reduced ammonia production in the African lungfish *Protopterus aethiopicus* during 46 days of aestivation in a mucus cocoon. J. Exp. Zool. A Comp. Exp. Biol. 303: 1054–1065.

Jordan, H.E. and C.C. Speidel. 1931. Blood formation in the African lungfish under normal conditions and under conditions of prolonged aestivation and recovery. J. Morphol. Physiol. 51: 319–371.

Joss, J. 2011. The lungfish endocrine system. pp. 369–391. *In*: J.M. Jorgensen and J. Joss (eds.). The Biology of Lungfishes. Science Publishers, Enfield.

Kardong, K.V. 2006. Vertebrates: Comparative Anatomy, Function, Evolution, 4th Edition. McGraw-Hill, New York.

Kind, P.K. 2011. The natural history of the Australian lungfish *Neoceratodus forsteri* (Krefft 1870). pp. 61–95. *In*: J.M. Jorgensen and J. Joss (eds.). The Biology of Lungfishes. Science Publishers, Enfield.

Mlewa, C.M., J.M. Gree and R.L. Dunbrack. 2011. The general natural history of the African lungfishes. pp. 97–127. *In*: J.M. Jorgensen and J. Joss (eds.). The Biology of Lungfishes. Science Publishers, Enfield.

Ostos-Garrido, M.V., J.I. Llorente, S. Camacho, M. García-Gallego, A. Sanz, A. Domezain and R. Carmona. 2009. Histological, histochemical and ultrastructural changes in the digestive tract of sturgeon *Acipenser naccarii* during early ontogeny. pp. 121–136. *In*: R. Carmona,

A. Domezain, M. García-Gallego, J.A. Hernando, F. Rodríguez and M. Ruiz-Rejón (eds.). Biology, Conservation and Sustainable Development of Sturgeons. Springer, New York.

Parker, W.N. 1892. On the anatomy and physiology of *Protopterus annectens*. Roy. Irish Acad. Trans. 30: 109–230.

Purkerson, M.L., J.U.M. Jarvis, S.A. Luse and E.W. Dempsey. 1975. Electron microscopy of the intestine of the African lungfish, *Protopterus aethiopicus*. Anat. Rec. 182: 71–90.

Radaelli, G., C. Domeneghini, S. Arrighi, M. Francolini and F. Mascarello. 2000. Ultrastructural features of the gut in the white sturgeon, *Acipenser transmontanus*. Histol. Histopathol. 15: 429–439.

Rafn, S. and K.G. Winstrand. 1981. Structure of intestine, pancreas, and spleen of the Australian lungfish, *Neoceratodus forsteri* (Krefft). Zool. Scripta. 10: 223–239.

Ross, M.H., G.I. Kaye and W. Pawlina. 2003. Histology. A Text and Atlas, with Cell and Molecular Biology, 4th Edition. Lippincott Williams & Wilkins, Inc., Baltimore.

Scheuermann, D.W., D. Adriaensen, J.-P. Timmermans and M.H. De Groot-Lasseel. 1991. Immunohistochemical localization of polypeptide hormones in pancreatic endocrine cells of a dipnoan fish, *Protopterus aethiopicus*. Acta Histochem. 91: 185–192.

Secor, S.M. 2005. Evolutionary and cellular mechanisms regulating intestinal performance of amphibians and reptiles. Integr. Comp. Biol. 45: 282–294.

Secor, S.M. 2008. Digestive physiology of the Burmese python: broad regulation of integrated performance. J. Exp. Biol. 211: 3767–3774.

Starck, J.M. and K. Beese. 2002. Structural flexibility of the small intestine and liver of garter snakes in response to feeding and fasting. J. Exp. Biol. 205: 1377–1388.

Vogel, V.O.P. and U. Mattheus. 1998. Lymphatic vessels in lungfishes (Dipnoi). I. The lymphatic vessel system in Lepidosireniformes. Zoomorphol. 117: 199–212.

Zapata, A.G. and E.L. Cooper. 1990. The Immune System: Comparative Histophysiology. Wiley, Chichester.

11

Morphology, Histology, and Functional Structure of the Alimentary Canal of Sturgeon

Ramón Carmona Martos,* Cristina E. Trenzado Romero
and *Ana Sanz Rus*

Introduction

Sturgeons are bony fish (superclass *Osteichthyes*, class *Actinopterygii,* subclass *Chondrostei*) of the order *Acipenseriformes*, which consists of two families: *Acipenseridae* or sturgeons, currently with 25 species; and *Poliodontidae* or paddlefish, with two species. In the evolutionary process, acipenserids lost part or all of their heavy ganoid scales and eliminated the ossification process of their cartilaginous skeleton.

Acipenserids constitute the most numerous group of 'living fossil fish' (Gardiner 1984), having existed since the Lower Jurassic (Bemis et al. 1997). They inhabit freshwater as well as brackish and marine waters, although in the latter case this is restricted to the continental platform. These fish are found primarily in cold and temperate regions of the Northern Hemisphere, i.e., North America, Europe, and Asia. Some species are found in rivers of the eastern and western coast of Canada and the United States as well as in the Mississippi River Basin. Others inhabit the rivers of Europe, particularly those that flow into the Atlantic Ocean, the Adriatic Sea, and the Baltic Sea.

Department of Cell Biology, University of Granada, Spain.
* Corresponding author: rcarmona@ugr.es

The most important sturgeon species are found in the Black Sea, the Azov Sea, the Caspian Sea, and the Aral Sea, as well as in the main rivers that flow into these seas. These fish are also found in the hydrographic basins of Asia, in the rivers flowing into the Okhotsk Sea, the Behring Sea, the Barents Sea, the Kara Sea, and the White Sea. This broad distribution reflects the importance that this group had in the past, whereas at present it is approaching extinction (Billard and Lecointre 2001).

In general, squaliforme sturgeons have a more or less elongated snout and four barbels on the underside, these being missing in paddlefish. The mouth is protractile (tubiforme) and appears in a ventral position. Embryonic and juvenile sturgeons show several rows of sharp teeth. However, teeth are lost in the adult stages. The taste buds are intraoral and extraoral on the barbels and lips but exclusively intraoral in paddlefish (Norris 1925). In addition to taste receptors, the conspicuous ampullary organs, electroreceptors, are distributed over the entire the head. Figure 1 shows the extraoral taste buds on the barbels and lips in a juvenile specimen (36 days post-hatching) of *Acipenser naccarii*. The caudal fin is heterocercal, with the upper lobe much larger than the inferior one. The body is massive and long, pentagonal in section or subcylindrical, covered by bare skin with five longitudinal series of bony scutes; one along the dorsal side, two on the flanks, and two more

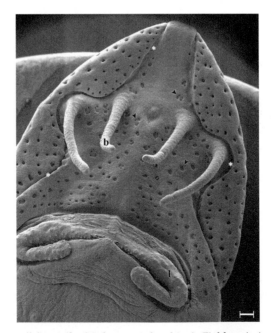

Fig. 1. *Acipenser naccarii* (juvenile, 36 days post-hatching). Field emission scanning electron microscopy. The head ventral surface shows the extraoral taste buds on the barbels (b) and lips (l). Numerous ampullary organs (arrowheads) and the lateral line (asterisks) are observed. Scale bar: 200 μm (From Camacho et al. 2011, Anat. Rec. 290: 1178–1189).

on the ventral side. Body length varies from 30 cm in *Pseudoscaphirhynchus hermanni* to 9 m in some specimens of *Huso huso*.

Regarding the digestive tract, all the species pass through a carnivorous life stage. The first phase is benthic zooplantivorous (Baranova and Miroshinichenko 1969), while the feeding habits of the juveniles and adults are broader, fluctuating from filtration, if paddlefish are included (*Polyodon spathula*; Weisel 1973), to ichthyophagy (*H. huso*). The majority of the adults are thought to be opportunistic benthic carnivores, their diet being composed of mollusks, crustaceans, insect larvae, benthic invertebrates, and occasionally fish (Buddington and Christofferson 1985). However, Soriguer et al. (1999) have shown that the adult sturgeon *A. naccarii* also consumes plant material in their natural environment (seeds and vegetal detritus), making it closer to omnivores than carnivores. This would explain a greater disposition of the digestive and metabolic machinery to use carbohydrates than in strictly carnivorous fish such as rainbow trout (*Onchorynchus mykiss*; Furné et al. 2005, 2008, 2012). Similarly, Sanz et al. (2005 patent) found this sturgeon species to have a better utilization of feeds that have a lower protein/energy relation than is adequate for a strict carnivore. In any case, studies on the digestive capacity of different sturgeon species (Medale et al. 1991; Kaushik et al. 1993; Sanz et al. 1997; Gisbert et al. 1999; Burel et al. 2000; Yetty et al. 2004) reveal a high digestibility coefficient very similar to that of other fish such as rainbow trout and a digestive enzyme system similar to that of other vertebrates.

Puberty usually occurs late in all the species, fluctuating between five and 30 years, and the reproductive cycle is not annual. Adults continue to grow throughout their life and some species such as beluga sturgeon (*H. huso*) can reach a 100 years of age and weigh more than 1000 kg. They reproduce in freshwater and most migrate for trophic or reproductive reasons with the exception of *A. ruthenus* and *A. baerii*, which are exclusively freshwater species.

In the last few decades, the natural populations have undergone enormous descents due mainly to human impact. Most species are in danger and need protective measures (Lelek 1987). Sturgeons are considered to be one of the most threatened groups of animals on the IUCN Red List of Threatened Species™ (IUCN 2013).

Functional Morphology of the Digestive Tract of Sturgeon

Gross Morphology

The first descriptions on the anatomy of the digestive tract in sturgeons date to the end of the 19th century (see Hopkins 1895) but in the last quarter of the 20th century, undoubtedly due to the rise in aquaculture, genuine interest arose in determining the anatomy, histology, physiology, and early development of sturgeons and in particular its digestive system (Weisel 1973, 1979; Buddington and Doroshov 1986; Dettlaff et al. 1993; Radaelli et al. 2000; Cataldi et al. 2002).

Anatomically, the structure of the digestive tube varies greatly in the different species. It is relative short (between 70 and 100% of the body length) and with a single loop that affects mainly the glandular stomach (Buddington and Christofferson 1985; Cataldi et al. 2002). The following parts can be distinguished: esophagus, stomach, and intestine with pyloric appendage (Fig. 2).

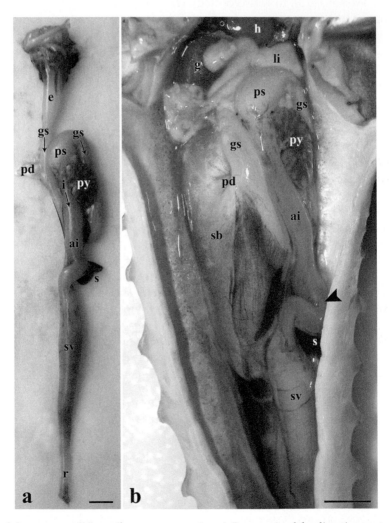

Fig. 2. *Acipenser naccarii* (juvenile, one year specimen). Panoramic of the digestive tract. **a**: note the short esophagus (e), the 'U'-shaped glandular stomach (gs) and the pyloric stomach (ps) with the form of a pouch. The pyloric appendage (py) locates between the intestine (i) and the stomach. **b**: abdominal cavity. Note the double bend (arrowhead) in the last part of the anterior intestine (ai) before connection to the spiral valve (sv). The pneumatic duct (pd) connects to the swim bladder (sb) in the first third of the glandular stomach. g, gallbladder; h, heart; li, liver; r, rectum; s, spleen. Scale bars: a–b, 5 mm.

The esophagus is short (Fig. 2a), approximately 5% the total length of the digestive tube and most of it is found before the thick septum of conjunctival tissue (transverse septum) which, in the manner of a diaphragm, separates the pericardial cavity from the abdominal one (Fig. 2b). Internally, the most notable feature is the presence of numerous pyramidal mucosal papillae throughout the cranial area. On the other hand, the esophagus is long and thin in *A. stellatus,* shorter and thicker in *A. gueldenstaedti,* and voluminous and slightly shorter in *H. huso* (Vajhi et al. 2013).

The stomach, with a caecum structure, is formed by two well-differentiated regions that comprise 35 to 50% of the total length of the digestive tract (Buddington and Christofferson 1985; Vajhi et al. 2013). The anterior region is the glandular stomach, which is 'U' shaped (Fig. 2a) and owes its name to the presence of numerous gastric or oxyntic-peptic glands. Internally, most noticeable is the presence of numerous longitudinal folds that affect the mucosa. In the first third of the glandular stomach is the pneumatic duct that connects with the swim bladder.

The posterior part of the stomach, with the form of a pouch or fist, corresponds to the pyloric regions and receives the name pyloric stomach or gizzard (Fig. 2). Rugae are absent in the gizzard and the inner lining has a velvety appearance. The muscle layers of the pyloric region are hypertrophied, forming what is often termed a 'gizzard', particularly in species which feed on hard food items. The gizzard terminates at the pyloric sphincter (Buddington and Christofferson 1985; Buddington and Doroshov 1986a; Cataldi et al. 2002). The main function of the glandular stomach is to start chemical digestion, while the pyloric stomach compensates for the lack of teeth in the adult stages of Acipenseriformes, grinding the food and permitting it pass in a much more fragmented form into the intestine (Gawlicka et al. 1995; Gisbert et al. 1998).

The pyloric appendage, a semicircular structure and more or less flattened, is located between the intestine and stomach (Fig. 2). The external appearance is lobed. This may represent the fusion of numerous pyloric caeca into a single organ (MacCallum 1886; Weisel 1979). A large number of internal channels within the pyloric appendage corroborate this hypothesis. The channels anastomose into a single opening, which communicates with the intestine. The bile duct enters at the tip of a papilla located within a few cm of the pyloric sphincter in adult fish (MacCallum 1886; Buddington and Doroshov 1986a) with the entrance to the pyloric caecum at approximately the same level.

The intestine is composed of three portions: the anterior intestine, the spiral valve, and the terminal intestine or the rectum (Fig. 2a). The anterior intestine, in its first two thirds is a straight tube but the last part, which some authors call the middle intestine (see Asgari et al. 2014), makes a double bend before connecting to the spiral valve (Fig. 2b). Backflow of the chyme into the intestine is prevented by a valve within the curvature which separates the intestine from the spiral valve (Buddington and Christofferson 1985).

The spiral valve (Fig. 2) together with the anterior intestine comprise up to 45 or 50% of the overall length of the digestive tube. The spiral valve is formed by folds of the mucosa in the form of Archimedes' screw, which run along the entire interior of this part of the digestive tube, with the resulting increase in the absorption surface area.

A short terminal intestine or rectum (2–3% of the gut length) extends from the end of the spiral valve to the anal vent. Longitudinal rugae are present internally. Some authors use the term 'posterior or terminal intestine' for the spiral valve and rectum together (Asgari et al. 2014).

In addition, at the level of the first bend of the anterior intestine and dorsal to it, lies the spleen. The liver constitutes a compact one-lobed organ that occupies most of the anterior region of the abdominal cavity. With respect to the pancreas, as in most Actinopterygii, it can be considered diffuse (Harder 1975), with nodules of variable size and disposition extending through the serous membrane of the intestine and the pyloric appendage. In Siberian sturgeon, Daprá et al. (2009) found a conspicuous part of the pancreas, which they called the 'pancreatic body', on the dorsal side of the pyloric appendage between the stomach and the small intestine, with three large lobes (long, left and right) constituting the rest of the pancreatic tissue that extends from this pancreatic body.

Developmental Histology of the Digestive Tract of Sturgeon

The histological structure of the wall of the digestive tube is very similar to that described for bony fish and far simpler than in terrestrial vertebrates (Buddington and Christofferson 1985). Although four basic layers have been established—mucosa, submucosa, muscle, and serous membrane—it should be taken into account that the *muscularis mucosae* is hardly developed in most of the wall and it is not always possible to distinguish the lamina propria of the submucosa. Consequently, the mucosa appears to continue without interruption with the muscle layer. The main difference in sturgeons with respect to other fish is the presence of a ciliated epithelium over a great part of their surface. In the transition area between the esophagus and stomach, the stratified epithelium gives way to a simple cylindrical epithelium with ciliated cells; the cilia continue towards the final digestive canal. These types of cells appear during embryonic life, first in the spiral valve, and then develop over time, becoming visible in the rest of the digestive tract at the onset of the exogenous phase of feeding in the sturgeon *A. naccarii*. According to Radaelli et al. (2000), the presence of these cells in the digestive tract, rather than being a primitive characteristic, could constitute a feeding adaptation and may play an important part in the mixture of different digestive secretions with food, transport of food particles, and the maintenance of homogeneous mucus layer throughout the digestive tube.

Esophagus

The ontogenic development of the esophagus is similar in most sturgeon species although it does not occur at the same speed, and thus Asgari et al. (2014) have reported that in *H. huso* the process is faster than in other species studied, all of them belonging to the genus *Acipenser*. In general, during the first week after the fish hatches, the esophagus ends up connecting to the stomach. At this time, portions can be differentiated by their cell characteristics: an anterior portion, showing an epithelium of cylindrical cells with microvilli and numerous mucous cells responsible for producing neutral and acid mucosubstances; and a posterior portion, with fewer mucous cells but a high concentration of ciliated cells (Fig. 3), which has a food-transport function

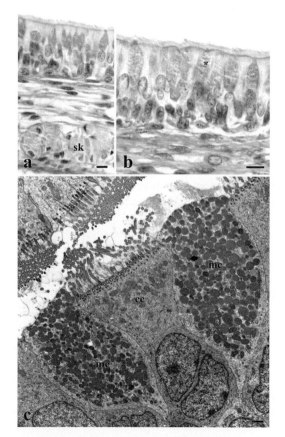

Fig. 3. *Acipenser naccarii* (7 days post-hatching). Esophagus. **a:** periodic acid-Schiff–Alcian Blue staining. Skeletal muscle (sk) associates to the wall of the posterior portion. **b:** Detail of the epithelium with numerous mucous (asterisk) cells and ciliated cells. **c:** transmission electron microscopy. Detail of apical region of ciliated (cc) and mucous (mc) cells. Scale bars: a–b, 10 μm; c, 1 μm.

(Ostos-Garrido et al. 2009; also see Gisbert et al. 1998 and Asgari et al. 2014). However, other authors (Cataldi et al. 2002) state that the posterior portion of the esophagus could in reality be part of the stomach and only the anterior portion should be considered as the esophagus.

We have observed that, in juvenile specimens of *A. naccarii* (about a month after hatching), the epithelium of the posterior portion of the esophagus shows a pseudostratified appearance and contains ciliated and mucous cells. The mucosa of this portion is surrounded by a thin layer of connective tissue (lamina propria-submucosa) and by a circular layer of skeletal muscle (Fig. 3a). Of note, skeletal muscle is a distinctive characteristic of the esophagus which is not shared by any other part of the digestive tract. It is not clear whether this portion of the juvenile esophagus is a transitional zone which will be incorporated into the stomach (Cataldi et al. 2002, see above) or it will remain as a structural component of the definitive esophagus. In adult specimens, at the esophagus-stomach junction, the stratified epithelium of the esophagus abruptly changes to single columnar while the stomach muscle layer is solely comprised of smooth muscle cells.

In juvenile specimens, the fold of the mucosa increases and is later arranged in primary longitudinal folds that are connected to other thinner and smaller secondary folds. In any case, the folding pattern, as in other freshwater fish, is simpler than in marine teleosts (see, e.g., Yamamoto and Hirano 1978). In two-year-old sturgeons (Cataldi et al. 2002) and five-year-old ones (pers. observation, R. Carmona), the epithelium of the esophagus mucosa is already, as in other fish (Buddington and Christofferson 1985), stratified with scattered PAS and Alcian-positive mucous cells (Fig. 4). The ultrastructural characteristics of the mucosa cells are very similar to those found in other vertebrates (Gisbert et al. 1999) and, in terms of function, it has been proposed that they facilitate food transport (Buddington and Christofferson 1985; Gisbert et al. 1998; Gisbert and Doroshov 2003). Furthermore, the neutral glycoconjugates of the esophagus are considered to cooperate in the digestion of substances such as disaccharides and short-chain fatty acids (Sarasquete et al. 2001). Similarly, it cannot be ruled out that sialic acid residues in mucus prevent viruses from recognizing their receptor determinants and also protect the mucosa from the attack of sialidase produced by bacteria (Zimmer et al. 1992).

Also, in adult sturgeons, the esophagus does not appear to fulfill the osmoregulatory function found in teleosts (Cataldi et al. 1995). Nevertheless, Allen et al. (2009), indicated that in the esophagus of juvenile *A. medirostris* can reduce the osmolality of the water, facilitating its absorption in the intestine and as a result the acclimation to hyperosmotic environments, which is of importance in this group of andromous species. Consistent with this finding, ultrastructural evidence in *A. transmontanus* reported by Radaelli et al. (2000) could be the presence of a large quantity of vesicles of various sizes in the adluminal cytoplasm of the surface cells of the esophagus epithelium, related

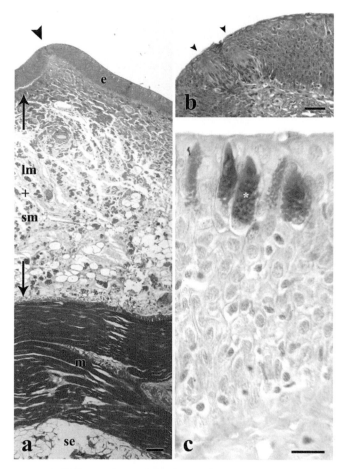

Fig. 4. *Acipenser naccarii* (five years specimen). Esophagus. **a**: Hematoxylin-Eosin staining. Histological structure of the wall. The stratified epithelium (e) is formed by squamous cells and scattered mucous cells. It is not possible to distinguish the lamina propria (lm) from the submucosa (sm). Hence, the mucosa appears in continuity with the skeletal muscle layer (arrows). Arrowhead indicates two taste buds situated in the epithelium. **b**: Detail of the taste buds (arrowheads) shown in **a**. **c**: Alcian Blue pH 2.5 staining. The epithelium contains Alcian-positive mucous cells (asterisk). m, muscle; se: serous membrane. Scale bars: a, 200 µm; b, 50 µm; c, 20 µm.

to osmoregulatory activity. Further in-depth research is needed on this issue to determine definitively whether the esophagus has some osmoregulatory function.

In the white sturgeon *A. transmontanus*, the most distinctive characteristic of the esophagus mucosa is the presence of large longitudinally-oriented myelinated nerve fiber bundles in the lamina propria-submucosa in close relationship with numerous lobules of multilocular adipose tissue, supplied by an extensive vasculature (Radaelli et al. 2000). In *A. naccarii*, myelinated nerve-fiber bundles are also visible at that location and even in the interior of the muscle layer, between the bundles of striated muscle fibers. The association between the lobes of adipose tissue, the vascular system and the nerve fibers suggests a high capacity of mobilization of esophagus adipose tissue in sturgeons (Hung et al. 1997).

Stomach

Sturgeons have a stomach with a caecal structure formed by two well-differentiated zones: the anterior or glandular portion, where the main characteristic is the presence of gastric glands; and a posterior or pyloric portion, which has a highly developed muscle sheath (Buddington and Christofferson 1985). This arrangement of the gastric compartment denotes a high degree of specialization so that the glandular stomach would be involved in the beginning of acid digestion of the food by the secretion of HCl and pepsin, while the pyloric stomach would facilitate the fragmentation of the ball of food by the contractions of its muscular wall (Figs. 5 and 6).

The ontogenesis of the stomach in sturgeons, as in the case of the esophagus, shows no differences other than temporal ones. In none of the species studied was a differentiated gastric compartment detected at the moment of hatching, but a large intraembryonic yolk sac appeared, delineated by an endodermic epithelium encompassing a large central mass of yolk. From this point on, the stomach undergoes a process of differentiation which begins with the compartmentalization of the yolk sac by the appearance of a fold in the dorsal-posterior region, which is divided into two well-defined portions: the gastric anlage and the intestinal anlage (Camacho et al. 2011). Next, the gastric anlage takes on a tubular shape and undergoes a process of torsion until adopting a 'U' shape. Then, the pyloric portion (ascending segment) and the glandular part of the stomach (descending segment) differentiate. Torsion of the gastric cavity occurs approximately 5 days post-hatching (dph) in *A. baerii* (Gisbert et al. 1998) and at 6 dph in *A. medirostris* (Gisbert and Dorsoshov 2003). In *A. naccarii*, the process concludes at roughly 7 dph with the acquisition of the caecal structure typical of juveniles and adults (Camacho et al. 2011).

The histological structure of the stomach appears abruptly in the transition zone of the esophagus, where there is a striking replacement of the stratified epithelium typical of the esophagus by the characteristic of the stomach, which is a simple cylinder. In addition, the striated musculature gives rise to smooth muscle. In the stomach epithelium it is possible to distinguish mucous and ciliated cells that in the glandular region form pits where gastric glands lead occupying most of the lamina propria (Fig. 6). The mucosecretory cells of the gastric epithelium present a cylindrical aspect, the irregular nucleus is located in a basal position, and the apical surface presents short and sparse microvilli without a prominent gycocalyx (Fig. 7). The cytoplasm contains

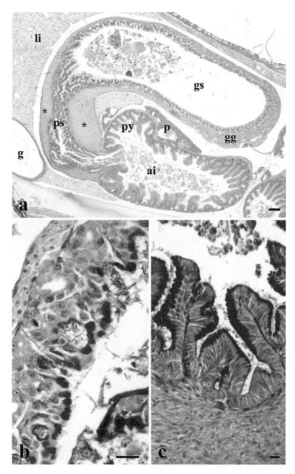

Fig. 5. *Acipenser naccarii* (juvenile, 36 days post-hatching). Proximal portion of the digestive tract. **a**: Alcian Blue pH 2.5 staining; dorsoventral section. Gastric glands (gg) in the glandular stomach (gs) are numerous. The pyloric stomach (ps) shows muscle layers (asterisk) thicker than in other portions. Note mucosal folds in the pyloric appendage (py) and in the anterior intestine (ai). **b**: detail of glandular stomach. Mucous cells are positive to periodic acid-Schiff (PAS) indicating the presence of neutral mucosubstances. **c**: detail of pyloric stomach wall. Mucous cells are PAS-positive. Note thickness of muscle layer. g, gallbladder; li: liver; p: pancreas. Scale bars: a, 200 μm; b, 50 μm; b–c, 20 μm.

abundant electrodense granulation related to the PAS-positive character of the optical microscope (see Fig. 5b). This feature, as in *A. transmontanus*, (Domeneghini et al. 1999), *A. baerii* (Gisbert et al. 1998, 1999), *A. medirostris* (Gisbert and Doroshov 2003), and different fish species with highly varied feeding habits indicate secretion of neutral mucosubstances (Reifel and Travill 1978). Gisbert et al. (1998) proposed that these mucosubstances protect the

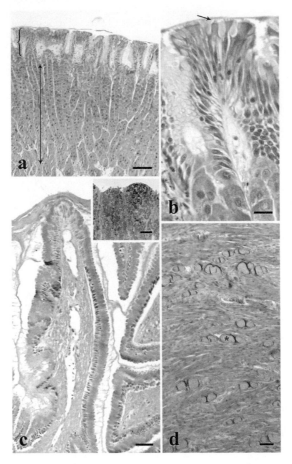

Fig. 6. *Acipenser naccarii* (five years specimen). Stomach. **a:** Hematoxylin-Eosin staining. Mucosa of the glandular region. Gastric pits (square bracket) and gastric glands occupy most of the lamina propria (double arrow). **b:** detail of the superficial mucosa of the glandular stomach. Note the mucous cells and cilia (arrow) in the epithelial surface. **c:** mucosa of the pyloric region. The epithelium of this portion of the stomach is composed predominantly of secretory cells with adluminal cytoplasm full of secretion granules (inset, transmission electron microscopy) that form a superficial mucus film. Note the absence of gastric glands in the lamina propria. **d:** detail of the smooth muscle layer of the pyloric stomach wall. Asterisk, connective tissue. Scale bars: a, 100 µm; b, 20 µm; c 50 µm; inset c, 2 µm; d, 100 µm.

gastric epithelium against the possible damage caused by the digestive process. Other authors suggested that these mucopolysaccharides act as stabilizers of the gastric pH (Scocco et al. 1997), or have related their presence to the absorption of disaccharides and short-chain fatty acids (Grau et al. 1992). It should be taken into account that most of the absorption of amino acids and carbohydrates occurs in the intestine. However, Buddington and Doroshov

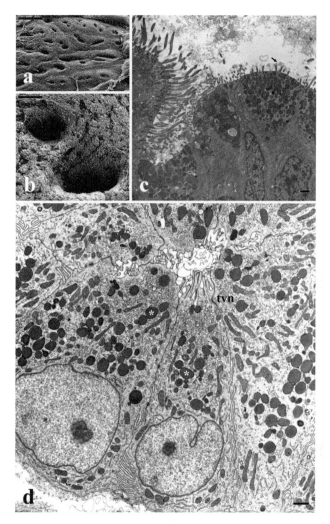

Fig. 7. *Acipenser nacarii.* Glandular stomach under scanning and transmission electron microscopy (SEM and TEM). **a**: SEM. Luminal surface showing the gastric pits. **b**: detail of gastric pits. Note the ciliated surface. **c**: TEM. Detail of the gastric epithelium. Mucosecretory cells show apical microvilli and associated surface mucus (arrow). **d**: Oxyntic-peptic cell showing a highly developed tubulovesicular network (tvn) and abundant zymogen granules (asterisks). Scale bars: a, 200 μm; b, 20 μm; c–d, 1 μm.

(1986) demonstrated that, in sturgeons, the uptake may also occur at the level of the gastric epithelium.

As mentioned above, in the gastric epithelium of sturgeons, both in their glandular portion as well as in the pyloric part, there is a high number of ciliated cells (see Fig. 7). These cells are present in the epithelium from the time at which exogenous feeding begins and remains there lifelong. Their function in the stomach would be the distribution of mucus, homogeneously generating a mucus film over the underlying epithelium to protect it against chemical and physical damage, which could occur during the digestive process. In addition, the presence of cilia facilitates the passage of fluids and food from the esophagus and its mixture with the secretions of the gastric glands themselves. Therefore, the presence of ciliated cells in the gastric epithelium in sturgeons, being a primitive characteristic, constitutes a feeding adaptation to facilitate the movement of food through the digestive system (Radaelli et al. 2000).

In Acipenseriformes and most teleosts, the gastric glands are restricted to the anterior region of the stomach (glandular portion) and are constituted by a single type of cell called the oxyntic-peptic cell (Gawlicka et al. 1995), which makes up tubular glands (Fig. 6a). The oxyntic-peptic cells are pyramidal or cuboid in shape and contain large quantities of granules of zymogen, rounded and strongly osmiophilic, that are generated by the rough endoplasmic reticulum and the Golgi body. Zymogen granules are discharged in the glandular tube by exocytosis. The apical tubulovesicular network of these cells (Fig. 7d) is involved in the production of HCl (Western and Jennings 1970).

Radaelli et al. (2000), studying the gastric glandular cells of adult white sturgeon *A. transmontanus* specimens, found a great quantity of zymogen granules together with a moderately developed tubulovesicular network. The moderate development of this network is consistent with the determinations of gastric pH in *A. transmontanus*, which was in no case lower than 4 (Buddington and Doroshov 1986b).

The appearance of gastric glands constitute one of the main events in the differentiation of the stomach in fishes and, according to Kolkowski (2001), indicates the point at which animals leave the embryonic period and acquire a digestive system capable of digesting and absorbing exogenous food.

In sturgeons, the time of the appearance of gastric glands also shows slight interspecific variations that should be caused by variations in the temperature of maintenance of eggs and embryos (Gisbert and Williot 2002). In *A. naccarii* the gastric glands appear at 6 dph (Camacho et al. 2011), while Gisbert et al. (1998) indicate 9 dph for *A. baerii*, and Gisbert and Doroshov (2003) suggest 10 dph for *A. medirostris*.

However, the presence of these glands does not signify that the stomach is fully functional. Thus, Buddington and Doroshov (1986b) detected pepsin activity in *A. transmontanus* long after the appearance of gastric glands, and Camacho et al. (2011) concluded that *A. naccarii* specimens are anatomically and functionally ready for gastric digestion approximately a month after hatching, when the oxyntic-peptic cells contain abundant zymogen granules and the protease activity is clearly detectable.

The pyloric stomach lacks gastric glands. The smooth muscle, which in fish may reach up to 80% of the thickness of the gastric wall, is the predominant tissue (Figs. 5 and 6). The epithelium of this portion of the stomach is composed predominantly of secretory cells with adluminal cytoplasm full of secretion granules (Fig. 6, inset) that form a superficial mucus film. Some ciliated cells appear intercalated between secretory cells.

The stomach wall contains large nerve-fiber bundles similar to those described for the esophagus. They are situated in the tunica propria-submucosa. These nerve bundles contain myelinated and unmyelinated nerve fibers as well as neuronal bodies (Domeneghini et al. 1999).

Pyloric Appendage

Many fish have developed a series of blind tubes around the pyloric sphincter called pyloric caeca. The fact that the number of caeca is not constant has been used as a character to differentiate species; thus, we can find from just a few (e.g., 7–12 in mullet) to several hundred (e.g., in tuna). Clearly, stomach-less fish also lack pyloric caeca. In Acipenseriformes, the pyloric caeca constitute a compact organ, the pyloric appendage, that takes on a flat, triangular shape and is situated between the stomach and the intestine (Buddington and Christofferson 1985). The pyloric appendage appears to represent the fusion of numerous individual caeca into a single organ. This is corroborated by the existence of numerous internal canals that end in a single aperture that communicates with the intestine (Figs. 8 and 9).

Around 5 dph to 6 dph, the anterior intestinal wall evaginates in the zone of the pyloric sphincter. This evagination self-organizes in the form of blind pouches to form the pyloric appendage (Sanz et al. 2011; Ostaszewska et al. 2011; Asgari et al. 2014). The mucosa of the pyloric appendage becomes differentiated before the onset of exogenous feeding and the epithelium acquires the typical columnar structure that is also found in the anterior and middle intestine.

Certainly, the histological characteristics of the pyloric appendage are similar to those of the anterior and middle intestine since each blind pouch is formed by mucosal folds limited by simple columnar epithelium rich in goblet cells (Figs. 8d and 9c). In sturgeons, the pyloric caeca perform similar functions to those described in teleosts (Gawlicka et al. 1995). Among these is the digestive function supplementary to the stomach and intestinal activity (Buddington et al. 1997), carbohydrate and fat absorption (Ahearn et al. 1992) and reabsorption of water and ions. Also, they increase the intestinal surface and, therefore, increase the transit time for several substrates (Kapoor 1975; Fänge and Grove 1979).

Filling of the pyloric caeca appears to be due to retroperistalsis. Ronnestad et al. (2000) suggested that the retrograde contractions of the intestine direct part of the intestinal content towards the pyloric sphincter in such a way that the intraluminal pressure of the duct increases and permits the filling of the pyloric caeca. Also, it has been proposed that the retroperistaltic movements could help in mixing the chyme with enzymatic secretions (such as bile and exocrine pancreatic enzymes). A similar mechanism has been observed in the small spotted dogfish *Scyliorhinus canicula* (Andrews and Young 1993).

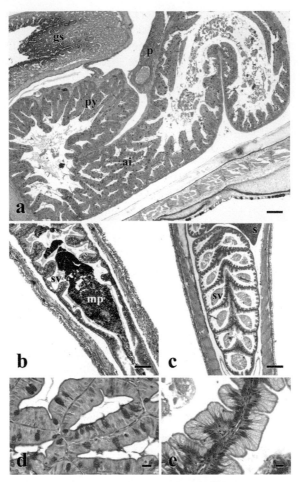

Fig. 8. *Acipenser naccarii* (**a**, **c**, **d**, **e**, 36 days post-hatching; **b**, 14 days post-hatching). Digestive tract of sturgeon. **a**: periodic acid-Schiff. The histological characteristics of the pyloric appendage (py) are similar to those of the anterior intestine (ai). **b**: Hematoxylin-Eosin staining. Final portion of spiral valve (sv) showing melanin plug (mp). **c**: panoramic of the spiral valve (sp) showing folds of the mucosa. **d**: periodic acid-Schiff. Histological characteristics of anterior intestine (ai). Note mucosal folds lined by simple columnar epithelium rich in goblet cells. **e**: Hematoxylin-Eosin. Detail of mucosal folds in the spiral valve. gs, glandular stomach; p, pancreas. s, spleen. Scale bars: a, 200 µm; b, 100 µm; c, 500 µm; d–e, 20 µm.

Intestine

The sturgeon intestine develops during the first few days post-hatching. The primitive intestine undergoes a process of specialization which leads to division into anterior intestine, spiral intestine (spiral valve) and rectum. The spiral valve is the first segment to develop (Buddington 1985). As it occurs

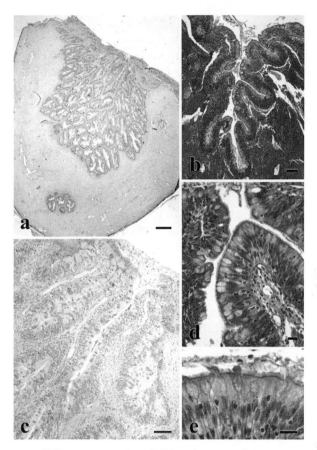

Fig. 9. *Acipenser naccarii* (five years specimen). Digestive tract of sturgeon. **a**: Alcian Blue pH 2.5. The pyloric appendage reunites a large number of pyloric caeca into a single struture. **b**: Hematoxylin-Eosin staining. Spiral valve. Note prominent mucosal folds. **c**: Alcian Blue pH 2.5. Mucosal folds in the pyloric appendage show a high number of goblet cells. **d** and **e**: Hematoxylin-Eosin staining. Mucosal folds in the spiral valve. Note, as in the pyloric caeca, the high number of goblet cells. Scale bars: a, 500 μm; b, 100 μm; c, 100 μm; d–e, 20 μm.

with other portions of the digestive tract, interspecific temporal differences have been reported. For instance, the appearance of the intestine in *H. huso* coincides with the hatching phase (Asgari et al. 2014); in *A. baerii* it occurs between days 1 and 2 post-hatching (Gisbert et al. 1998); in *A. medirostris* (Gisbert and Doroshov 2003) and *A. naccarii* (Boglione et al. 1999; Camacho et al. 2011; Cataldi et al. 2002) around day 3 and, in *A. ruthenus*, towards day 4 (Wegner et al. 2009).

Towards the end of the endogenous feeding period in *A. naccarii* embryos, the epithelium lining the anterior intestine and the spiral valve is a columnar epithelium formed by enterocytes, ciliated cells and interposed goblet cells.

In the rectum, however, the epithelium is stratified, contains a large quantity of goblet cells and lacks ciliated cells (Radaelli et al. 2000). In the intestine of *A. medirostris*, the first goblet cells appear around 6 dph, but are inactive at this stage. It is only later (from 7 dph to 8 dph) that these cells begin to produce neutral and acidic mucosubstances (carboxylated and sulfated; Gisbert and Doroshov 2003).

As a sub-product of yolk-reserve utilization, an accumulation of melanin is produced in the final portion of the spiral valve. The continuous deposit of melanin granules forms a mass that ultimately occupies the entire lumen of the posterior portion of the spiral valve, i.e., 'the melanin plug'. The plug is expelled immediately after the first exogenous feeding (Fig. 8b). This occurs, depending on the species, between 7 dph in *A. gueldenstaedti*, and 14 dph in *H. huso* (Asgari et al. 2014).

From the moment at which exogenous feeding begins, the posterior portion of the gastrointestinal tract is anatomically similar to that of adults (see Figs. 8 and 9). The most notable feature is the general increase in length, as well as the increase in the number and height of the mucosal folds (Gisbert et al. 1998).

The intestinal mucosa shows, after the developmental period, a certain degree of folding. The folds are lined by a simple columnar epithelium (Figs. 9–12) formed by brush-border cells or enterocytes, ciliated cells and intercalated goblet cells (Detlaff et al. 1993).

The enterocytes have a free border which exhibits closely packed microvilli that form a typical brush border (Radaelli et al. 2000). The most important cytoplasmic characteristic to emphasize is the presence of apical vacuoles related to the absorption of lipids, amino acids, and proteins (Caceci 1984). Supranuclear vacuoles have also been implicated in hydromineral exchange (Noaillac-Depeyre and Gas 1973). The enterocytes of the pyloric portion are involved in the absorption of lipids and amino acids while those of the middle region of the intestine maintain the ability to absorb macromolecules from the initial stages of development until the adult phase (Nose 1989).

Ciliated cells are characterized by the presence of numerous, long, 9+2 cilia that form distinct ciliated tufts (Fig. 10). Their function, as stated above, is to help in the transport of food particles and in the maintenance of a constant mucus layer on the intestinal adluminal surface. This occurs not only in the embryonic stages but throughout the entire life of the fish (Gisbert et al. 1998).

Goblet cells show similar histological characteristics to those observed in other osteichthyes, or in vertebrates in general (Specian and Oliver 1991; Tibbetts 1997; Gisbert et al. 1999). However, the composition of the glycoconjugates produced by these cells varies between species (Domeneghini et al. 1999; Gisbert et al. 1999; Gisbert and Dorshov 2003). This is probably due to differences in feeding habits and to the characteristics of the habitats (Scocco et al. 1997; Tibbetts 1997; Domeneghini et al. 1998).

In addition to the basic three cell types, the intestine epithelium also contains blood cells of the white series and ramified cells resembling interdigitating dendritic cells. However, they could also be M-cells. Nevertheless, it would

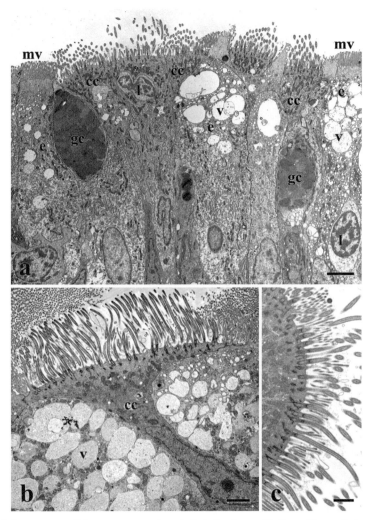

Fig. 10. *Acipenser naccarii* (two years specimen). Spiral valve epithelium. Transmission electron microscopy. **a:** the simple columnar epithelium contains enterocytes (e), ciliated cells (cc) and intercalated goblet cells (gc). Note infiltrated white cell (l). **b:** ciliated cells (cc). **c:** detail of the apical pole of the ciliated cells. The tubular structure of the cilia and the ciliary roots are evident. mv, microvilli; v, enterocite vacuoles. Scale bars: a, 5 μm; b, 2 μm; c, 1 μm.

be necessary to use immunocytochemistry to confirm the identity of the latter, which we have found in *A. naccarii* by electron microscopy (Fig. 11). Several types of endocrine cells also appear distributed between the cells of the gastric and the intestinal epithelium. These cells are abundant (Salimova and Fehér 1982) and have the same ultrastructural characteristics than those described in higher vertebrates and have been classified according to size, texture, and electron density of their granulations (Radaelli et al. 2000). In

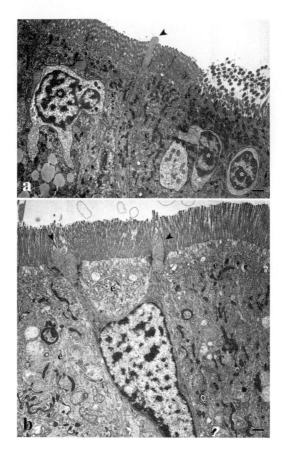

Fig. 11. *Acipenser naccarii* (one year juvenile specimen). Anterior intestine epithelium. Transmission electron microscopy. **a**: the columnar epithelium is mostly formed by enterocytes (e) and ciliated cells (cc). Note the presence of a cytoplasmic process (arrowhead) possibly associated to a dendritic cell or to a M-cell, Asterisks, intraepithelial white cells. **b**: Note cellular processes (arrowheads) from dendritic cells or M-cells. e, enterocytes. Scale bars: a–b, 1 μm.

fishes, enteroendocrine cells are involved in important functions regulating gut physiology such as motility (even complex functions like peristalsis), secretion, regional blood flow, and possibly also other functions which are less well elucidated (Holmgren and Olsson 2009).

The lamina propria of the intestine wall, especially in the spiral valve, contains large numbers of lymphocytes. Several follicle-like accumulations of lymphocytes are found in histological sections through the thickened margin of the spiral valve (Fänge 1986). The lymphoid tissues of the intestine, especially the extensive lymphocytic accumulations in the spiral valve, are probably involved in immune reactions against bacteria and/or parasites (Weisel 1973).

The presence of lymphoid tissue in the intestinal mucosa in different species of fish has been related to the so-termed Gut-Associated Lymphoid Tissue (GALT) described in the intestinal mucosa of higher vertebrates (Rombout et al. 1985, 1986; Georgopoulou et al. 1986; Hart et al. 1988; Abelli et al. 1997). GALT, although a lymphoid organ as such, is found to be associated with major defensive functions. As in the skin and in the gills, the presence in the intestinal mucosa of lymphocytes capable of secreting antibodies under immunization has been demonstrated (Georgopoulou and Vernier 1986). A study by Davidson et al. (1991) demonstrates that lymphocytes isolated from lymphocyte accumulations in the gut of the rainbow trout are capable of producing the Macrophage-Activating Factor (MAF), a function assigned to the T lymphocytes in higher vertebrates (Graham and Secombes 1990).

The muscle layer of the intestine wall is formed by two layers of smooth muscle cells arranged in different directions. It is surrounded by loose connective tissue and abundant blood vessels. Externally, the gut wall is limited by the visceral membrane of the serosa. The peristaltic intestinal movements are found to be, as in other fish, under the control of the nervous and endocrine systems, and to present characteristics very similar to those of mammals (Olsson and Holmgren 2001). As in higher vertebrates, the presence of food in the stomach excites the vagus nerve which, in turn, regulates the contraction of the gastric wall and the passage of food to the intestine. The latter induces the synthesis of a secretine-type molecule which favors the release of pancreatic juices. Although cholecystokinin also triggers the release of pancreatic enzymes in several species, the best-known role of this hormone in fish is the stimulation of the contraction of the bile vesicle (Aldman and Holmgren 1987).

Morpho-functional Specialization of the Intestine

Anterior or Pyloric Intestine

The enterocytes of this portion of the intestine (Fig. 12a) appear to be involved in the absorption of lipids and amino acids. They are columnar cells with brush border and a great quantity of vacuoles as well as multivesicular bodies and lysosomes in diverse degrees of functioning (Gawlicka et al. 1995). The cytoplasm of the enterocytes also contains lipoprotein vesicles that appear to be the morphological equivalent of the lipid bodies related to the absorption of the fatty component of the diet, described in the intestinal tract of teleosts (Radaelli et al. 2000). These ultrastructural characteristics suggest that the columnar cells of the anterior intestine show a maximum lipid-absorption capacity (Buddington and Christofferson 1985).

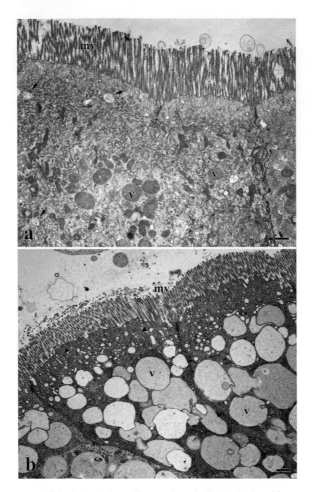

Fig. 12. *Acipenser naccarii* (one year juvenile specimen). Transmission electron microscopy. **a:** enterocytes in the anterior intestine show long microvilli (mv), numerous vacuoles and vesicles (v) and multivesicular bodies (arrows). **b:** supranuclear cytoplasm and apical surface of spiral valve enterocytes showing microvilli (mv), numerous vesicles (v) and small pinocytotic vacuoles (arrowheads). Scale bars: a–b, 1 μm.

Spiral Valve

In Chondrichthyes, Acipenseriformes and Holostei, the inner surface of the short middle intestine is increased by the formation of a spiral valve whose morphology varies from several screw turns to a series of interfitting cones (Stevens and Hume 1995).

In Acipenseriformes, the spiral valve represents between 60 and 70% of the post-gastric alimentary canal and constitutes an anatomical adaptation to the type of feeding (Buddington and Christofferson 1985). The number of turns of the spiral valve reflects interspecific variations, ranging from four to five for the genus *Scaphirhynchus* (Weisel 1979) to six to eight for other Acipenserids (Gisbert et al. 1998; Gisbert and Doroshov 2003). The histological organization of the wall of this portion is a continuation of that found in the anterior intestine, although the mucosal folds are shorter in this zone (see Fig. 8).

Gisbert et al. (1998) confirmed that the differential characteristic of the enterocytes of this region is the presence in the cytoplasm of a great number of small vacuoles in charge of gathering proteins by pinocytosis (Fig. 12b). This ability to absorb protein macromolecules, which is preserved over the lifespan of the animal, has nutritional significance (Buddington 1991) and, as in the genus *Polyodon*, may be related to the capture of antigens (Petrie-Hanson and Peterman 2005).

Rectum

The distal intestine or rectum is a short structure that can make up between 2 and 3% of the total length of the digestive tube (Buddington and Christofferson 1985). It extends from the end of the spiral valve to the anus. According to Cataldi et al. (2002), the stratified epithelium of the rectum shows no ciliated cells but there is a large quantity of goblet cells. The epithelial cells of this intestinal portion have shorter microvilli than those of the enterocytes of the anterior intestine or of the spiral valve (Radaelli et al. 2000).

In summary, the anatomical, histological and functional organization of the digestive tract of sturgeons can be said to be similar, in general terms, to that of the teleosts. However, we can list several striking particularities such as: the disappearance of teeth in the adult stage; a short, well-developed esophagus without an osmoregulatory function; the presence of pyloric caeca that associate to form a single structure termed the pyloric appendage; and a spiral valve as in chondrichthyes and holostei. Also, in our opinion, the most characteristic feature in relation to other fish is the presence of ciliated cells distributed over a large part of the digestive epithelium. The presence of ciliated cells, despite of a primitive trait, appears to constitute an alimentary adaptation that plays an important role in mixing different digestive secretions with the food, in the transport of food particles, and in the maintenance of a homogeneous mucus layer throughout the digestive tube that protects the underlying epithelium against chemical or physical damage during digestion. Finally, the presence of an adequate digestive enzyme pool, comparable to other monogastric vertebrates, ensures sturgeons an appropriate digestive use of food. The digestive structure and the functional characteristic of the adult stage are reached in sturgeons at around one month after hatching.

References

Abelli, L., S. Picchietti, N. Romano, L. Mastrolia and G. Scapigliati. 1997. Immunohistochemistry of gut-associated lymphoid tissue of the sea bass *Dicentrarchus labrax* (L.). Fish Shellfish Immunol. 7: 235–245.

Ahearn, G.A., R.D. Behnke, V. Zonno and C. Storelli. 1992. Kinetic heterogeneity of Na-D-glucose cotransport in teleost gastrointestinal tract. Am. J. Physiol. 263: 1018–1023.

Aldman, G. and S. Holmgren. 1987. Control of gallbladder motility in the rainbow trout, *Salmo gairdneri*. Fish Physiol. Biochem. 4: 143–155.

Allen, P.A., J.J. Cech and D. Kültz. 2009. Mechanisms of seawater acclimation in a primitive, anadromous fish, the green sturgeon. J. Comp. Physiol. B 179: 903–920.

Andrews, P.L.R. and J.Z. Young. 1993. Gastric motility patterns for digestion and vomiting evoked by sympathethic nerve stimulation and 5-hydroxytryptamine in the dogfish *Scyliorhinus canicula*. Phil. Trans. R. Soc. Lond. B 342: 363–380.

Asgari, R., G. Rafiee, S. Eagderi, R. Shahrooz, H. Poorbagher, N. Agh and E. Gisbert. 2014. Ontogeny of the digestive system in hatchery produce Beluga (*Huso huso* Linnaeus 1758); a comparative study between *Belulga* and genus *Acipenser*. Aquaculture Nutrition 20: 595–608.

Baranova, V.P. and M.P. Miroshnichenko. 1969. Conditions and prospects for culturing sturgeon fry in the Volgograd sturgeon nursery. Hydrobiol. J. 5: 63–67.

Bemis, W.E., E.K. Findeis and L. Grande. 1997. An overview of Acipenseriformes. Environ. Biol. Fish. 48: 25–71.

Billard, R. and G. Lecointre. 2001. Biology and conservation of sturgeon and paddlefish. Rev. Fish Biol. Fisheries 10: 355–392.

Boglione, C., P. Bronzi, E. Cataldi, S. Serra, F. Gagliardi and S. Cataudella. 1999. Aspects of early development in the Adriatic sturgeon, *Acipenser naccarii*. J. Appl. Ichthyol. 15: 207–213.

Buddington, R.K. 1985. Digestive secretion of lake sturgeon (*Acipenser fulvescens*) during early development. J. Fish Biol. 26: 715–723.

Buddington, R.K. 1991. Ontogenic development of sturgeons: selected physiological examples. pp. 53–63. *In*: P. Williot (ed.). Proceedings of the First International Symposium on the Sturgeon, CEMAGREF-DICOVA, Bordeaux.

Buddington, R.K. and J.P. Christofferson. 1985. Digestive and feeding characteristics of the chondrosteans. Environ. Biol. Fish. 14: 31–41.

Buddington, R.K. and S.I. Doroshov. 1986a. Structural and functional relations of the white sturgeon alimentary canal (*Acipenser transmontanus*). J. Morphol. 190: 201–213.

Buddington, R.K. and S.I. Doroshov. 1986b. Development of digestive secretions in white sturgeon juveniles (*Acipenser transmontanus*). Comp. Biochem. Physiol. 83A: 233–238.

Buddington, R.K., A. Krogdahl and A.M. Bakke-Mckellep. 1997. The intestines of carnivorous fish: structure and functions and the relations with diet. Acta Physiol. Scand. Suppl. 638: 67–80.

Burel, C., T. Boujard, F. Tulli and S.J. Kaushik. 2000. Digestibility of extruded peas, extruded lupin, and rapeseed meal in rainbow trout (*Oncorhynchus mykiss*) and turbot (*Psetta maxima*). Aquaculture 112A: 167–177.

Caceci, T. 1984. Scanning electron microscopy of goldfish, *Carassius auratus*, intestinal mucosa. J. Fish Biol. 25: 1–12.

Camacho, S., R. Carmona, J.I. Llorente, A. Sanz, M. García-Gallego, A. Domezain, N. Dominguez and M.V. Ostos-Garrido. 2011. Stomach development in the sturgeon *Acipenser naccarii*: histoenzymatic and ultrastructural analysis. J. Appl. Ichthyol. 27: 693–700.

Cataldi, E., E. Ciccotti, P. Di Marco, O. Di Santo, P. Bronzi and S. Cataudella. 1995. Acclimation trials of juvenile Italian sturgeon to different salinities: morpho-physiological descriptors. J. Fish Biol. 47: 609–618.

Cataldi, E., C. Albano, C. Boglione, L. Dini, G. Monaco, P. Bronzi and S. Cataudella. 2002. *Acipenser naccarii*: fine structure of the alimentary canal with references to its ontogenesis. J. Appl. Ichthyol. 18: 329–337.

Daprà, F., F. Gai, G.B. Palmegiano, B. Sicuro, M. Falzone, K. Cabiale and M. Galloni. 2009. Siberian sturgeon (*Acipenser baeri*, Brandt JF 1869) gut: anatomic description. Int. Aquat. Res. 1: 45–60.

Davidson, G.A., A.E. Ellis and C.J. Secombes. 1991. Cellular responses of leucocytes isolated from the gut of rainbow trout, *Oncorhynchus mykiss* (Walbaum). J. Fish Dis. 14: 651–659.

Dettlaff, T.A., A.S. Ginsburg and O.I. Schmalhausen. 1993. Sturgeon Fishes. Developmental Biology and Aquaculture. Springer-Verlag, Berlin.

Domeneghini, C., R. Pannelli Straini and A. Veggetti. 1998. Gut glyconjugates in *Sparus aurata* L. (Pisces, Teleostei). A comparative histochemical study in larval and adult ages. Histol. Histopathol. 13: 359–372.

Domeneghini, C., S. Arrighim, G. Radaelli, G. Bosi and F. Mascarello. 1999. Morphological and histochemical peculiarities of the gut in the white sturgeon, *Acipenser transmontanus*. Eur. J. Histochem. 43: 135–145.

Fänge, R. 1986. Lymphoid organs in sturgeons (Acipenseridae). Vet. Immunol. Immunopathol. 12: 153–161.

Fänge, R. and D. Grove. 1979. 4. Digestion. pp. 161–260. *In*: W.S. Hoar, D.J. Randall and J.R. Brett (eds.). Fish Physiology, Vol. VIII, Bioenergetics and Growth. Academic Press, New York.

Furné, M., M.C. Hidalgo, A. López, M. García Gallego, A.E. Morales, A. Domezain and A. Sanz. 2005. Digestive enzyme activities in Adriatic sturgeon (*Acipenser naccarii*) and rainbow trout (*Oncorhynchus mykiss*). A comparative study. Aquaculture 250: 391–398.

Furné, M., M. García-Gallego, M.C. Hidalgo, A.E. Morales, A. Domezain, J. Domezain and A. Sanz. 2008. Effect of starvation and refeeding on digestive enzyme activities in sturgeon (*Acipenser naccarii*) and trout (*Oncorhynchus mykiss*). Comp. Biochem. Physiol. A 149: 420–425.

Furné, M., A.E. Morales, C.E. Trenzado, M. García-Gallego, M.C. Hidalgo, A. Domezain and A. Sanz. 2012. The metabolic effects of prolonged starvation and refeeding in sturgeon and rainbow trout. J. Comp. Physiol. B 182: 63–76.

Gardiner, B.G. 1984. The relationships of the palaeoniscid fishes, a review based on new specimens of *Mimia* and *Moythomasia* from the Upper Devonian of Western Australia. Bull. Br. Mus. Nat. Hist. Geol. 37: 173–248.

Gawlicka, A., S.J. The, S. Hung, D. Hinton and J. De La Noüe. 1995. Histological and histochemical changes in the intestine tract of white sturgeon larvae during ontogeny. Fish Physiol. Biochem. 14: 357–371.

Georgopoulou, V. and J.M. Vernier. 1986. Local immunological response in the posterior intestinal segment of the rainbow trout after oral administration of macromolecules. Dev. Comp. Immunol. 10: 529–537.

Georgopoulou, V., M.F. Sire and J.M. Vernier. 1986. Immunological demonstration of intestinal absorption and digestion of protein macromolecules in the trout (*Salmo gairdneri*). Cell Tiss. Res. 245: 387–395.

Gisbert, E. and P. Williot. 2002. Advances in larval rearing of Siberian sturgeon. J. Fish Biol. 60: 1071–1092.

Gisbert, E. and S. Doroshov. 2003. Histology of the developing digestive system and the effect of food deprivation in larval green sturgeon (*Acipenser medirostris*). Aquat. Liv. Res. 16: 77–89.

Gisbert, E., A. Rodríguez, F. Castelló-Orvay and P. Williot. 1998. A histological study of the development of the digestive tract of Siberian sturgeon (*Acipenser baerii*) during early ontogeny. Aquaculture 167: 195–209.

Gisbert, E., M.C. Sarasquete, P. Williot and F. Castello-Orvay. 1999. Histochemistry of the development of the digestive system of Siberian sturgeon during early ontogeny. J. Fish Biol. 55: 596–616.

Graham, S. and C.J. Secombes. 1990. Cellular requirements for lymphokine secretion by rainbow trout *Salmo gairdneri* leucocytes. Dev. Comp. Immunol. 14: 59–68.

Grau, A., S. Crespo, M. Sarasquete and M. González de Canales. 1992. The digestive system of the amberjack *Seriola dumerili*, Risso: a light and scanning microscopic study. J. Fish Biol. 41: 287–303.

Harder, W. 1975. *In*: Hans Richardz Publications editor. Anatomy of Fishes. Parts I and II. Schweizerbart'sche Verlagsbuchhandlung, Stuttgart, West Germany.

Hart, S., A.B. Wrathmell, J.E. Harris and T.H. Garyson. 1988. Gut immunology in fish: a review. Dev. Comp. Immunol. 12: 453–480.

Holmgren, S. and C. Olsson. 2009. The neuronal and endocrine regulation of gut function. pp. 467–512. *In*: N.J. Bernier, G. van der Kraak, A.P. Farrell and C.J. Brauner (eds.). Fish Neuroendocrinology. Academic Press, Amsterdam.

Hopkins, C.S. 1895. On the enteron of American ganoids. J. Morph. 11: 411–422.

Hung, S.S.O., W. Liu, H. Li, T. Storebakken and Y. Cui. 1997. Effect of starvation on some morphological and biochemical parameters in white sturgeon, *Acipenser transmontanus*. Aquaculture 151: 357–363.

Kapoor, B.C., H. Smit and I.A. Verighira. 1975. The alimentary canal and digestion in teleosts. pp. 102–219. *In*: F.S. Russell and C.M. Yonge (eds.). Advances in Marine Biology, Vol. 13. Academic Press, London.

Kaushik, S.J., J. Breque and D. Blanc. 1993. Apparent amino acid availability and plasma-free amino acid levels in Siberian sturgeon (*Acipenser baeri*). Comp. Biochem. Physiol. 107A: 433–438.

Kolkowski, S. 2001. Digestive enzymes in fish larvae and juveniles-implications and applications to formulated diets. Aquaculture 200: 181–201.

Lelek, A. 1987. The Freshwater Fishes of Europe, Volume 9: Threatened Fishes of Europe Aula-Verlag. Wiesbaden, 343 p.

MacCallum, A.B. 1886. The alimentary canal and pancreas of *Acipenser*, *Amia* and *Lepidosteus*. J. Anat. Physiol. 20: 604–636.

Medale, F., D. Blanc and S.J. Kaushik. 1991. Studies on the nutrition of Siberian sturgeon, *Acipenser baeri*. II. Utilization of dietary non-protein energy by sturgeon. Aquaculture 93: 143–154.

Noaillac-Depeyre, J. and N. Gas. 1978. Absorption of protein macromolecules by the enterocytes of the carp (*Cyprinus carpio* L.). Z Zellforsch 146: 525–541.

Norris, H.W. 1925. Observations upon the peripheral distribution of the cranial nerves of certain ganoid fishes (*Amia*, *Lepidosteus*, *Polyodon*, *Scaphirhynchus*, and *Acipenser*). J. Comp. Neurol. 39: 345–416.

Nose, T. 1989. Protein and Aminoacid Nutrition in Fish. Roche Research Prize for Animal Nutrition, to Takeshi Nose. November 9th, 1989. Switzerland : F. Hoffmann-La Roche pp. 51–69.

Olsson, C. and S. Holmgren. 2001. The control of gut motility. Comp. Biochem. Physiol. 128A: 481–503.

Ostaszewska, T., R. Kolman, M. Kamaszewski, G. Wiszniewski, D. Adamek and A. Duda. 2011. Morphological changes in digestive tract of Atlantic sturgeon (*Acipenser oxyrinchus*) during organogenesis. Int. Aquacult. Res. 3: 101–105.

Ostos-Garrido, M.V., J.I. LLorente, S. Camacho, M. García-Gallego, A. Sanz, A. Domezain and R. Carmona. 2009. Histological, histochemical and ultrastructural changes in the digestive tract of sturgeon *Acipenser naccarii* during early ontogeny. pp. 121–136. *In*: R. Carmona, A. Domezain, M. Garcia Gallego, J.A. Hernando, F. Rodriguez and M. Ruiz Rejón (eds.). Biology, Conservation and Sustainable Development of Sturgeon. Springer Science Business Media B.V., Heidelberg.

Petrie-Hanson, L. and A.E. Peterman. 2005. American paddlefish leukocytes demonstrate mammalian-like cytochemical staining characteristics in lymphoid tissues. J. Fish Biol. 66: 1101–1115.

Radaelli, G., C. Domeneghini, S. Arrighi, M. Francolini and F. Mascarello. 2000. Ultrastructural features of the gut in the white sturgeon, *Acipenser transmontanus*. Histol. Histopathol. 15: 429–439.

Reifel, C.W. and A.A. Travill. 1978. Structure and carbohydrate histochemistry of the stomach in eight species of teleosts. J. Morphol. 158: 155–168.

Rombout, J.H.W.M., C.H.J. Lamers, M.H. Helfrich, A. Dekker and J.J. Taverne-Thiele. 1985. Uptake and transport of intact macromolecules in the intestinal epithelium of carp (*Cyprinus carpio* L.) and the possible immunological implications. Cell Tiss. Res. 239: 519–530.

Rombout, J.H.W.M., L.J. Blok, C.H.J. Lamers and E. Egberts. 1986. Immunization of carp (*Cyprinus carpio*) with a *Vibrio anguillarum* bacterin: indications for a common mucosal immune system. Dev. Comp. Immunol. 10: 341–351.

Rønnestad, I., C.R. Rojas-García and J. Skadal. 2000. Retrograde peristalsis; a possible mechanism for filling the pyloric caeca? J. Fish Biol. 56: 216–218.

Salimova, N. and E. Fehér. 1982. Innervation of the alimentary tract in chondrostean fish (Acipenseridae). A histochemical, microspectrofluorimetric and ultrastructural study. Acta Morphol. Hung. 30: 213–222.

Sanz, A., A. Domezain, C. Trenzado and M. García-Gallego. 1997. Primera aproximación al conocimiento de la utilización digestiva de los macronutrientes por el esturión Acipenser naccarii. Actas del VI Congreso Nacional de Acuicultura. 9–11 Julio, 1997, Cartagena, Spain pp. 653–659.

Sanz, A., M. García Gallego, M. de la Higuera, J. Domezain and A. Domezain. 2005. *Piensos para la alimentación del esturión desde los seis meses de edad hasta tamaño comercial.* Universidad de Granada. N° de solicitud Patente.: 200100776. Concesión 18.02.2005.

Sanz, A., J.I. Llorente, M. Furné, M.V. Ostos-Garrido, R. Carmona, A. Domezain and M.C. Hidalgo. 2011. Digestive enzymes during ontogeny of the sturgeon *Acipenser naccarii*: intestine and pancreas development. J. Appl. Ichthyol. 27: 1139–1146.

Sarasquete, C., E. Gisbert, L. Ribeiro, L. Vieira and M.T. Dinis. 2001. Glycoconjugates in epidermal, branchial and digestive mucous cells and gastric glands of gilthead sea bream, *Sparus aurata*, *Senegal sole*, *Solea senegalensis* and Siberian sturgeon, *Acipenser baeri* development. Eur. J. Histochem. 45: 267–278.

Scocco, P., G. Menghi and P. Ceccarelli. 1997. Histochemical differentiation of glycoconjugates occurring in the tilapine intestine. J. Fish Biol. 51: 848–857.

Soriguer, M.C., J. Domezain, A. Domezain, M. Bernal, C. Esteban, J.C. Pumar and J.A. Hernando. 1999. An approximation of the feeding habits of *Acipenser naccarii* (Bonaparte 1836) in an artificial river. J. Appl. Ichthyol. 15: 348–349.

Specian, R.D. and M.G. Oliver. 1991. Functional biology of intestinal goblet cells. Am. J. Physiol. 260: 183–193.

Stevens, C.E. and I.D. Hume. 1995. Comparative Physiology of the Vertebrate Digestive System, 2nd ed. Cambridge University Press, Cambridge, 400 p.

Tibbetts, I.R. 1997. The distribution and function of mucous cells and their secretions in the alimentary tract of *Arrhamphus sclerolepis krefftii*. J. Fish Biol. 50: 809–820.

Vajhi, A.R., O. Zehtabvar, M. Masoudifard, M. Moghim and M. Akhtarzade. 2013. Digestive system anatomy of the *Acipenser persicus*: New features. Iranian J. Fish. Sci. 12: 939–946.

Wegner, A., T. Ostaszewska and W. Rozek. 2009. The ontogenetic development of the digestive tract and accessory glands of sterlet (*Acipenser ruthenus* L.) larvae during endogenous feeding. Rev. Fish Biol. Fish. 19: 431–444.

Weisel, G.F. 1973. Anatomy and histology of the digestive system of the Paddlefish (*Polyodon spathula*). J. Morphol. 140: 243–256.

Weisel, G.F. 1979. Histology of the feeding and digestive organs of the shovelnose sturgeon, *Scaphirhynchus platorynchus*. Copeia 1979: 518–525.

Western, J. and J. Jennings. 1970. Histochemical demostration of the hydrochloric acid in the gastric tubules of teleost using *in vivo* Prusian blue technique. Comp. Biochem. Physiol. 35: 879–884.

Yamamoto, M. and T. Hirano. 1978. Morphological changes in the esophageal epithelium of the eel, *Anguilla japonica*, during adaptation to seawater. Cell Tiss. Res. 192: 25–38.

Yetty, N., H. Roshada, A. Ahyaudin and A. Chong. 2004. Characterization of digestive enzymes in a carnivorous ornamental fish, the Asian bony tongue *Scleropages formosus* (Osteoglossidae). Aquaculture 233: 305–320.

Zimmer, G., G. Reuter and R. Schauer. 1992. Use of influenza c-virus for detection of 9-o-acetylated sialic acids on immobilised conjugates by esterase activity. Eur. J. Biochem. 204: 209–215.

12

The Structural Organization in the Olfactory System of the Teleosts and Garfishes

Michal Kuciel,[1,] Krystyna Żuwała,[2] Eugenia Rita Lauriano,[3,a] Leszek Satora[4] and Giacomo Zaccone[3,b]*

Introduction

Olfaction plays a more or less important role in the life of various fish species. This may be the dominant sense (e.g., catfish, *Silurus glanis* (Jakubowski 1981)), regressed (e.g., Tetraodontidae (Yamamoto and Ueda 1979)), or co-dominant with other senses (most Cypriniformes (Burne 1909)). Most construction plans of the olfactory organ and its role are interdependent and characteristic for the species. Olfaction can be used in various/pivotal fields of life. Its most common function is to find and identify food, and during the mating season—a partner for reproduction. A special case is salmon and eels taking a distant journey to the spawning areas, which are found most likely due to olfaction

[1] Poison Information Centre, Jagiellonian University Medical Collage, Śniadeckich 10, 31-531 Crackow, Poland.
[2] Department of Comparative Anatomy, Institute of Zoology, Jagiellonian University, Gronostajowa 9, 30-387 Crackow, Poland.
  E-mail: krystyna.zuwala@uj.edu.pl
[3] Department of Food and Environmental Science, Faculty of Science, Messina University, Viale Stagno d'Alcontres 31, I-98166 Messina, Italy.
[a] E-mail: elauriano@unime.it
[b] E-mail: gzaccone@unime.it
[4] Institute of Applied Biotechnology and Basic Sciences, University of Rzeszow, Werynia 502, 36-100 Kolbuszowa, Poland.
  E-mail: satora@wp.pl
* Corresponding author: michalkuciel@gmail.com

(Creutzberg 1961; Hassler 1966). Another fairly common olfactory organ role is to detect a serum substance from serous glands of the skin which are released as a result of injury to the skin by a predator (for review see: Pfeifer 1962, 1963). The presence of these substances in water is identified as a threat and calls for appropriate behavior (e.g., seeking shelter, escape).

Stimulation of the Olfactory Sensory Neurons (OSNs) is followed by a specific trans-membrane receptor protein complex reaction with a specific chemical molecule (ligand). This combination results in a series of reactions within the cell which causes the disorder of resting potential. Formed in this way, the nerve impulse is propagated to the telencephalon (Ma 2007; Kato and Touhara 2009; Chig-Ying et al. 2009).

Gross Morphology

The structure of fish olfactory organ has evolved and developed along with the evolution of this group of vertebrates. Today, thanks to research of many different species for over 100 years we have a slightly wider, but still very fragmented representation showing the enormous diversity of the olfactory organ structure (for review see: Burne 1909; Kleerekoper 1969; Doving et al. 1977; Hara 1975; Zeiske et al. 1992; Hansen and Zielinski 2005). Cyclostomata's olfactory organ is a single cavity located on the dorsal side of the head in the area between the end of the snout and eyes. Its bottom lines the olfactory epithelium (Doving and Holmberg 1974). In Elasmobranchians the olfactory organ is symmetrically located on the dorsal side of the rostrum (Thiesen et al. 1986; Zeiske et al. 1986). The exception are species of Chlamydoselachiidae family, in which the olfactory organ is located on the dorsal side of rostrum (Jasinski 1982). In Telosteans, the olfactory organ is always located on the dorsal, preorbital part of the head (Figs. 1A, 2A). It consists of olfactory chambers where usually at the bottom there is an olfactory rosette. The olfactory rosette's surface develops olfactory epithelium. The most typical olfactory rosette has a longitudinal fold (Figs. 3A, 3B, 3C), the so-called median strip from which on both sides transverse olfactory lamellas arise. Most of new olfactory lamellas are formed in the frontal part of olfactory rosette, and then its surface area increases as well as fish growth. Less typical structure of the olfactory rosettes has been described in numerous fish species, and as it turns out, differences concerning the location of the median strip, the number and size of olfactory lamellas, the way of olfactory lamellas attachment in the olfactory chamber and their orientation in accordance to the fish long axis (Bannister 1965; Yamamoto 1982). The number of olfactory lamellas varies depending on the species, the most common increases with the individual's size (Figs. 3A, 3B) and can be a kind of indicator of a particular species olfactory ability, e.g., 9–18 in *Esox lucius*, Esociformes (Wunder 1957), or 100–150 in *Silurus glanis*, Siluriformes (Jakubowski and Kunysz 1979; Jakubowski 1981). Special examples are: *Tetraodon fluviatilis*, Tetradontiformes (Doroshenko and Motavkin

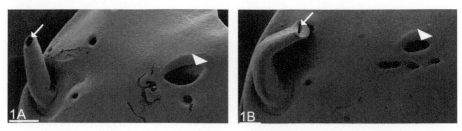

Fig. 1. Dorso-rostral part of the head of 21-days old (A), and 46-days old individual (B) of *Amia calva*. Arrow – inlet nostril, arrowhead – outlet nostril. SEM. Scale bar: 1 mm.

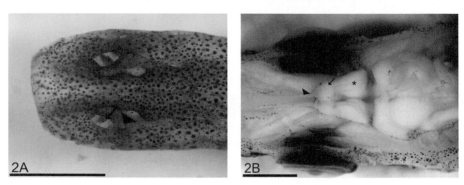

Fig. 2. *Lepisosteus oculatus* (A) rostral part with visible inlet (arrow) and outlet (arrowhead) nostrils. Star – olfactory lamella of right sided olfactory rosette. (B) Bare brain with visible telencephalon (star), adjacent olfactory bulbus (arrow) and olfactory nerve I (arrowhead). Scale bar: 5 mm.

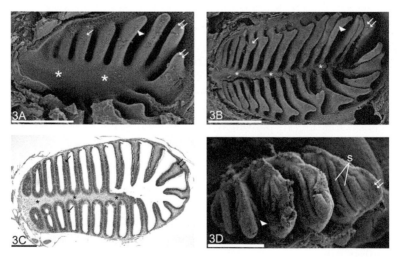

Fig. 3. Olfactory rosette of 21-days old (A), and 46–days old individual (B) and cross-section of 46-days old individual (C) of *Amia calva*. (D) olfactory rosette of *Lepisosteus oculatus*. Star – central strip (lacking in *L. oculatus*), arrow – single olfactory lamella in frontal part of olfactory rosette, double arrow - single olfactory lamella in caudal part of olfactory rosette, arrowhead – bottom of the olfactory chamber, s – secondary olfactory lamella (in *L. oculatus*). A, B – SEM; C – LM. Scale bar: A – 2000 µm; B – 5000 µm; C – 20 µm.

1986); *Conidens laticephalus*, Gobiesociformes; *Gadus macrocephalus*, Gadiformes; *Cololabis saira*, Athereniformes; *Dietyosoma burger, Acanthogobius flawimanus*, Perciformes (Yamamoto 1982), in which the only one olfactory lamella was observed and *Holopagrus guentheri*, Perciformes, where 230 lamellas was observed (Pfeifer 1964). In some species of the Gobiidae family olfactory rosette does not occur—olfactory epithelium covers the olfactory chamber's wall (Belanger et al. 2003) or the olfactory canal (Kuciel et al. 2011; Kuciel et al. 2014).

Diversity of olfactory organ's gross morphology is not the only variation of olfactory organ structure. There are also differences in olfactory epithelium location. It can be arranged in several ways within the single olfactory lamellae: continuous except for borders, e.g., in *Oncorhynchus masou*, (Pfeiffer 1963), zonal, e.g., in *Silurus glanis* (Jakubowski 1981), irregular, e.g., in Gasterosteiformes (Yamamoto and Ueda 1978), or islet form, e.g., in Tetraodontiformes (Yamamoto and Ueda 1979). Lacking the olfactory rosette causes several different types of olfactosensory epithelium location in gobiids: (i) evenly at the bottom, lateral walls, and on the roof of the olfactory chamber as in *Neogobius melanostomus* (Belanger et al. 2003) and *Neogobius gymnotrachelus*, Żuwała et al. (unpubl. data), (ii) evenly at the bottom and on lateral walls of the olfactory chamber as in *Neogobius fluviatilis*, Żuwała et al. (unpubl. data), (iii) evenly at the bottom and on lateral walls of the olfactory canal as in *Boleophthalmus boddarti, Scartelaos. histophorus* (Kuciel et al. 2013) and *Parapocryptes rictuosus, Proterorchinus marmoratus*, Żuwała et al. (unpubl. data), (iv) fascicular with finger-like ramifications as in *Periophthalmus argentilineatus* and *Periophthalmus minutus* (Kuciel et al. 2013) and finally (v) islet form, as in *Periophthalmus barbarus* (Kuciel et al. 2011), *Periophthalmus chrysospilos* and *Periophthalmus variabilis* (Kuciel et al. 2013).

Access of water with fragrance particles to olfactory epithelium is possible due to inlet and outlet nostrils (Figs. 1A, 1B, 2A). These are of a different structure depending on the species as a result of its life style and the level of olfaction use. The most common nostrils are just a kind of slots divided by a skin flap which guides the water into the olfactory chamber (Burne 1909). More sophisticated structure of pipe-like inlet nostrils are in *Polypterus senegalus* or *Amia calva* (Figs. 1A, 1B).

In a quite large number of Teleostei, the olfactory organ is equipped with so-called accessory nasal sacs. This is a non-sensory structure actively assisting (by it's compression and depression) in olfactory epithelium ventilation. Until now the occurrence of one-two accessory nasal sacs have been described. Doving et al. (1977) named species having an active mechanism for ventilation of olfactory organ as cyclosmates. Other species, wherein the organ is ventilated passively while fish swimming by water current or by ciliated epithelium—isosmates. Depending on the sensitivity of olfaction, three main types have been distinguished by Burne (1909): good, average and weak smellers, respectively: macrosmatic, mesosmatic and microsmatic.

Lepisosteus oculatus (Osteichthyes, Holostei) olfactory organ is a paired structure localized at the end of much elongated rostrum (Fig. 2A). A single organ is built from a relatively shallow chamber with an olfactory rosette at the bottom. There is a large slit-like inlet nostril and a round-shaped small outlet nostril at a very close distance (Fig. 2A). Olfactory rosette has olfactory lamellas without connection by a central strip, arranged transversely in accordance to fish long axis. On the surface of olfactory lamellas, there are secondary lamellas (Fig. 3D). Accessory nasal sacs were not observed. At the surface of olfactory lamellas there is a dense, evenly distributed non-sensory ciliated epithelium. Numerous microvillar OSNs, rare ciliated OSN and no crypt OSNs were found in TEM study.

Amia calva (Osteichthyes, Holostei) olfactory organ is a paired structure localized on the dorsal preorbital part of the head (Figs. 1A, 1B). A single olfactory chamber is quite deep and spacious. There is a pipe-like inlet nostril and round-shaped outlet nostril. Accessory nasal sac was not observed. At the bottom, there is an olfactory rosette with olfactory lamellas connected by a central strip (Figs. 3A, 3B, 3C). The sides of olfactory lamellas are not fused with the olfactory chamber edge. As the fish grow, the number and size of olfactory lamellas increase. There are two ways of olfactory lamellas increase in numbers in *Amia*: (i) new olfactory lamellas appear in the frontal side of olfactory rosette and then enlarge the surface, (ii) longitudinal division of olfactory lamellas at the central and terminal olfactory rosettes section, Kuciel and Żuwała (unpubl. data). This additional way of multiplying olfactory lamellas number was not described previously and needs more study.

Polypterus senegalus (Osteichthyes, Osteostei) olfactory organ is paired and localized on the dorsal preorbital part of the head. A single organ is quite large and it takes all rostral volume between the eye and the end of the snout. Pipe-like inlet nostril which exhibits the ability to move leads to the frontal part of olfactory chamber where a unusual olfactory rosette occurs. The rosette is extended spheric-like with olfactory lamellas arranged around this sphere and oriented longitudinally in accordance to fish long axis. The rear portion of olfactory rosette is connected to olfactory chamber. Lack of an outlet nostril suggests an active mechanism of olfactory organ ventilation which is probably generated by jaw movements. Accessory nasal sacs were not observed.

Histology

Olfactory Sensory Neurons (OSNs) are specialized nerve cells. They are classified as primary type receptors, which have peripheral outgrowths (dendrites) and centripetal outgrowth (axon). The central part is mostly a spindle-shaped nerve cell body (perikaryon). At the apical side (facing the olfactory chamber), olfactory sensory neurons form dendritic protrusions, where specific trans-membrane protein receptors are localized. On the OSNs cell body opposite side, facing the basal lamina, the axon is formed. Together with axons of the other OSNs olfactory nerve I is created, running towards the olfactory bulb. According Gemne and Doving (1969) in *Lota lota* there are about 10^7 axons. Depending on the species, the olfactory bulb can be placed right on the telencephalon (e.g., most of Teleosteans, *Anguilla, Esox, Salmo* (Hara 1975)), half-way between the brain and rosette (e.g., *Raniceps raninus* (Doving 1967) and *Gymnothorax kidako, Coryphaena hippurus* (Uchihashi 1953)), or at the olfactory rosette (e.g., *Lepisosteus* (Fig. 2B), *Polypterus, Amia, Carassius, Ictalurus,* cobitids, silurids). So there are species either with elongated olfactory nerve (olf. rosette-olf. bulb), or with elongated olfactory pathway (olf. bulb-telencephalon). In the olfactory bulb there are synaptic connections with mitral cells's dendritic appendages forming the second link for the olfactory signal.

Olfactory Epithelium Ultrastructure

The longitudinal-section (TEM) of the olfactory epithelium (Fig. 4A) allows to distinguish two groups of cells: sensory cells—olfactory sensory neurons (OSNs) and non-sensory cells. The first group includes three basic types: ciliated OSNs (Figs. 4B, 4C), microvillar OSNs (Figs. 4B, 4D) and crypt OSNs. The latter include: non-sensory ciliated cells (Figs. 5A, 5B, 5C), supporting cells (Fig. 4D) and basal cells (Fig. 4A).

Ciliated OSNs

The body of ciliated OSNs (perikarions) has an elongated shape with a spindle-like ending at the basal side of the cell, and at about 2/3 of olfactory epithelium depth, form centripetally signaling axon. At the apical side of the cell, there are dendritic protrusions penetrating the olfactory cavity formed around the olfactory knob. Cilias protrude quite significantly above the surface of the epithelium. The number and length of cilias are different and depend on the species, size, age of the individual, and typically range from three to 10. Its length varies and is in the range of 2 to 3 μm in *Danio rerio* (Hansen and Zeiske 1998) to about 10 μm in *Acipenser baeri* (Hansen and Zielinski 2005). Cilias are built in a typical aksonemal manner, with 9+2 microtubule dublets in cross-section. At the base of each of cilia, basal bodies are present, which are composed with two centrioles perpendicular to each other. In ciliated OSNs, basal bodies are located just under the apical membrane of olfactory knob. In some species of teleosts, ciliated OSNs have additional short microvillar-like appendages, such as *Acipenser baeri* (Hansen and Zielinski 2005). The nucleus is usually in the basal part of the cell. Above the nucleus, numerous mitochondria have been observed.

A specific group of ciliated OSNs are those with a single thick knob (rod-like/giant cell) (Figs. 4B, 4E). These cells are rarely observed in fish and their function is not fully understood. Their structure is similar to ciliated OSNs. The differences are visible in the apical part, where every cilium is surrounded by an outer membrane, thereby forming a thick vertical single protrusion (Hansen and Zielinski 2005) which is raised considerably above the surface of olfactory epithelium.

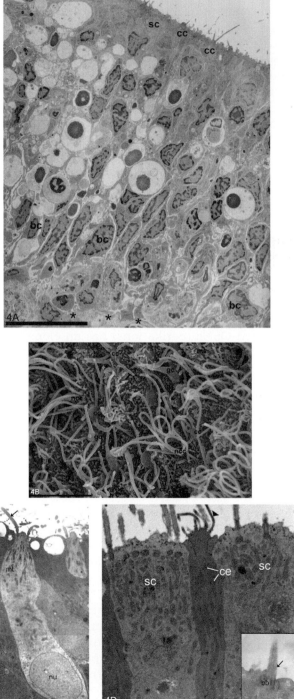

Fig. 4. contd....

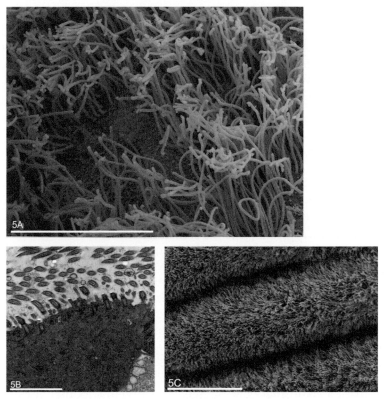

Fig. 5. Apical parts of non-sensory ciliated epithelium with dense ciliation in *Amia calva* (A). Longitudinal section of apical part of non-sensory ciliated cell showing dense ciliation in *Lepisosteus oculatus* (B), and it's olfactory rosette secondary lamella's surface ciliation (C). A, C – SEM; B – TEM. Scale bar: A – 10 μm; B – 2 μm; C – 1 mm.

Microvillar OSNs

Perikarion of microvillar OSNs is elongated and shaped like a spindle at the basal part, where the axon is formed. Perikarion extends to about 2/3 of the olfactory epithelium's depth. The apical part is slightly enlarged and extends a little above the surface of olfactory epithelium. The olfactory knob of microvillar OSNs is generally much less emphasized than in ciliated OSNs. On

---

Fig. 4. Longitudinal section of *Boleophthalmus boddarti* olfactory epithelium with ciliated OSN (cc), supporting cell (sc), basal cells (bc) and basal lamina (star) (A). Apical parts of olfactory epithelium in *Salmo trutta m. trutta* with: giant cell (gc), ciliated OSN (cc), microvillar OSN (mc), and non-sensory ciliated cells (ncc) (B). Longitudinal section of *Lepisosteus oculatus* ciliated OSN with cilias (arrow), mitochondria (mt) and nucleus (nu) (C). Longitudinal section of *Lepisosteus oculatus* microvillar OSN with micrivilli (arrowhead) and centrioles (ce) and supporting ciliated cells (scc) on both sides (D). Longitudinal section of giant cell apical part with one thick cilia with common membrane (arrow), and basal bodies (bb) (E-insert). A, C, D, E – TEM; B – SEM. Scale bar: A – 10 μm; B, D – 5 μm; C, E-insert – 2 μm.

the surface of olfactory knob, there are small microvillar protrusions in different numbers depending on the species, e.g., 10–30 in *Dicentrarchus labrax* (Diaz et al. 2002), 30–60 in *Anguilla anguilla* (Schulte 1972), 40–70 in *Carassius auratus* (Hansen et al. 1999). Centrioles are always located below the constriction and may be more numerous (Hansen and Zielinski 2005). Nucleus and other organelles in the cytoplasm are located similarly as in ciliated OSNs.

Crypt OSNs

The most striking feature of crypt OSNs is the 'crypt', found at the apical part of the cell reaching up to about 1/3 of the cell depth. Inside the crypt, a small number of microvilli, which do not reach the surface of sensory epithelium are found. The total number of microvilli depends on the species (usually 3–10) and corresponds approximately to the number of cilia usually observed in ciliated OSNs. Crypt OSNs are surrounded by the supporting cells, which are equipped with numerous microvilli (Hansen and Finger 2000). The shape is different from ciliated and microvillous OSNs. Most are stocky, ovoid to oval cells, extending to about 1/3 of the depth of the olfactory epithelium. As with the previous two types of cells, the nucleus is situated in the basal part of perikarion. Above the nucleus, cytoplasm contains organelles such as the Golgi apparatus, mitochondria, lysosomes and rough endoplasmic reticulum cisterns.

OSNs are equipped with specific trans-membrane receptors related to G-protein (Nagai et al. 1996; Hansen et al. 2003). Reaction with a specific ligand causes a conformational change of G-protein subunits. It causes the disorder of the resting potential at the cell membrane which produces a signal. Signal propagation along the axon's membrane is followed by its analysis and recognition in the telencephalon. Particular types of OSNs are sensitive to various groups of chemical compounds. This is related to the presence of individual type of OSNs membrane of different subunits of G proteins. Hara (1992), Laberge and Hara (2001) and Hansen et al. (2003) reported that the ciliated OSN are equipped with subunit G α olf/s which bind specifically to the amino acids, and bile acids. Microvillar OSNs have subunit G α q/11 and bind specifically to the amino acids and nucleotides. Crypt OSNs have subunit G αo, but it is not fully understood what kind of substances it is linked with. It is believed that these cells are responsible for identifying sex hormones and possibly occur in the olfactory epithelium in greater numbers during the reproductional season. Previous studies did not specify the nature of the substances identified by rod-like OSNs (Hansen and Zielinski 2005).

Non-sensory Ciliated Cells

Non-sensory ciliated cells are cylindrical in shape extending from the basal lamina to the surface of the epithelium. They are present in all Teleostean's olfactory epithelium and are responsible for movement of mucus and water within the epithelial surface (Zeiske et al. 1992). There are numerous mitochondria in the cytoplasm at the cell tip portion. At the basal cell part, a relatively small nucleus has been observed surrounded by a dense endoplasmic reticulum. The nucleus may have a different shape and typically do not contain condensed chromatin. Numerous cilia grow on a wide, flat apical part of the cell. The number and size of cilia depend on the species, e.g., 40 in *Danio rerio* (Hansen and Zeiske 1998), and to 140 in *Anguilla anguilla* (Schulte 1972). Cilias of those cells are usually slightly longer than a cilias of OSNs (Hansen and Zielinski 2005). The presence of a very dense ciliation of non-sensory ciliated epithelium was specific for both *Lepisosteus* and *Polypterus*. Their location depends on the species and can be of various types: (i) on the periphery of the olfactory lamellas, (ii) dispersed, (iii) islets/bands/zones (Jakubowski 1981; Hansen et al. 1999).

Supporting Cells

Supporting cells are a kind of scaffold filling the space between the OSNs and keep them in a correct position. A single supporting cell often accompanies a few OSNs. They have a cylindrical shape, often with their basal part clearly adhering to the basal lamina. The apical membrane of this cell forms characteristic microridges. They were described as supporting cells with short protruding cilia-like in *Salmo trutta m. trutta* (Bertmar 1973). There is no data in the literature whether supporting cells with cilia-like protrudings are the same as non-sensory ciliated cells. Desmosomes at the apical part of the epithelium of the olfactory receptor engage adjacent to the supporting cells. The most important feature is the presence of tonofilaments long tufts in the cytoplasm, typically perpendicular to the long axis of the cell. At the longitudinal-section of the olfactory epithelium, nucleus of supporting cells were observed below the nuclei of OSNs (Zeiske et al. 1992).

Basal Cells

Basal cells are highly diversified in shape with relatively large nucleus in respect to the size of the cell. They are located in the deepest area of the olfactory epithelium below the base of the supporting cells, on basal lamina. A new sensory and non-sensory cells are formed from them (Breipohl et al. 1976; Evans et al. 1982; Hansen et al. 1999; Thornhill 1967; Zippel et al. 1997). To date, it is not clear whether supporting cells that have an mitosis ability can also be formed from basal cells (Hansen and Zielinski 2005).

Rodlet Cell

So far, in many fish species the so-called rodlet cells have been observed in various tissues (brain, gill, kidney, gonads). They are most likely parasitic organisms (Thelohan 1892), rather than the tissue cells of fish. They are separated by a thick cuticle from other cells in the tissue. There are projections at the apical, which are probably involved in the subsequent cells infections.

References

Bannister, L.H. 1965. The fine structure of the olfactory surface of teleostean fishes. Q. ft. Microsc. Sci. 106: 333–342.

Belanger, R.M., C.M. Smith, L.D. Corkum and B.S. Zielinski. 2003. Morphology and histochemistry of the peripheral olfactory organ in the round goby, *Neogobius melanostomus* (Teleostei; Gobiidae). J. Morphol. 257: 62–71.

Bertmar, G. 1973. Ultrastructure of the olfactory mucosa in the homing Baltic sea trout *Salmo trutta trutta*. Mar. Biol. 19: 74–88.

Breipohl, W., H.P. Zippel, K. Ruckert and H. Oggolter. 1976. Morphologische und elektrophysiologische Studien zur Struktur und Funktion des olfaktorischen Systems beim Goldfisch unter normalen und experimentellen Bedingungen. Beitr. Elektronenmikrosk. Direkt. Oberfl. 9: 561–584.

Burne, R.H. 1909. The anatomy of the olfactory organ of teleostean fishes. Proc. Zool. Soc. London 2: 610–663.

Chig-Ying, S., K. Menus and J.R. Carlson. 2009. Olfactory perception: receptors. Cells and circuits. Cell 139: 45–59.

Diaz, J.P., M. Prie-grane, T. Noell and R. Connes. 2002. Ultrastructural study of the olfactory organ in adult and developing European sea bass, *Dicentrarchus labrax*. Can. J. Zool. 80: 1610–1622.

Doroshenko, M.A. and P.A. Motavkin. 1986. Olfactory epithelium of marine fishes in scanning electron microscopy. Acta Morphol. Hungarica 34: 143–155.

Doving, K.B. 1967. Comparative electrophysiological studies on the olfactory tract of some teleost. J. Comp. Neurol. 131: 365–370.

Doving, K.B. and K.A. Holmberg. 1974. A note on the function of the olfactory organ of the hagfish *Myxine glutinosa*. Acta Physiol. Scand. 91: 430–432.

Doving, K.B., M. Dubois-Dauphin, A. Holley and F. Jourdan. 1977. Functional anatomy of the olfactory organ of fish and ciliary mechanism of water transport. Acta Zool. 58: 245–255.

Evans, R.E., B. Zielinski and T.J. Hara. 1982. Development and regeneration of the olfactory organ in rainbow trout. pp. 15–37. *In*: T.J. Hara (ed.). Chemoreception in Fishes. Elsevier Scientific Publishing Co., Amsterdam, Oxford, New York.

Gemne, G. and K.B. Doving. 1969. Ultrastructural properties of primary olfactory neurons in fish (*Lota lota* L.) Am. J. Anat. 126: 457–476.

Hansen, A. and E. Zeiske. 1998. The peripheral olfactory organ of the zebrafish, *Danio rerio*: an ultrastructural study. Chem. Senses 23: 39–48.

Hansen, A. and T.E. Finger. 2000. Phyletic distribution of crypt-type olfactory receptor neurons in fishes. Brain Behav. Evol. 55: 100–110.

Hansen, A. and B.S. Zeilinski. 2005. Diversity of the olfactory epithelium of bony fishes: development, lamellar arrangement, sensory neuron cell types and transduction components. J. Neurocytol. 34: 183–208.

Hansen, A., H.P. Zippel, P.W. Sorensen and J. Caprio. 1999. Ultrastructure of the olfactory epithelium in intact, axotomized, and bulbectomized goldfish, *Carassius auratus*. Micr. Res. Techn. 45: 325–338.

Hansen, A., S.H. Rolen, K. Anderson, Y. Morita, J. Caprio and T.E. Finger. 2003. Correlation between olfactory receptor cell type and function in the channel catfish. J. Neuros. 23(28): 9328–9339.

Hara, T.J. 1975. Olfaction in fish. pp. 271–335. Progress in Neurobiology, Vol. 5, Part 4. Pergamon Press, Great Britain.

Hara, T.J. 1992. Fish Chemoreception. Chapman & Hall, London.

Hassler, A.D. 1966. Underwater Guideposts—Homing of Salmon. University of Wisconsin Press, Madison.

Jakubowski, M. 1981. Ultrastructure (SEM, TEM) of the olfactory epithelium in the Wels, *Silurus glanis* L. (Siluridae, Pisces). Z. mikrosk.—anat. Forsch. 93: 728–735.

Jakubowski, M. and E. Kunysz. 1979. Anatomy and morphometry of the olfactory organ of the wels *Silurus glanis* L. (Siluridae, Pisces). Z. mikrosk.—anat. Forsch. 93: 728–735.

Jasiński, A. 1982. Narządy zmysłów. pp. 428–442. *In*: H. Szarski (ed.). Anatomia porównawcza kręgowców. PWN, Warszawa.

Jones, F.R.H. 1968. Fish Migration. Edward Arnold, London.

Kato, A. and K. Touhara. 2009. Mammalian olfactory receptors: pharmacology, G protein coupling and desensitization. Cell. Mol. Life Sci. 66: 3743–3753.

Kleerkoper, H. 1969. Olfaction in Fishes. Indiana University Press, Bloomington.

Kuciel, M., K. Żuwała and M. Jakubowski. 2011. A new type of fish olfactory organ structure in *Periophthalmus barbarus* (Gobiidae, Oxudercinae). Acta Zool. Stockholm 92: 276–280.

Kuciel, M., K. Żuwała and U. Satapoomin. 2013. Comparative morphology (SEM) of the peripheral olfactory organ in the Oxudercinae subfamily (Gogiidea, Perciformes). Zool. Anz. 252: 424–430.

Kuciel, M., E.R. Lauriano, G. Silvestri, K. Żuwała, S. Pergolizzi and D. Zaccone. 2014. The structural organization and immunohistochemistry of G-protein alpha subunits in the olfactory system of the air-breathing musdskipper, *Periophthalmus barbarus* (Linneaus, 1766) (Gobiidae, Oxudercinae). Acta Histochem. 116: 70–78.

Laberge, F. and T.J. Hara. 2001. Neurobiology of fish olfaction: a review. Brain Res. Rev. 36: 46–59.

Ma, M. 2007. Encoding olfactory signals via multiple chemosensory systems. Crit. Rev. Biochem. Mol. Biol. 42: 463–480.

Nagai, T., D.J. Kim, R.J. Delay and S.D. Roper. 1996. Neuromodulation of transduction and signal processing in the end organs of taste. Chemosens. Sens. 21: 353–365.

Pfeifer, W. 1962. The fright reaction of fish. Biol. Rev. Camb. Phil. Soc. 37: 495–511.

Pfeifer, W. 1963. Alarm substances. Experientia 19: 11–123.

Pfeifer, W. 1964. The morphology of the olfactory organ of *Holopagrus guentheri* Gill, 1862. Can. J. Zool. 42: 235–237.

Schulte, E. 1972. Untersuchungen an der Regio olfactoria des Aals, *Anguilla anguilla* L. Zeit. Zellforsch. Mikrosk. Anat. 125: 210–228.

Thelohan, P. 1892. Sur des sporozoaires indetermines parasites des poissons. J. d'Anat. Physiol. Paris 28: 163–171.

Thiesen, B., E. Zeiske and H. Breucker. 1986. Functional morphology of the olfactory organs in the spiny dogfish (*Squalus acanthius* L.) and the small-spotted catshark (*Scyliorhinus canicula* (L.)). Acta Zool. 67: 73–86.

Thornhill, R.A. 1967. The ultrastructure of the olfactory epithelium of the lamprey *Lampetra fluviatilis*. J. Cell. Sci. 2: 591–602.

Uchihashi, K. 1953. Ecological study of the Japanese teleost in relation to the brain morphology. Bull. Jap. Reg. Fish. Res. Lab. 2: 1–66.

Wunder, W. 1957. Die Sinnesorange der Fische. Allgem Fichereizeit. 82: 171–173.

Yamamoto, M. 1982. Comparative morphology of the peripheral olfactory organ in teleost. pp. 39–59. *In*: T.J. Hara (ed.). Chemoreception in Fishes. Elsevier, Amsterdam.

Yamamoto, M. and K. Ueda. 1978. Comparative morphology of fish olfactory epithelium. V. Gasterosteiformes, Channiformes and Synbranchiformes. Bull. Japan. Soc. Sci. Fish. 44: 1309–1314.

Yamamoto, M. and K. Ueda. 1979. Comparative morphology of fish olfactory epithelium. IX. Tetraodontiformes. Zool. Mag. (Tokyo) 88: 210–8.

Zeiske, E., J. Caprio and S.H. Gruber. 1986. Morphological and electrophysiological studies on the olfactory organ of the lemon shark, *Negaprion brevirostris*. pp. 381–391. *In*: T. Uyeno, R. Arai, T. Taniuchi and K. Matsuura (eds.). Indo-pacific Fish Biology: Proceedings of the Second International Conference on Indo-Pacific Fishes. Ichthyological Society of Japan, Tokyo.

Zeiske, E., B. Thiesen and H. Breucker. 1992. Structure, development and evolutionary aspects of the peripheral olfactory system. pp. 13–39. *In*: T.J. Hara (ed.). Fish Chemoreception. Chapman and Hall, London.

Zippel, H.P., A. Hansen and J. Caprio. 1997. Renewing olfactory receptor neurons do not require contact with the olfactory bulb to develop normal responsiveness. J. Comp. Physiol. A-Sensory Neural and Behavioral Physiology 181: 425–437.

13

Hagfish Slime and Slime Glands

Douglas S. Fudge, Timothy M. Winegard*[a]*
and *Julia E. Herr*[b]

Introduction

Hagfishes possess many unique traits, but are best known for their ability to produce large volumes of slime when they are provoked (Fig. 1). Slime is so essential to the identity of hagfishes that two of their common names contain the word slime ('slime eel' and 'slime hag'). The Atlantic hagfish (*Myxine glutinosa*) was twice named for its slime (myx = slime; glutin = glue) by Linnaeus, who concisely summed up its biology with the following: "Intrat et devorat pisces; aquam in gluten mutat" ("enters into and devours fishes; transforms water into glue") (Linnaeus 1758). Charles Darwin also knew about hagfish slime and even witnessed the phenomenon himself. In the fish section of The Zoology of the Voyage of the H.M.S. Beagle, Leonard Jenyns relates Darwin's encounter with a hagfish caught near Tierra del Fuego in which he "observed a milky fluid transuding through the row of lateral pores" (Jenyns 1842). The great German anatomist Johannes Müller published an anatomical description of hagfishes in the mid 19th century, which included detailed descriptions and drawings of the slime glands as well as individual thread skeins that he had observed under the microscope (Müller 1845).

In the early 1940s, the polymer physicist J.D. Ferry published a paper on hagfish slime in which he cautioned that "the heterogeneity of the original slime and its irreversible contraction render it unsuitable for study of mechanical

Dept. of Integrative Biology, University of Guelph, Guelph ON N1G-2W1 Canada.
[a] E-mail: winegart@uoguelph.ca
[b] E-mail: julia.herr@gmail.com
* Corresponding author: dfudge@uoguelph.ca

Fig. 1. Hagfishes are well known for their ability to make large volumes of slime when they are attacked or stressed. Here a Pacific hagfish *Eptaretus stoutii* has produced several liters of slime in response to being handled in an aquarium (photo by Andra Zommers).

properties in relation to its composition and structure" (Ferry 1941). A few years later, Newby investigated the structure of hagfish slime glands and further probed the enigmatic Gland Thread Cells (GTCs) (Newby 1946). The late 1970s and early 1980s marked the beginning of a fertile period of research on hagfish slime, with papers from Downing, Spitzer, and Koch as well as Terakado and Fernholm probing ever deeper into the structure and function of cells within the slime glands as well as their secretory products. Downing, Salo, Luchtel and colleagues published early studies on the mucous vesicles from hagfish slime exudate (Downing et al. 1981; Salo 1983; Luchtel et al. 1991). By the mid 1990s, Downing, Spitzer, Koch and colleagues had isolated the proteins that make up slime threads, identified them as intermediate filament (IF) proteins, and reported the sequence of their genes.

More recent work on the biophysics of hagfish slime began with D.S. Fudge's PhD research (Fudge 2002) under the supervision of J.M. Gosline, whose expertise in mucus, silk, and keratin biomaterials was an ideal combination for tackling several questions about the biophysics of hagfish slime (in spite of Ferry's ominous warnings!). Recent work by Zintzen and colleagues has filled in significant gaps about how hagfishes use slime in the wild (Zintzen et al. 2011). Our research on slime biophysics and this chapter in general owe a profound debt to the researchers listed above, and in particular to the authors of the chapters on slime from the two previous hagfish texts. Blackstad's chapter in the *Biology of Myxine* laid the groundwork for much of what was to come in the next few decades (Blackstad 1963), and Spitzer and Koch's chapter in *The Biology of Hagfishes* did the same for the next wave of interest in hagfishes and their slime (Spitzer and Koch 1998).

Hagfish Slime Glands

In their chapter on slime and slime glands, Spitzer and Koch (1998) make the distinction between mucus and slime, and it is worth repeating their definitions here. Mucus is defined as the glycoprotein product from secretory cells of the skin, whereas hagfish slime refers specifically to the material arising from secretory products made within the slime glands. It is this slime that sets the hagfishes apart from other marine animals, most of which are capable of producing epidermal mucus of some kind. Hagfish slime is produced within the highly specialized slime glands, which exist in segmental pairs down both sides of the body (Fig. 2). The slime gland pores are approximately equi-spaced such that there are no regions of the body that could be attacked without eliciting a local release of slime exudate from nearby glands. The number of slime glands varies among the hagfishes, with common species such as *Eptatretus stoutii* having 158 glands and *Myxine glutinosa* having 194. Out of all the known species, *E. taiwanae* has the smallest number glands (120) while *Nemamyxine elongata* has the most (400) (Fernholm 1998). While the number of slime glands is a convenient character for taxonomic studies, little is known about whether the differences among species have any functional significance. Our work on captive hagfishes suggests that *E. stoutii* slime glands contain larger volumes of slime than *M. glutinosa*, although the functional significance of this observation is not yet clear.

Slime Gland Morphology

The slime glands contain two main types of secretory cells, Gland Mucous Cells (GMCs) and GTCs (Figs. 2, 3), which produce the mucous and fibrous components of the slime, respectively. Although they are highly specialized, the slime glands are best understood as having evolved as invaginations of the skin, with subsequent (or concurrent) specializations allowing for rapid and copious holocrine ejection of secretory products. The gland is encased by a thin collagenous capsule, which itself is surrounded by a thin layer of striated muscle, the *musculus decussatus* (Lametschwandtner et al. 1986). These layers of muscle fibers surely contribute to the forceful ejection of slime exudate from the gland, although contributions from the more powerful surrounding myotomal musculature are also possible. GMCs and GTCs arise from stem cells in an epithelial layer underlying the capsule, and get pushed by dividing and growing cells below toward the center of the gland as they themselves grow and mature (Newby 1946). The result is a gradient of cell size and maturity, with the oldest, most mature cells in the center of the gland, and the youngest cells closest to the epithelium. GTCs develop with their pointed, apical ends toward the gland pore, which likely facilitates their passage through the gland duct. Based on the dimensions of the duct and the GTCs, the cells likely pass through it single file, where it is also likely that the plasma membranes of GTCs

A

slime gland pore

B

gland pore

gland mucous cell

collagenous capsule

gland thread cell

striated muscle

500 µm

epithelial layer

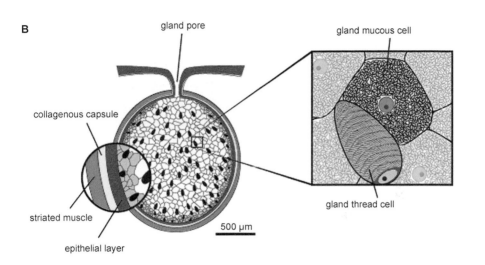

C

thread skein   mucous vesicle

gland pore

seawater

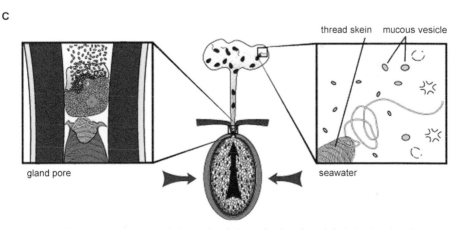

Fig. 2. Hagfishes (A) have a row of slime gland down both sides of their body. The gland (B) is surrounded by a thin layer of striated muscle, which ejects a bolus of slime exudate into the surrounding seawater when it contracts. Most of the gland volume is occupied by two kinds of cells, gland thread cells and gland mucous cells, which produce the fibrous and mucous portions of the slime, respectively. (C) Exudate release involves these cells passing through the narrow gland duct, which strips the plasma membrane from both kinds of cells. In seawater, the coiled slime threads unravel and mucous vesicles swell and rupture (modified from Herr 2012).

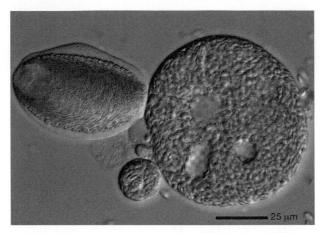

Fig. 3. Immature GMCs and GTCs in culture. These cells were obtained by digesting *M. glutinosa* slime glands and maintaining the cells in modified Leibovitz L-15 media. From Winegard (2012).

and GMCs are disrupted, resulting in the release of naked thread skeins and countless mucous vesicles.

In spite of the great variability that exists in the number of slime glands, little is known about how the slime glands themselves differ among the greater than 80 species of known hagfishes (Martini and Beulig 2013). Leppi (1968) carried out histological studies of skin and slime glands from three species of hagfishes, but found no large differences in the staining properties of the glands. One salient difference between the morphology of slime glands from *M. glutinosa* and *E. stoutii* has to do with their vascularization. Vascular corrosion casts and scanning electron microscopy reveal a basket-like vascular pattern that extends throughout the gland capsule of both species; however, capillary loops penetrate into the gland lumen of *E. stoutii* but not *M. glutinosa* (Lametschwandtner et al. 1986). These differences may correspond to differences in the recharge rate of the gland after sliming, but at this point there is no evidence to support this claim.

Gland Mucous Cells

GMCs are large (~100–150 µm in diameter), specialized cells that are responsible for producing the mucous component of hagfish slime (Figs. 2, 3). GMCs are distributed throughout the slime gland with GTCs in a pattern that appears to minimize the occurrence of adjacent GTC/GTC or GMC/GMC pairs. Such a pattern likely contributes to an even distribution of thread skeins and mucous vesicles in slime exudate released from the gland. GMCs are packed full of disk-shaped 7 µm diameter mucous vesicles, which are released from the gland with their membranes intact (Luchtel et al. 1991). The glycoproteins within the mucous vesicles are comprised of 77% protein (rich in proline, threonine and valine), 12–18% carbohydrate, and 5% lipid (Salo et al. 1983; Lehtonen et al. 1966). Such a composition is unusual for mucous secretions, which typically have carbohydrate contents closer to 85% (Gum 1985). GMCs are similar to the LMCs (Large Mucous Cells) of the epidermis both in size and their mucous secretion products. While it is clear that GMCs release mucous vesicles via the holocrine mode from the slime gland, it is not known if LMCs of the epidermis release mucus via holocrine or merosecretion, although some have suggested it is the latter based on the similarities between LMCs and teleost goblet cells (Whitear 1986). While we are beginning to understand the stages of GTC growth and maturation, little is currently known about GMC development.

Gland Thread Cells

GTCs represent the most specialized epidermal cell type among lampreys and hagfishes (Quay 1963) and are responsible for making the fibrous component of hagfish slime (Figs. 2, 3). Undifferentiated epithelial cells adjacent to the basement membrane of the gland are believed to give rise to both the thread and mucus producing cells of the gland (Newby 1946). Although a complete developmental series does not exist for GTCs, it is known that these cells undergo substantial morphological changes during development. Following differentiation, GTCs increase approximately three-fold in diameter (Newby 1946). This initial increase in size is followed by a uniaxial lengthening of the cell and polar repositioning of the nucleus with its prominent nucleolus within the GTC (Newby 1946; Terakado et al. 1975; Downing et al. 1984). The polar position of the GTC nucleus, which is maintained throughout development, in combination with the primarily uniaxial lengthening of the cell, gives GTCs their characteristic ellipsoidal shape. The nucleated or basal side of the developing GTC remains blunted in comparison to the pointed apical side of the cell. When developing GTCs reach approximately four times their original diameter, the slime thread begins to appear (Newby 1946; Downing et al. 1984; Winegard and Fudge 2010). Subsequent maturation involves the precision coiling of the GTC protein thread into a skein, which at maturity occupies over 70% of the cell's volume and contains a single thread that can reach 15 cm in length (Fig. 4). At maturity, the thread has a bi-directional taper, with a diameter of about 3 µm in the middle and about 0.5 µm at the ends (Fudge et al. 2005; Winegard 2012).

Slime Thread Synthesis Within GTCs

Work by Downing, Spitzer, Koch, and colleagues in the 1980s and 1990s characterized the two proteins that make up the slime thread in GTCs and identified them as part of the IF family of proteins, which are a diverse group of fibrous proteins that form networks in the cytoplasm and nucleus (Downing et al. 1981; Aebi et al. 1988; Steinert and Roop 1988; Koch et al. 1994, 1995).

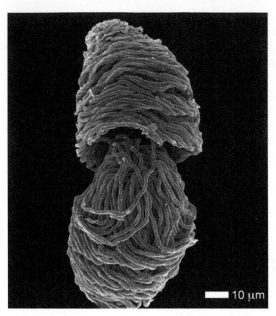

Fig. 4. SEM of isolated thread skein from *M. glutinosa* showing the pattern of staggered thread loops that spiral around the long axis of the skein. Mature slime threads are about 15 cm long and can stretch to about 34 cm before they break. Image by Mark Bernards.

Hagfish slime thread IF proteins can be divided into two groups; (1) alpha (643 residues, 66.6 kDa) and (2) gamma (603 residues, 62.7 kDa) (Spitzer et al. 1988). Another protein, beta, is believed to be a post-translational form of gamma that occurs during late stages of GTC development (Spitzer et al. 1984; Spitzer et al. 1988). Although they clearly belong to the IF family, alpha and gamma proteins share less than a 30% homology with their most closely related type II and type I IF proteins (Koch et al. 1994). They also exhibit a high threonine content (alpha, 13%; gamma, 10%) when compared to other IF proteins (2–5% on average) (Koch et al. 1994). The high threonine content may be involved in the post-translational modifications of gamma mentioned above, as hydroxyl groups of threonine residues are common sites of phosphorylation.

Although many of the features of thread development in GTCs have been described by careful TEM studies (Terakado et al. 1975; Downing et al. 1984), the exact mechanism of thread assembly and coiling in GTCs is not well understood. The thinnest regions of thread are located adjacent to the apical side of the GTC nucleus. This region, termed the Mitochondrial Rich Zone (MRZ), contains large numbers of mitochondria as well as cytoplasmic ribosomes and polysomes (Downing et al. 1984) (Fig. 5), which all point to the MRZ as the most likely site of slime thread synthesis. At some point in thread growth and maturation, microtubules (MTs) appear in thread cross-sections, although their function is not clear. Possibilities include increasing the flexural

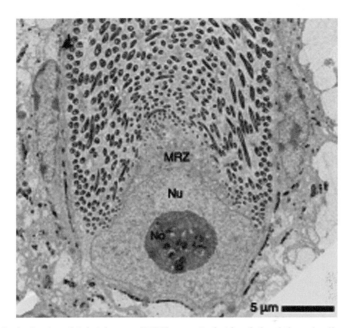

Fig. 5. TEM of mitochondrial rich zone (MRZ) on apical side of gland thread cell nucleus (Nu), which contains a large and prominent nucleolus (No). The MRZ is devoid of growing slime thread, contains a high concentration of mitochondria and ribosomes, and is believed to be involved in the process of slime thread elongation, although the mechanism is currently unknown (Winegard 2012).

stiffness of the developing thread to facilitate coiling, and the delivery of IFs or IF subunits to the interior of the growing thread. The growing thread also appears to be wrapped with a helical filament about 12 nm in diameter, but it is unclear whether this filament exists in a low-pitch helix or as separate rings (Downing et al. 1984) (Fig. 6). The function of this wrapping filament is also not clear, but one possibility is that it prevents the merging of adjacent portions of the developing thread. At late stages of GTC maturation, the IFs in the thread appear to rearrange into a single, electron dense thread with no discernable ultrastructure (Fan 1965; Terakado et al. 1975; Downing et al. 1984). After this condensation of IF proteins, MTs in the thread are surrounded by an electron lucent halo, and eventually the MTs disappear and the spaces they occupied are filled in. IF condensation is also accompanied by the appearance of a fluffy rind on the surface of the thread (Downing et al. 1984). This feature may coincide with the addition of IF proteins or subunits directly to the thread, which at this stage is still increasing in diameter.

Gland Interstitial Cells

It was previously thought that GMCs and GTCs are the only cells found within the slime gland interior (Newby 1946; Fernholm 1981; Spitzer and Koch 1998; Blackstad 1963). However, Winegard (2012) recently discovered and described

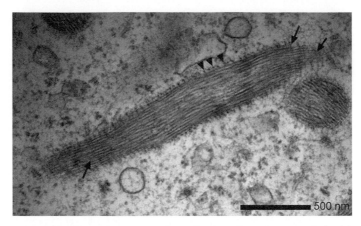

Fig. 6. Immature slime threads consist of parallel bundles of intermediate filaments and microtubules, and appear to be wrapped with another filament of unknown function. In this longitudinal section of a growing TEM, the wrapping filament can be seen crossing the surface of the thread (arrows) and appearing as regularly spaced dots at the thread periphery (arrowheads). From Winegard (2012).

another cell type, gland interstitial cells (GICs), which exist between GMCs and GTCs. The cells were first discovered via fluorescent nuclear staining of paraffin embedded hagfish slime gland sections, and their presence was confirmed by TEM (Fig. 7). GICs were likely overlooked for so long due to their spindle-like morphology and their vanishingly small volume compared to their GMC and GTC neighbors. GICs possess a prominent nucleus and long bifurcating processes that form connections with other GICs, GTCs, and GMCs. From TEM, it appears that each GTC and GMC borders one to three GICs, suggesting

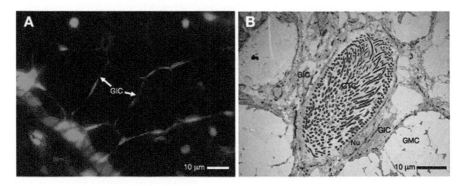

Fig. 7. Fluorescence and TEM images of gland interstitial cells (GICs). In the fluorescence image (A), nuclei were stained blue with DAPI and intermediate filaments were stained green with a pan-cytokeratin antibody. In the lower right of the TEM image (B) a slender GIC with an elongate nucleus (Nu) can be seen wedged between a gland thread cell (GTC) and a gland mucous cell (GMC). In the upper left is likely another GIC sandwiched between the GTC and another GMC (Winegard 2012).

that GICs may be the most abundant cells in the slime glands. GICs exhibit large numbers of mitochondria, Golgi, and vesicles, with some GIC vesicles appearing in TEM to be in the process of fusing with neighboring GTCs and GMCs. The function of GICs is not known, but possibilities include acting as nurse cells for developing GTCs and GMCs and retaining immature GTCs and GMCs during sliming when mature cells are expelled from the gland.

Slime Function

Several functions have been proposed for hagfish slime, but the most obvious is defense against fish predators. Fernholm (1981) relates observations of captive fishes suffocating after attacking hagfish in aquaria. Lim et al. (2006) tested the hypothesis that the slime functions specifically to adhere to fish gills and disrupt respiratory water flow, and found that the slime is exceptionally good at doing this in isolated fish heads. In 2011, Zintzen et al. published observations of wild hagfishes being attacked by 14 different fish predators (Fig. 8). In every case, the hagfishes escaped almost entirely unharmed, but the plight of the predators with mouths full of slime was unclear. Other proposed functions for the slime include: localization of eggs (Koch et al. 1991a), denying competitors access to scavenged food sources (Isaacs and Schwartzlose 1975; Zintzen et al. 2011), and predation (Zintzen et al. 2011). Hagfish slime may also provide some defense against microbial pathogens, as higher levels of innate immune substances such as alkaline phosphatase, lysozyme, and cathepsin B are found in the hagfish slime compared to hagfish epidermal mucus (Subramanian et al. 2008).

Fig. 8. Recent work by Zintzen et al. (2011) showed that hagfish slime is a remarkably good defense against fish predators. Here a seal shark *Daliatis licha* is shown attacking a New Zealand hagfish *Eptatretus cirrhatus*. The arrow indicates slime that has been released by the hagfish. In subsequent frames of the video, the shark aborts the attack and swims away with large volumes of slime trailing from its mouth (from Zintzen et al. (2011)).

Slime Composition and Deployment

Fudge et al. (2005) measured the concentration of mucus and slime threads in naturally produced slime and found them to be present in remarkably small amounts. A liter of slime produced by *E. stoutii* (0.9 L) requires only 20 mg of slime threads and 15 mg of mucus (dry weight). Such economical use of threads and mucus explains how a 150 g Pacific hagfish can produce 25 L of slime. If the mucus in hagfish slime were at the same concentration as those found in secretions such as gastric mucus (more than three orders of magnitude greater), the same animal would only be able to make about 12 mL of slime.

Using high-speed video observations of hagfishes sliming in aquaria, Lim et al. (2006) demonstrated that exudate is released from the slime glands in a forceful jet with a velocity of about 0.17 m/s. The same study showed that proper deployment of the slime requires convective mixing with seawater, which could be supplied by the thrashing hagfish, the movements of an attacking predator, or flow caused by suction feeding teleosts. They also showed that exudate release is local and only occurs from glands in the immediate vicinity of where the hagfish is physically attacked (Lim et al. 2006; Zintzen et al. 2011) (Fig. 9). These experiments demonstrated that hagfishes

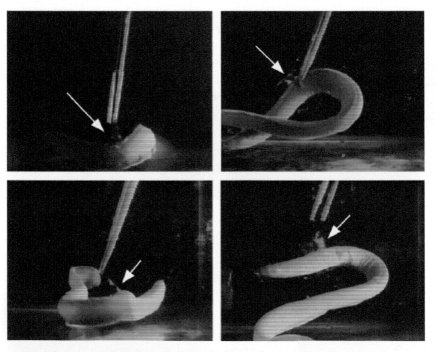

Fig. 9. Hagfish slime is released locally. These high speed video stills of hagfish being attacked with forceps in an aquarium show that slime exudate is only released from glands close to where a hagfish is attacked (arrows). Each panel shows a different hagfish. From Lim et al. (2006).

do not hide within their slime, but instead immediately try to escape after exudate is released. Videos of hagfishes being attacked by a variety of fish predators in the wild corroborate this conclusion (Zintzen et al. 2011). Koch et al. (1991b) first tackled the question of how the thick exudate produced by the slime glands interacts with seawater to transform into mature slime. Using *in vitro* experiments in which exudate stabilized in sodium citrate buffer was progressively diluted with distilled water, they observed intermediate stages of skein unraveling, mucous vesicle rupture, as well as mucus thread interactions. These experiments provided early glimpses into the mechanisms of slime deployment, but the use of stabilized slime and distilled water limited the generalizability of these observations to more natural conditions.

Mucous Vesicles

Downing et al. (1981) published some of the first observations of the mucous component of the slime along with methods for slime collection and stabilization, which would facilitate later research on hagfish slime mucus by them and others. Luchtel et al. (1991) explored mechanisms of mucous vesicle rupture by stirring slime exudate into a variety of solutions and assessing the degree of slime formation. From these experiments, they concluded that the interior of the vesicles has an osmolarity of about 900 mOsm, or close to that of seawater, which is consistent with the fact that hagfishes are marine osmoconformers. Luchtel et al. also showed that the vesicles rupture in all hyperosmotic salt solutions except ones made from di- and trivalent anions such as sulfate and citrate. Herr et al. (2010) developed a rupture assay that allowed them to observe the behavior of single immobilized mucous vesicles as they are exposed to solutions of interest (Fig. 10). They also measured the ionic composition of the fluid component of slime exudate, as well as the abundance of organic osmolytes. The most common osmolytes are the methylamines betaine, TMAO, and dimethylglycine, which have a cumulative concentration of 390 mM in exudate fluid. The distribution of ions in exudate fluid is similar to that of plasma, but the concentrations of the various ions are generally lower, presumably due to the higher concentrations of organic osmolytes in the slime gland. The discovery of high levels of methylamines in the slime exudate raised the possibility that these compounds are involved in the stabilization of mucous vesicles within intact GMCs in the slime gland. However, functional studies revealed that only about half of all vesicles could be stabilized in even very concentrated solutions of TMAO or betaine, suggesting that these compounds may be involved in another function. These studies also revealed significant heterogeneity in the swelling behavior of mucous vesicles exposed to seawater, with some vesicles swelling slowly and immediately and others swelling very quickly, but only after a substantial delay (Herr et al. 2010; Herr 2012).

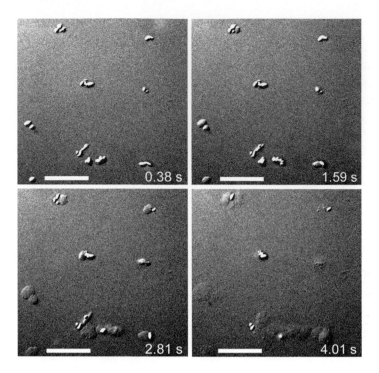

Fig. 10. Swelling and rupture of hagfish slime mucous vesicles from *M. glutinosa*. Timestamps in the lower right corner indicate time following contact with seawater. Scale bars are 20 μm (from Herr et al. 2010).

Skein Deployment

Recent work has attempted to understand how the coiled thread within a thread skein unravels in the fraction of a second that it takes for the slime to deploy in seawater. Newby (1946) characterized skein unraveling as 'explosive' and 'spontaneous' but more recent work has demonstrated the importance of other factors such as convective mixing and the presence of mucus. Koch et al. (1991b) showed that the 'removable mass' (an indication of slime cohesion) of the slime goes down with increasing concentrations of the disulfide-cleaving compound DTT, which is known to disrupt mucus networks. Subsequent work showed that proper setup of the slime requires convective mixing (Fudge et al. 2005; Lim et al. 2006), and a study of *M. glutinosa* slime showed that mixing forces and the presence of mucus are both important for skein unraveling (Winegard et al. 2010) (Fig. 11).

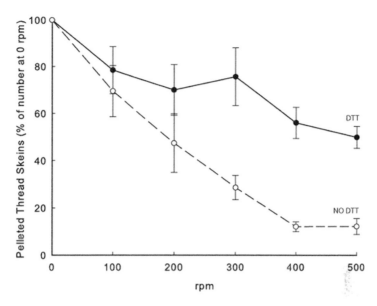

Fig. 11. Unraveling of thread skeins in *M. glutinosa* requires both hydrodynamic mixing as well as a functional mucous component. The number of skeins that pellet (and therefore do not deploy) decreases as the stirring rate of the slime in seawater increases, which demonstrates that mixing is important for proper slime setup. The strong effect of DTT, which is a reducing agent that cleaves disulphide bonds and disperses mucous networks, suggests that the mucous component of the slime is required for proper skein deployment. Figure from Winegard et al. (2010).

Evolution of Sliming in Hagfishes

All extant hagfishes possess slime glands, but it is not clear when this trait appeared in the evolutionary history of the lineage. One hagfish fossil (*Myxinikela siroka*) (Bardack 1991) that was described, dates from over 300 Ma, and shows no evidence of slime gland pores or slime glands. Absence of evidence is not necessarily evidence of absence, but given the quality of the specimen and some of the fine features that were preserved (branchial blood vessels and branchial pouches, for example), it is quite possible that *Myxinikela* lacked slime glands. If this is true, then it is tempting to speculate that the slime glands evolved in response to increasing predation pressure from gnathostomes. Indeed, modern chondrichthyans and osteichthyans are both successfully repelled by hagfish slime (Zintzen et al. 2011). While answering the question of why the slime glands evolved is difficult, imagining how they evolved may be more straightforward. The two main secretory cell types in the slime glands, GMCs and GTCs both have analogues in the epidermis—LMCs and Epidermal Thread Cells (ETCs). The presence of these cells is consistent with the hypothesis that the slime glands evolved via invagination of the skin into a specialized gland that could release large amounts of mucus and threads much more rapidly and in a more directed manner than the skin alone could. LMCs are present in low numbers in the epidermis (they are far outnumbered by the Small Mucous Cells (SMCs)), but they are similar to GMCs in size and general morphology. Epidermal thread cells (ETCs) are large, enigmatic cells of unknown function in the skin that, like GTCs, produce a keratin-like polymer within their cytoplasm (Schreiner 1916; Blackstad 1963). Interestingly, they also resemble 'skein cells' found in lamprey skin, which also produce a coiled cytoskeletal polymer, also of unknown function (Downing and Novales 1971; Lane and Whitear 1980; Spitzer and Koch 1998). The presence of skein cells in lamprey skin suggests that the cellular precursors of GTCs existed before the hagfishes diverged from the common ancestor they share with the lampreys, and only after this divergence became more specialized within the slime glands.

Slime Mechanics

Defensive sliming by hagfishes is astounding to naive observers because of the vast volumes of slime that are produced, but there is clearly something unique about the material properties of the slime that is also startling. As stated earlier, the slime is remarkably dilute, with mucus concentrations a full three orders of magnitude lower than materials like gastric mucus. On their own, mucus at this concentration have almost no effect on the viscosity of seawater (Fudge et al. 2005), but in concert with slime threads, mucus acts synergistically to entrain large volumes of water with very little secreted material (Koch et al. 1991b; Fudge et al. 2005). Ewoldt et al. (2010) quantified the mechanical properties of whole slime and concluded that hagfish slime is one of the softest biomaterials known, with an elastic modulus of 0.02 Pa, which is more than three orders of magnitude softer than gelatin. The slime also exhibits strain softening at large strains and simultaneous strain stiffening locally, which may correspond with the breaking of weak mucus-thread cross-links and the stretching of slime threads, respectively (Ewoldt et al. 2010).

Slime threads consist of IF proteins, yet their tensile mechanical properties differ radically from mammalian keratins like wool, hair, and nail, which consist of IFs embedded in an amorphous keratin protein matrix (Fraser et al. 1972). Slime threads are remarkably extensible, stretching to strains of 220% (more than three times their original length) before breaking (Fudge et al. 2003). As a comparison, wool breaks at a strain of about 50% (Szewciw et al. 2010). The elastic modulus of the threads is similar to that of rubber (about 6 MPa), which is about 400 times softer than hydrated wool (Fudge et al. 2003). At strains up to about 35%, the threads exhibit rubberlike mechanics, but at higher strains, they strain plastically and they exhibit dramatic strain stiffening at strains greater than 100%. Synchrotron x-ray diffraction experiments reveal that the transition between elastic and plastic behavior corresponds with protein chains in an alpha-helical conformation being pulled apart and reannealing into beta-sheets (Fudge et al. 2003). It is the formation of these stable beta-sheet structures that impart slime threads with their impressively high breaking stress. The study of slime thread mechanics has provided insight not only into the function of hagfish slime, but also has led to new insights into the behavior and function of IFs in living cells (Fudge et al. 2003; Fudge et al. 2009), in mammalian keratins (Fudge and Gosline 2004; Szewciw et al. 2010; Greenberg and Fudge, 2012), and in the quest to manufacture high performance protein materials (Fudge et al. 2010; Negishi et al. 2012) (Fig. 12).

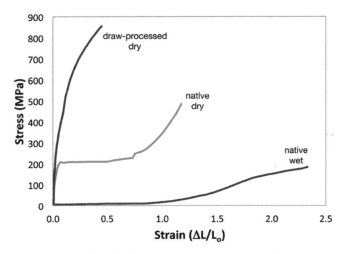

Fig. 12. Stress-strain curves for hagfish slime threads under a variety of conditions. In their native, hydrated state, the threads are soft and highly extensible, whereas in air they are far stiffer and less extensible. Slime threads stretched in water and then tested in air are even stiffer and less extensible, and far stronger. The wet mechanics of slime threads have provided important insights into the mechanical function of intermediate filament networks in living cells. The mechanics of dry threads have illuminated the biomechanics of keratins, and draw-processed threads are inspiring biomimetic efforts to produce high performance protein materials. Data from Fudge et al. (2003), Fudge and Gosline (2004), and Fudge et al. (2010).

Future Directions

Although we have made steady progress in understanding hagfish slime and slime glands, several interesting and fundamental questions remain. Future work will focus on understanding the function of GICs as well as quantifying the kinetics of gland refilling after a sliming event. Little is known about the innervation of the slime glands, although a recent study reports the presence of 5HT-, ChAT- and TH-positive neurons in the slime gland capsule (Zaccone et al. 2014), and future work will further characterize the function of these neurons in mechanoreception and slime release. Several questions remain regarding how the slime thread is manufactured within GTCs, and insight into this problem will be useful for biomimetic efforts to make high performance protein materials inspired by slime threads. The biophysics of mucous vesicle swelling are also not completely understood, and studies of the isolated mucous gel as well as the properties of the vesicle membranes are required to close this gap. Lastly, we know very little about the diversity of slime function among the more than 80 species of hagfishes that have been described, and we suspect that factors such as predation pressure and lifestyle (e.g., burrowing vs. epibenthic) may have had interesting effects on the structure and function of the slime.

References

Aebi, U., M. Haner, J. Troncosca, R. Eichner and A. Engel. 1988. Unifying principles in intermediate filament (IF) structure and assembly. Protoplasma 145: 73–81.

Bardack, D. 1991. First fossil hagfish (myxinoidea): a record from the Pennsylvanian of Illinois. Science 254: 701–3.

Blackstad, T.W. 1968. The skin and the slime gland. pp. 195–230. In: A. Brodal and R. Fänge (eds.). The Biology of Myxine. Universitetsforlaget, Oslo, Norway.

Downing, S. and R. Novales. 1971. The fine structure of lamprey epidermis III. Granular cells. J. Ultrastruct. Res. 35: 304–313.

Downing, S.W., R.H. Spitzer, W.L. Salo, J.T. Downing, L.J. Saidel and E. Koch. 1981. Threads in the hagfish slime gland thread cells: Organization, biochemical features, and length. Science 212: 326–328.

Downing, S.W., R.H. Spitzer, E.A. Koch and W.L. Salo. 1984. The hagfish slime gland thread cell. I. A unique cellular system for the study of intermediate filaments and intermediate filament-microtubule interactions. J. Cell Biol. 98: 653–669.

Ewoldt, R.H., T.M. Winegard and D.S. Fudge. 2010. Non-linear viscoelasticity of hagfish slime. Int. J. Nonlinear Mech. 46: 627–36.

Fan, W.J.W. 1965. The fine structure of thread cell differentiation in the slime glands of the Pacific hagfish, *Polistotrema stouti*. Anat. Rec. 151: 348.

Fernholm, B. 1981. Thread cells from the slime glands of hagfish (Myxinidae). Acta Zool. 62: 137–145.

Fernholm, B. 1998. Hagfish systematics. pp. 33–44. In: J.M.J. Jorgensen, J.P. Lumholt, R.E. Weber and H. Malte (eds.). The Biology of Hagfishes. Chapman and Hall, New York.

Ferry, J.D. 1941. A fibrous protein from the slime of the hagfish. J. Biol. Chem. 138: 263–268.

Fraser, R.D., T.P. MacRae and G.E. Rogers. 1972. Keratins: Their Composition, Structure, and Biosynthesis. Charles C. Thomas, Springfield, IL, 304 pp.

Fudge, D.S. 2002. The biomechanics of intermediate filament-based materials—insights from hagfish slime threads. Ph.D. Thesis, University of British Columbia, Vancouver, BC, Canada.

Fudge, D.S. and J.M. Gosline. 2004. Molecular design of the α-keratin composite: insights from a matrix-free model, hagfish slime threads. Proc. Roy. Soc. B 271: 291–299.

Fudge, D.S., K.H. Gardner, V.T. Forsyth, C. Riekel and J.M. Gosline. 2003. The mechanical properties of hydrated intermediate filaments: Insights from hagfish slime threads. Biophys. J. 85: 2015–2027.

Fudge, D.S., N. Levy, S. Chiu and J.M. Gosline. 2005. Composition, morphology and mechanics of hagfish slime. J. Exp. Biol. 208: 4613–4625.

Fudge, D.S., T.M. Winegard, R.H. Ewoldt, D.R. Beriault, L.J. Szewciw and G.H. McKinley. 2009. From ultra-soft slime to hard α-keratins: the many lives of intermediate filaments. Integr. Comp. Biol. 49: 32–39.

Fudge, D.S., S. Hillis, N. Levy and J.M. Gosline. 2010. Hagfish slime threads as a biomimetic model for high performance protein fibres. Bioinspir. Biomim. doi:10.1088/1748-3182/5/3/035002.

Greenberg, D.A. and D.S. Fudge. 2012. Regulation of hard α-keratin mechanics via control of intermediate filament hydration: Matrix squeeze revisited. Proc. Roy. Soc. B doi: 10.1098/rspb.2012.2158.

Gum, J.R. 1995. Mucins: their structure and biology. Biochem. Soc. Trans. 23: 795–799.

Herr, J.E. 2012. Mechanisms of rupture of mucin vesicles from the slime of Pacific hagfish *Eptatretus stoutii*: roles of inorganic ions and aquaporin water channels. M.Sc. Thesis, University of Guelph, Guelph, ON, Canada.

Herr, J.E., T.M. Winegard and D.S. Fudge. 2010. Stabilization and swelling of hagfish slime mucin vesicles. J. Exp. Biol. 213: 1092–1099.

Isaacs, J.D. and R.A. Schwartzlose. 1975. Active animals of the deep-sea floor. Sci. Am. October: 84–91.

Jenyns, L. 1842. Fish. *In*: C. Darwin (ed.). The Zoology of the Voyage of H.M.S. Beagle, under the command of Captain Fitzroy, R.N., during the years 1832–1836. Smith, Elder & Co., London.

Koch, E.A., R.H. Spitzer and R.B. Pithawalla. 1991a. Structural forms and possible roles of aligned cytoskeletal biopolymers in hagfish (slime eel) mucus. J. Struct. Biol. 106: 205–210.

Koch, E.A., R.H. Spitzer, R.B. Pithawalla and S.W. Downing. 1991b. Keratin-like components of gland thread cells modulate the properties of mucus from hagfish (Eptatretus stouti). Cell Tissue Res. 264: 79–86.

Koch, E.A., R.H. Spitzer, R.B. Pithawalla and D.A.D. Parry. 1994. An unusual intermediate filament subunit from the cytoskeletal biopolymer released extracellularly into seawater by the primitive hagfish (Eptatretus stouti). J. Cell Sci. 107: 3133–3144.

Koch, E.A., R.H. Spitzer, R.B. Pithawalla, F.A. Castillos 3rd and D.A. Parry. 1995. Hagfish biopolymer: a type I/type II homologue of epidermal keratin intermediate filaments. Int. J. Biol. Macromol. 17: 283–292.

Lametschwandtner, A., U. Lametschwandtner and A.R. Patzner. 1986. The different vascular patterns of slime glands in the hagfishes, *Myxine glutinosa* Linnaeus and *Eptatretus stoutii* Lockington. A scanning electron microscopy study of vascular corrosion casts. Acta Zool. 67: 243–248.

Lane, E.B. and M. Whitear. 1980. Skein cells in lamprey epidermis. Can. J. Zool. 58: 450–455.

Lehtonin, A., J. Karkkainen and E. Haahti. 1966. Carbohydrate components in the epithelial mucin of hagfish, *Myxine glutinosa*. Acta Chem. Scand. 20: 1456–1462.

Leppi, T.J. 1968. Morphochemical analysis of mucous cells in the skin and slime glands of hagfishes. Histochemie 15: 68–78.

Lim, J., D.S. Fudge, N. Levy and J.M. Gosline. 2006. Hagfish slime ecomechanics: testing the gill-clogging hypothesis. J. Exp. Biol. 209: 702–710.

Linnaeus, C. 1758. Systema Naturae (10th ed.). Holmiae: Laurentii Salvii.

Luchtel, D.L., I. Deyrup-Olsen and A.W. Martin. 1991. Ultrastructure and lysis of mucin-containing granules in epidermal secretions of the terrestrial slug *Ariolimax columbianus* (mollusca: Gastropoda:Pulmonata). Cell Tissue Res. 26. 375–383.

Martini, F.H. and A. Beulig 2013. Morphometics and Gonadal Development of the Hagfish Eptatretus cirrhatus in New Zealand. PLoS ONE doi:10.1371/journal.pone.0078740.

Müller, J. 1845. Vergleichende Anatomie der Myxinoiden. Untersuchungen uber die Eingeweide der Fische. Abh. K. Akad. Wissensch. Berlin 1843: 109–170.

Negishi, A., C.L. Armstrong, L. Kreplak, M.C. Rheinstadter, L.-T. Lim, T.E. Gillis and D.S. Fudge. 2012. The production of fibers and films from solubilized hagfish slime thread proteins. Biomacromolecules doi: 10.1021/bm3011837.

Newby, W.W. 1946. The slime glands and thread cells of hagfish, *Polistrotrema stouti*. J. Morphol. 78: 397–409.

Quay, W.B. 1972. Integument and the environment: glandular composition, function and evolution. American Zoologist 12: 95–108.

Salo, W.L., S.W. Downing, W.A. Lidinsky, W.H. Gallagher, R.H. Spitzer and E.A. Koch. 1983. Fractionation of hagfish slime gland secretions: partial characterization of the mucous vesicle fraction. Prep. Biochem. 13: 103–135.

Schreiner, K.E. 1916. Zur Kenntnis der Zellgranula. Untersuchungen uber den feineren Bau der Haut von *Myxine glutinosa*. Arch. Mikrosk. Anat. Entw. Mech. 87: 79–188.

Spitzer, R.H. and E.A. Koch. 1998. Hagfish skin and slime glands. pp. 109–32. *In*: J.M.J. Jorgensen, J.P. Lumholt, R.E. Weber and H. Malte (eds.). The Biology of Hagfishes, Chapman and Hall, New York.

Spitzer, R.H., S.W. Downing, E.A. Koch, W.L. Salo and L.J. Saidel. 1984. The hagfish slime gland thread cell. II. Isolation and characterisation of intermediate filament components associated with the thread. J. Cell Biol. 98: 670–677.

Spitzer, R.H., E.A. Koch and S.W. Downing. 1988. Maturation of hagfish gland thread cells—composition and characterization of intermediate filament polypeptides. Cell Motility Cytoskeleton 11: 31–45.

Steinert, P.M. and D.R. Roop. 1988. Molecular and cellular biology of intermediate filaments. Ann. Rev. Biochem. 57: 593–625.

Subramanian, S., N.W. Ross and S.L. MacKinnon. 2008. Comparison of the biochemical composition of normal epidermal mucus and extruded slime of hagfish (*Myxine glutinosa* L.) Fish Shellfish Immunol. 25: 625–632.

Szewciw, L.J., D. de Kerckhove, G. Grime and D.S. Fudge. 2010. Calcification provides mechanical reinforcement to whale baleen alpha-keratin. Proc. Roy. Soc. B 277: 2597–605.

Terakado, K., M. Ogawa and Y. Hashimoto. 1975. Ultrastructure of the thread cells in the slime gland of Japanese hagfishes, *Paramyxine atami* and *Eptatretus burgeri*. Cell Tissue Res. 159: 311–323.

Winegard, T.M. 2012. Slime gland cytology and mechanisms of slime thread production in the Atlantic hagfish *Myxine glutinosa*. M.Sc. Thesis, University of Guelph, Guelph, ON, Canada.

Winegard, T.M. and D.S. Fudge. 2010. Deployment of hagfish thread skeins requires the transmission of mixing forces via mucin strands. J. Exp. Biol. 213: 1235–1240.

Whitear, M. 1986. The skin of fishes including cyclostomes. pp. 8–38. *In*: J. Bereiter-Hahn, A.G. Matoltsy and K.S. Richards (eds.). Biology of the Integument, Vol. 1. Springer Verlag Berlin, Heidelberg, New York, Tokyo.

Zaccone, G., D.S. Fudge, T.M. Winegard, G. Capillo, M. Kuciel, K. Funakoshi and E.R. Lauriano. 2014. Confocal imaging and phylogenetic considerations of the subcutaneous neurons in the Atlantic hagfish *Myxine glutinosa*. Acta Zool. doi 10.1111/azo.12068.

Zintzen, V., C.D. Roberts, M.J. Anderson, A.L. Stewart, C.D. Struthers and E.S. Harvey. 2011. Hagfish predatory behaviour and slime defense mechanism. Sci. Rep. 1: 1–6.

Index

A

A. baerii 235, 242, 243, 246, 249
A. gueldenstaedti 237, 250
A. japonica 26
A. medirostris 240, 242, 243, 246, 249, 250
A. naccarii 235, 238, 240–242, 246, 249, 251
A. ruthenus 235, 249
A. stellatus 237
A. transmontanus 240, 241, 243, 246
acanthodians 4, 7, 8
Acetylcholine 154
ACh 25–29
Acipenser 234, 236, 239, 241, 243–245, 248, 249, 251, 252, 254
Acipenser ruthenus 203
Acipenseridae 203
Acipenseriformes 233, 237, 246, 247, 254
Actinistia 8–10, 13
actinistians 8, 13
actinopterians 209
actinopterygians 1, 7–9, 11–14, 152, 153, 161, 162, 174
Actinopterygii 11, 161, 162, 180, 183, 188, 233, 238
adipose tissue 241, 242
adrenaline 27
Adrenergic 164, 168, 173
adrenergic control 27
adult sturgeons 240
Aestivation 20, 22, 30, 31, 43–49, 52, 81–105, 107–123, 133, 134, 140–148, 213, 216, 220–225, 229
aethiopicus, Protopterus annectens, Protopterus dolloi 83
Africa 38, 39, 46
African 39, 44–47, 49
African lungfish *Protopterus* 20
Agnatha 1, 2, 4
agnathans 1, 2, 4, 154, 156, 157, 171, 174
air-breathing 133, 134, 143, 169, 179–190, 193–196
alpha-helical 286
Amia 180–191, 194–196

Amia calva 153, 169, 170
ammonia production 87, 90, 92, 93, 97–99, 102, 113–117, 122
ammonia toxicity 96–99, 114–116
Amniota 203
amphibians 39, 40, 42, 209
Amphiuma 182
anaerobic glycolysis 43
anaerobic metabolism 144, 145
Andrias japonicus 203
Aneural heart 176
AngII 24, 27–29
Anguilla anguilla 20
Anguilla rostrata 28
Anlagen 205, 208, 209
anterior gut 4
anterior intestine 236–238, 243, 248, 249, 252–255
anus 255
apomorphic 203, 207
ARC 31
arousal 225
arousal phases 83, 90, 97, 120, 123
arterial pO$_2$ 144, 145
AT$_1$ 27, 28
AT$_2$ 27
Atrial Natriuretic Peptides 20
atrioventricular 135–137, 139
atrium 135–137, 139, 151, 153, 154, 158, 163, 165, 168, 170–172
Australia 38–40
Australian 39, 40, 46, 47, 50
autapomorphic 203
AV plug 135, 137

B

β3-AR 25, 27, 29
β-ARs 27
Backflow 159, 166
betaine 283
beta-sheets 287
bichirs 153, 161, 162, 164, 167–169, 180
bile duct 237

Bimodal breathing 169
birds 203
Blackstad's 273
Blood O_2 content 191
bony fishes 38, 233, 238
bowfin 153, 161, 169, 180, 181
branchial arches 139, 140, 143
brush-border 250
bulbar folds 138–140
Bulbus 152, 153, 155–171, 173, 174
Bulbus arteriosus 153, 155–171, 173, 174
bullfrog 190
buoyancy 179–181, 183, 184, 187, 188, 196
burrow 47–49

C

Ca^{2+} reuptake 24, 25
caeca 237, 242, 247, 249, 255
caecilians 203
caecum 237
calcium 25, 29
capillary 276
carbamoyl phosphate synthetase 83
carbohydrate 277
cardiac performance 23
Cardiogenesis 153, 166, 174
Cardiovascular system 164, 167
CAs 24–28
Catecholamine 154, 155, 161
Catestatin 29
Caudata 203
cell death 142, 144, 146, 148
Cell survival 146
cellular stress 146
Ceratodontidae 58
cerebrospinal fluid 184
Chaenocephalus aceratus 22
chemoreceptors 180, 184–186, 194
Chionodraco hamatus 22
Cholinergic 160, 161, 164, 168
chondrichthyans 5–8, 13, 14, 152, 153,
 157–162, 164, 166, 167, 169, 171, 174
Chondrichthyes 4, 6, 7
Chondrostei 14
chromaffin cells 24, 26, 154, 155, 161, 173
chronotropism 27
cilia 238, 244, 246, 250, 251
ciliated cells 215–220, 225, 238, 239, 242,
 245–247, 249–252, 255
ciliated epithelium 238
ciliated OSNs 265, 267, 268
cloaca 212–214, 216, 217, 224, 225
CO 23, 24, 28
cocoon 44, 48, 49, 82–88, 95, 96, 98, 99,
 102–105, 109, 116, 119

coelacanth 38, 40, 41, 45, 170, 171
Coelacanthiformes 203
coelacanths 9, 10, 203
Collagen 157, 159, 162, 164, 167, 173
Colossoma 188
columnar cells 253
Compact myocardium 160, 164, 166, 167,
 169, 172
compacta 20, 23, 168, 171
Composition 273, 277, 282, 283
conductance 64, 67, 68
Control of breathing 179, 180, 185, 186
conus 137–139, 152, 153, 155–174
Conus arteriosus 152, 153, 155–174
conus valves 137, 138, 159
Coronary vessels 157, 160, 161
Cps I 83, 90, 91, 98, 99, 114
Cps III 83, 90, 91, 98, 99, 114
cranial fissure 7, 8
cranial nerves 186
Craniata 1
crocodiles 203
crypt OSNs 264, 265, 268
Cyclostomata 2
cylindrical cells 239

D

Darwin 272
defense 281
dehydration 84, 85, 87, 91–93, 102, 109,
 118
deoxyhemoglobin 29
deoxymyoglobin 29
Deployment 282–285
desiccation 83, 85, 95, 96, 117
development 57–66, 69, 72–77
Developmental Histology 238
Devonian period 39, 40
diffusion 64, 67, 183, 187, 190
digestive canal 238
digestive tract 235–238, 240, 243, 248, 249, 255
digestive tube 236–238, 255
Dipnoi 9, 10, 13, 38–40, 151, 162, 170, 172,
 173, 203
divergence 39–41
diversity 41, 51
DNA 41, 42
Dogfish 160
Downing 273, 277–279, 283, 286
ductus pneumaticus 204

E

E. taiwanae 274
ecophysiology 59
Eel 19, 20, 22–25, 27–30

egg 57, 59, 61–65, 67–73, 76–78
egg capsule 59, 63–65, 67–70
Elasmobranchii 157
Elastin 156, 162, 164
electric eel 189
Electrophorus 189, 193
elopomorphs, osteoglossomorphs 14
Elpistostegalia 8–10
Embryo 166
embryonic 57–59, 62–67, 72, 76, 77, 152, 153, 164, 171
embryonic life 238
endocardium 138, 147, 154, 159, 161–163, 167, 168
endocrine cells 251, 252
endodermic epithelium 242
energy use 58, 64, 72, 76
eNOS 22–25, 27–29, 31
enterocytes 217–220, 223, 225, 249–255
enteroendocrine cells 252
Enzyme 43, 44, 48, 49
Epidermal Thread Cells 285, 286
epithelial 274, 275, 277
Eptatretus stoutii 274
Erpetoichthys calabaricus 203
esophagus 7, 13, 14, 236–242, 246, 247, 255
euteleosts 14
Evo-Devo 201
Evolution 38–40, 42, 45, 52, 285
evolutionary novelties 201
evolutionary scenario 202
Ewoldt 286
excretion 39, 45, 46, 48
exogenous feeding 245, 247, 250
extant 38, 39, 46
Extrinsic Regulation 25
eye stalk 7

F

fasting 213, 220
feedback 180, 181, 185, 190, 194
feeding 83, 84, 95, 104, 113, 122, 123
Fernholm 273, 274, 279, 281
Ferry 272, 273
fissures 6, 8
fossils 39, 40
four-stroke mechanism 182
Frank–Starling mechanism 24
Frank-Starling response 24, 29, 30
freshwater 212–214, 216, 218, 221, 223–226, 229
frogs 203
Fudge's 273
Function 273, 278–281, 283, 286–288

G

Galeaspidida 4
GALT 253
Ganglion cells 154, 157, 164
gars 153, 161, 167, 169, 173, 174, 180
gas bladder 14, 167
gas exchange 64, 65, 67, 69, 70, 76, 77, 151
gastric anlage 242
gastric cavity 242
gastric compartment 242
gastric glands 242–244, 246
gene 41, 42
gene expression 206
genetic cascade 207, 209
genetic changes 40
genetic code 205
genome 39, 41, 42
genome size 39, 41, 42
gills 1, 4, 10, 133, 134, 138, 143, 151, 152, 157, 159, 162, 169, 172
gizzard 237
Gland Interstitial Cells 279, 280
Gland Mucous Cells (GMCs) 274–277, 279–281, 283, 285, 286
Gland Thread Cells 273, 275, 277
glands 272–274, 276, 281–283, 285–287
glandular stomach 236, 237, 242–245, 248
gluconeogenesis 111–113, 116
glucose 144, 145, 147
glutamate dehydrogenase 113, 115, 116
glutamine 90, 91, 97–99
glutamine synthetase 98, 99
glycoconjugates 240, 250
glycogen 43, 44, 86, 87, 99, 110–113, 141, 142, 144
glycoprotein 274, 276
Gnathostomata 1, 2, 4
gnathostomes 1, 2, 4–7, 10, 14
goblet cells 217–221, 223–225, 247–251, 255, 277
goldfish 184
Golgi 281
Gondwana 39
Gosline 273, 287
granulocytes 219, 222, 226–229
growth 59, 64, 72, 73, 76
gut 212–220, 222–230
gycocalyx 242
Gymnotus 182

H

Habitat 40, 46
Hagfish 154–157
hair 286

hatching 60–67, 69, 70, 72–74
heart 133–140, 142–148, 151–158, 160,
    162–168, 170–174
heart rate 140, 144, 147
heart structure 134
hierarchical 202, 205, 206
histological structure 238, 241, 242
Holocephali 157
holocrine 274, 277
Holostei 14, 169
homeostasis 179
homologized 201
homology 201, 202, 205–209
Hoplerythrinus 182
HR 22–24, 26, 30
Hsp-90 31
Huso huso 235
hydrostatic 183
hypercarbia 184
hypoxia 63, 65, 73, 77, 83, 87, 88, 99, 102, 103,
    110, 111, 114–117, 144, 185–187, 196

I

identity 207
immunolocalization 23–24
induction phase 84, 85, 87, 91–93, 102, 109,
    118
Innervation 23, 25, 26, 160, 164, 167, 171, 173
iNOS 23, 24, 28
inotropism 24, 26–28
integrate and fire' models 189
intestinal anlage 242
intestinal mucosa 250, 253
intestinal vestibule 214–217, 219, 224, 225,
    227
intestine 5, 12, 236–238, 240, 243, 244,
    247–250, 252–255
Intrapericardial 153, 155, 156, 160, 173
ions 283
ischemia/reperfusion 29
islets of Langerhans 227, 228

J

Jenyns 272
juvenile 234–236, 240, 242, 243, 252, 254

K

keratin 273, 280, 286, 287
Koch 273, 274, 277–279, 281, 283, 284, 286
Krebs cycle 145

L

L. menadoensis 203
lactate dehydrogenase 107, 111

lamina propria 238, 240–242, 244, 252
lamprey 154–157, 277, 286
Large Mucous Cells 277
Latimeria 170, 171
Latimeria chalumnae 203
Lepidosiren 38–40, 42–44, 46–52, 152, 170, 173,
    180, 184, 185, 212, 226, 228, 230
Lepidosiren paradoxa 133, 203
Lepidosireniformes 39, 40
Lepisosteus 167, 180, 182, 184–187, 189, 190
Lim 281, 282, 284
lineages 39, 41
Linnaeus 272
lipid 146, 277
Lissamphibia 203
liver 236, 238, 243
living fossils 40
lizards 203
lobed fins 38
longevity 40
Luchtel 273, 276, 283
lung 4, 7, 8, 10, 12–14, 179–194, 196, 201–205,
    207–209
lung diffusion capacity 190
Lungfish 9, 10, 14, 19, 20, 22, 23, 30, 31, 38–52,
    81–114, 116–123, 133–135, 138, 139,
    141–146, 148, 151, 152, 169, 170, 172–174,
    180, 185, 212–217, 219, 222–226, 228
lymphatic micropumps 218, 221, 229, 230
lymphocytes 252, 253
lympho-granulocytic tissue 213
lymphoid tissues 252
lysosomes 253

M

macromeric 5, 8
maintenance phase 84–87, 89, 90, 97, 100–102,
    104, 105, 110–113, 115–122
mammals 203, 209
Mechanics 286, 287
mechanoreceptors 180, 184, 186, 187
melanin 248, 250
melanin plug 248, 250
melanomacrophage centres 226–228
merosecretion 277
Mesozoic 39, 40
metabolic depression 82, 88, 98, 103, 117
metabolic rate 64, 74, 75
metabolism 39, 43–45, 58, 69, 72, 75–77
methylamines 283
micrometric 6, 7
microtubules 278, 280
microvillar OSNs 264, 265, 267, 268
middle intestine 237, 247, 254
mitochondria 43, 278, 279, 281

Mitochondrial Rich Zone 278, 279
Modeling 188, 191, 196
molecular functional constraint 207
monophyletic 2, 4
mucosa 237, 238, 240, 241, 244, 247, 248, 250, 253
mucosal folds 243, 247–250, 255
mucosubstances 239, 243, 250
mucous cells 239–241, 243, 244
mucus 238, 240, 244–247, 250, 255, 273, 274, 277, 281–286
mucus layer 238, 250, 255
Müller 272
multivesicular bodies 253, 254
muscle 237–244, 246, 253
muscle atrophy 89, 105
muscle layers 237, 243
muscularis mucosae 238
musculus decussatus 274
Myllokunmingiida 2, 4
myocardium 138, 139, 142, 148, 152, 154–156, 158–160, 162–164, 166–169, 172–174
Myxine glutinosa 272, 274
Myxinikela siroka 285
Myxinoida 1, 2

N

NADPH-diaphorase 22
nail 286
NANC 26, 31
nasal capsules 1, 2, 4
naso-hypophysal duct 4
Nemamyxine elongata 274
Neoceratodus 38–40, 42, 46, 47, 50, 152, 170, 173, 180, 185, 212, 225, 228, 230
Neoceratodus forsteri 133, 203
nerve bundles 247
nerve-fiber bundles 242, 247
Nerve fibers 154, 155, 157, 161, 167
Nerves 164, 168
neuroepithelial cells 186
neuronal bodies 247
neuropeptide Y 26
Newby 273, 274, 277, 279, 284
newts 203
nitric 101, 105, 108, 118
Nitric Oxide 19, 147
Nitric Oxide Synthase 19
Nitrite 29, 30
nitrogen metabolism 45
nitrogenous 39, 45
nNOS 22, 24, 26
noradrenaline 27
NOS/NO System 19, 20, 30–32

O

O_2 dissociation curves 191
oesophagus 212, 214, 215, 229
OFT 151–174
Olfaction 260, 263
olfactory organ 260, 261, 263, 264
olfactory rosette 261–265, 267
Olfactory Sensory Neurons (OSNs) 261, 264, 265
Oncorhynchus mykiss 28
ontogenesis 242
ontogenetic 201, 207, 209
ontogeny 203
ornithine-urea cycle 83, 98, 114, 116
osmolarity 283
osmoregulatory 240, 241, 255
Osteichthyans 5–8, 11, 13, 14
Osteichthyes 1, 4–6, 8, 38, 40, 161
osteichthyians 202
Osteognathostomata 1, 204, 208
osteognathostomes 202, 203
Osteolepiformes 9
Osteostraci 2
ostracoderms 2
Outflow Tract 137, 138, 151, 155, 160, 166, 173
Outflow valves 153, 156, 161
oxide 101, 105, 108, 118
oxygen 39, 43, 45–48, 50, 58, 62–74, 76, 77
Oxygen Consumption 64–66
oxygen partial pressures 62
oxyntic 237, 245, 246

P

Palaeozoic 40
Paleozoic 188
pancreas 213, 214, 217, 218, 225–228, 238, 243, 248
paraphyletic 2, 4
parasites 224
Parsimonious 204, 209
parsimony 205, 209
pathogens 281
Pericardial cavity 152, 155, 156, 158, 159, 161–163, 166, 169, 171
periodicity 194
Permian 40
Petromizon marinus 22
Petromyzontida 1, 2
pharyngeal 207
pharynx 202–205, 207–209
physiology 57, 75–77
physoclist 204
physoclistous 183
physostomous 183, 184
pirambóia 39

Placodermi, Chondrichthyes, Acanthodii 4
placoderms 5, 6, 14
plasma 274, 275, 283
plasma cells 226–229
plesiomorphic 207, 209
PLN 25, 29
pneumatic duct 236, 237
PO$_2$ threshold 193, 194, 196
Polyodon spathula 235
Polypterids 162, 173
Polypteriformes 14, 203
polypteriforms 209
Polypterus 162, 163, 180–182, 186, 193, 203
Portal heart 154
positive inotropism 26, 27
positive-pressure 181
predation 281, 285, 288
primitive characteristics 38
protein 88–90, 92–96, 98–100, 102, 105, 108,
    112, 113, 116–118, 122, 123, 276–278,
    286–288
protein degradation 89, 90, 98, 100, 105, 117,
    123
protein kinase B 24
Protopterus 38–40, 42, 46–51, 83, 84, 103, 105,
    133, 134, 138, 144, 147, 152, 170, 172, 173,
    180, 185–187, 203, 212, 213, 215, 225–230
Protopterus aethiopicus 30
Protopterus amphibious 30
Protopterus amphibius 83
Protopterus annectens 30, 212, 213
Pseudoscaphirhynchus 235
pseudostratified 240
Pteraspidomorphi 4
pulmonalis fold 135, 136
Pulmonary arteries 162
pulmonary channel 135, 137, 139
Pulmonary circulation 151
pulmonary vein 134–137, 140
pulmonary veins and arteries 10
pyloric 212, 214–217, 225–227, 236–238,
    242–250, 253, 255
pylorus 216, 217

R

Rapidly-Adapting Receptors 186
Ray-finned 152, 161
rectum 236–238, 248, 250, 255
red pulp 226, 229
Reedfish 153, 162
regionalization 1
renin-angiotensin system 27
reptiles 209
respiratory pharynx 208, 209
Respiratory rhythm generation 184

S

salamanders 203
Salo 273, 277
sarcoplasmic reticulum 43, 44
sarcopterygians 1, 9–11, 13, 209
Sarcopterygii 9, 38, 40, 161, 170, 180, 183, 188
Scenario 202, 207, 208
Secondary heart field 152, 166
secretory cells 242, 244–247
sequence 205–207
SERCA2a 24, 25, 29
serosa 213, 214
serotonin 26
serous membrane 238, 241
shallow 40, 46, 49, 50
Shark 157, 159–161, 171
Siberian sturgeon 238
silk 273
sinus venosus 135–137
Skate 157, 159, 160, 169, 171
skein cells 286
skeletal muscle 239–241
skin 274, 276, 285, 286
Slime 272–288
slime glands 272–274, 276, 281–283, 285–287
Slowly-Adapting Receptors 186
smooth muscle 152, 153, 155–158, 160–164,
    166–169, 171, 173, 174, 240, 242, 244, 246,
    253
snakes 203
S-nitrosylation 24, 25, 29, 30
South America 38, 39, 43, 46
South American 38, 39, 42, 44–51
Spawning 59, 62
spiral valve 5, 12, 172, 212–225, 228, 229,
    236–238, 248–252, 254, 255
Spitzer 273, 274, 277–279, 286
spleen 213, 214, 216, 217, 224–230, 236, 238,
    248
spongiosa 20, 23, 163, 166
Spongy myocardium 160
stabilization 283
stomach 212, 215, 236–240, 242–248, 253
stomach epithelium 242
striated muscle fibers 242
Sturgeons 153, 161, 164–167, 169, 171, 172,
    233–235, 238, 240, 242, 245–247, 255
Subendocardium 162, 163, 167, 168
Subepicardium 162, 164, 167
submucosa 238, 240, 241
substance P 26
SV 23, 24
swim bladder 4, 7, 8, 12–15, 202–205, 207–209
    236, 237

synthesis 83, 89, 90, 93, 97–100, 102, 103, 105,
111, 112, 114–118, 122, 123
Systemic circulation 151, 152, 154, 172

T

tambaqui 188
taste buds 234, 241
teeth 5, 6, 8–11, 234, 237, 255
teleosts 1, 11, 12, 14, 20, 25, 30, 152, 153, 156,
157, 161, 162, 164, 169, 174, 203, 204
temperature 58, 60, 61, 63–65, 67, 69, 70,
72–76
Terakado 273, 277–279
terminal intestine 237, 238
Testudines 203
Tetrapoda 13
tetrapodomorph 180
tetrapods 1, 8–10, 14, 39–42, 45
time series analysis 194
TMAO, and dimethylglycine 283
toads 203
tooth plates 6, 10
trabecular 20, 22
Trachemys scripta 203
traits 201
Trematomus bernacchii 22
tuatara 203
tubular glands 246
TUNEL assay 31
Tunica media 155, 156
turtles 203, 207
two-stroke mechanism 181, 182, 188
type I breath 182, 188, 191, 194–196
type II breath 182, 187, 188–196

U

unraveling 283–285
urea 45, 46, 48, 82, 83, 85, 87, 90, 92–94, 97–99,
102, 110, 112–117, 120–122
urea accumulation 93, 116, 117

urea excretion 92, 93, 98, 120–122
urea synthesis 83, 90, 93, 97–99, 102, 114–117

V

vacuoles 250, 251, 253–255
vagotomy 187
Valve tier 169
vascularization 276
vasointestinal peptide 26
Vasostatin-1 29
Ventilation 179–182, 184, 185, 187, 195
ventral aorta 139, 152, 155, 156, 159, 163,
165–167, 169, 171
ventricle 136–141, 143, 147
ventricular relaxation 24
ventricular septum 135–138, 142, 144
Vertebrata 1
vertebrate 39–45, 51, 151–157, 160, 162, 166,
171, 174, 202, 205, 207
vesicles 273, 275–277, 281, 283, 284
vitelline membrane 59, 65, 67, 69
volume threshold 187, 194, 196

W

white pulp 226
wool 286

X

xanthine-oxidoreductase 29
Xenopus 182, 189

Y

yolk 242, 250
yolk sac 242

Z

Zintzen 273, 281–283, 285
Zebrafish 152, 153, 157, 162, 166